School Publishing

PROGRAM AUTHORS

Malcolm B. Butler, Ph.D.

Judith S. Lederman, Ph.D.

Randy Bell, Ph.D.

Kathy Cabe Trundle, Ph.D.

David W. Moore, Ph.D.

NATIONAL GEOGRAPHIC
SCIENCE

Built to target your state standards, *National Geographic Science* is a research-based program that brings science learning to life through the lens of National Geographic.

Promote Science success as you share

Grade 3

The National Geographic Experience . . .

- **Immerses Students in the Nature of Science and Inquiry.**

- **Unlocks the Big Ideas in Science for All Learners.**

- **Builds Scientific and Content Literacy.**

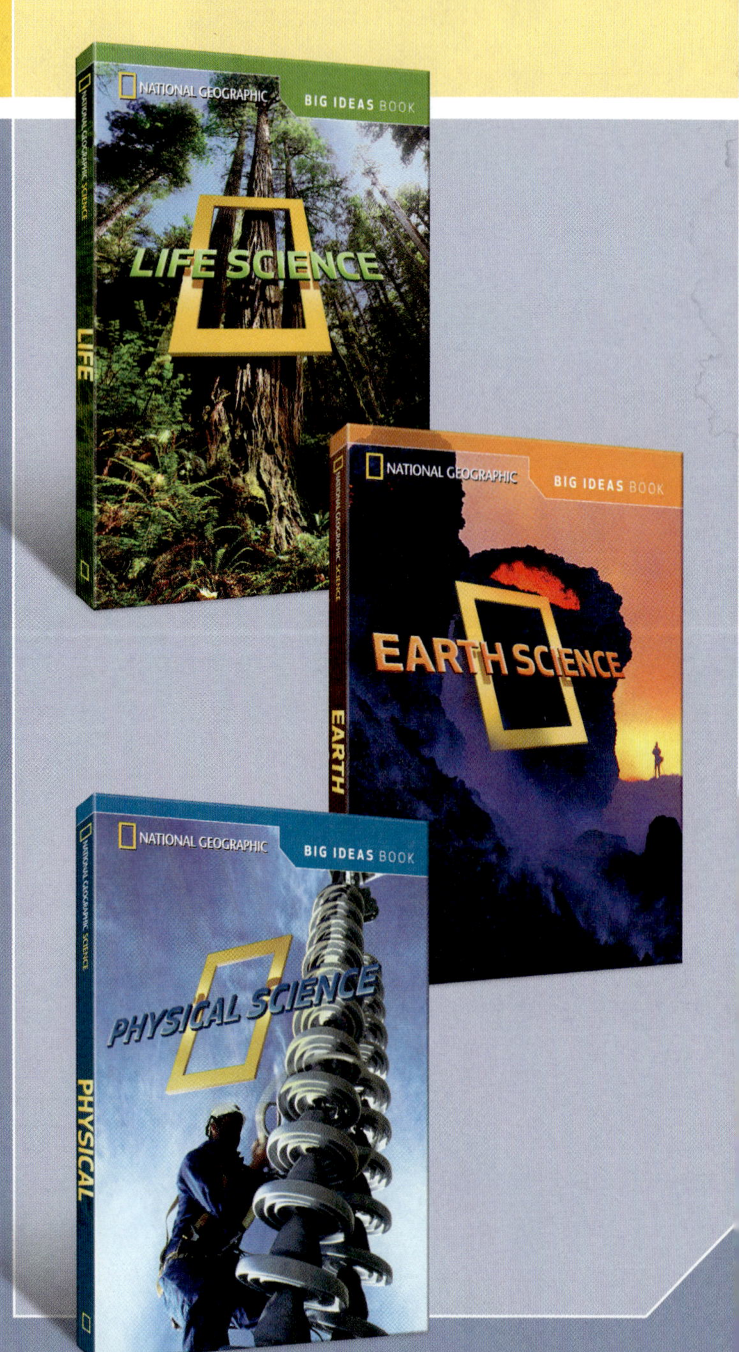

Grade 4

Grade 5

Immerse Students in . . . The Nature of Science.

In *National Geographic Science,* process skills build at each grade level to ensure a complete understanding of the Nature of Science. This chart shows how process skills and the Nature of Science work together to help students think and act like scientists.

PROCESS SKILLS

Nature of Science

Grade K

OBSERVE

- Science knowledge is based on evidence.
- Science knowledge can change based on new evidence.

Grades 1 & 2

OBSERVE & INFER

- Science conclusions are based on observation and inference.
- Science theories are based partly on things that cannot be observed.

Immerse Students in . . . Leveled, Hands-on Inquiry.

National Geographic Science provides students with abundant and relevant hands-on explorations to facilitate a thorough understanding of key Science concepts. The four levels of inquiry in *National Geographic Science* are designed to help students build confidence and competence in scientific thought and inquiry.

Explore Activity

The *Explore Activity* builds background for the unit. This activity *Engages* students as they *Explore.*

Explore Activity

Investigate Star Positions

Question How do some stars appear to move across the sky?

Science Process Vocabulary

observe verb

Scientists often use tools to take a close look at, or **observe,** objects and events.

Materials

safety goggles

Star Finder

What to Do

1. Put on your safety goggles. Cut out both parts of the Star Finder. Cut out the viewing window in the top piece of the Star Finder.

Directed Inquiry

In *Directed Inquiry,* the teacher gives direct instruction throughout the activity. Students are given opportunities to *Explain* what they have done, *Elaborate* by asking further questions, and *Evaluate* by using a self-reflection rubric.

Directed Inquiry

Investigate Weathering

Question How does physical weathering compare with chemical weathering?

Science Process Vocabulary

model noun

A **model** can show how a process, such as weathering, works.

predict verb

When you want to explain what you think will happen,

Materials

safety goggles

sandstone

paper towel

water

graduated cylinder

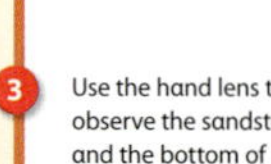

2 jars with lids

What to Do

1. Put on your safety goggles. **Predict** what will happen if you rub 2 pieces of sandstone together. Hold the pieces of sandstone over the paper towel and rub them together for a few seconds. **Observe** what happens. Record your observations in your science notebook.

2. **Measure** 250 mL of water into a jar. Place 5 pieces of sandstone in the jar. Put the lid on the jar securely.

3. 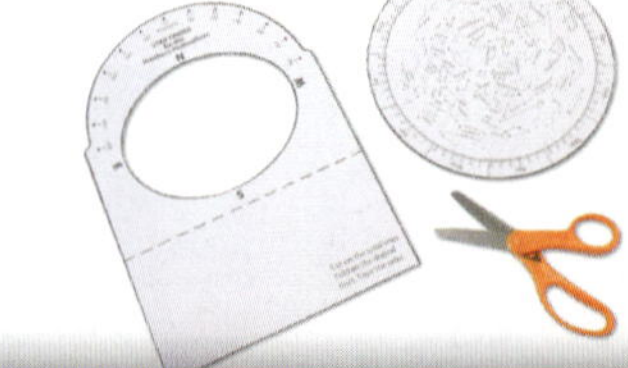 Use the hand lens to observe the sandstone and the bottom of the jar. Record your observations.

- **There is often no single "right" answer in science.**

- **Scientific theories provide the base upon which predictions and hypotheses are built.**

- **There is no single, scientific method that all scientists follow.**
- **There are a number of ways to do science.**

Also includes

Science in a Snap!

Offers quick investigations to activate understanding of science concepts

Guided Inquiry

In *Guided Inquiry*, students become independent learners with guidance from the teacher. Students manipulate variables, provide *Explanations, Elaborate* by asking further questions, and *Evaluate* by using a self-reflection rubric.

Open Inquiry

In *Open Inquiry*, students choose their own questions, create their own plans, carry out their plans, collect and record their own data, look for patterns, and share that data. Students *Explain* their results, *Elaborate* by asking further questions, and *Evaluate* by using a self-reflection rubric.

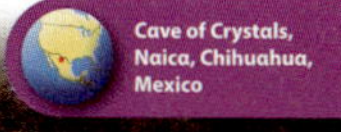

The Big Ideas in Science Unlocked and In Depth

The Big Ideas in Science should be targeted and focused for the teacher and the student. Therefore, *National Geographic Science* was created to be targeted and focused on these Big Ideas in Science as well.

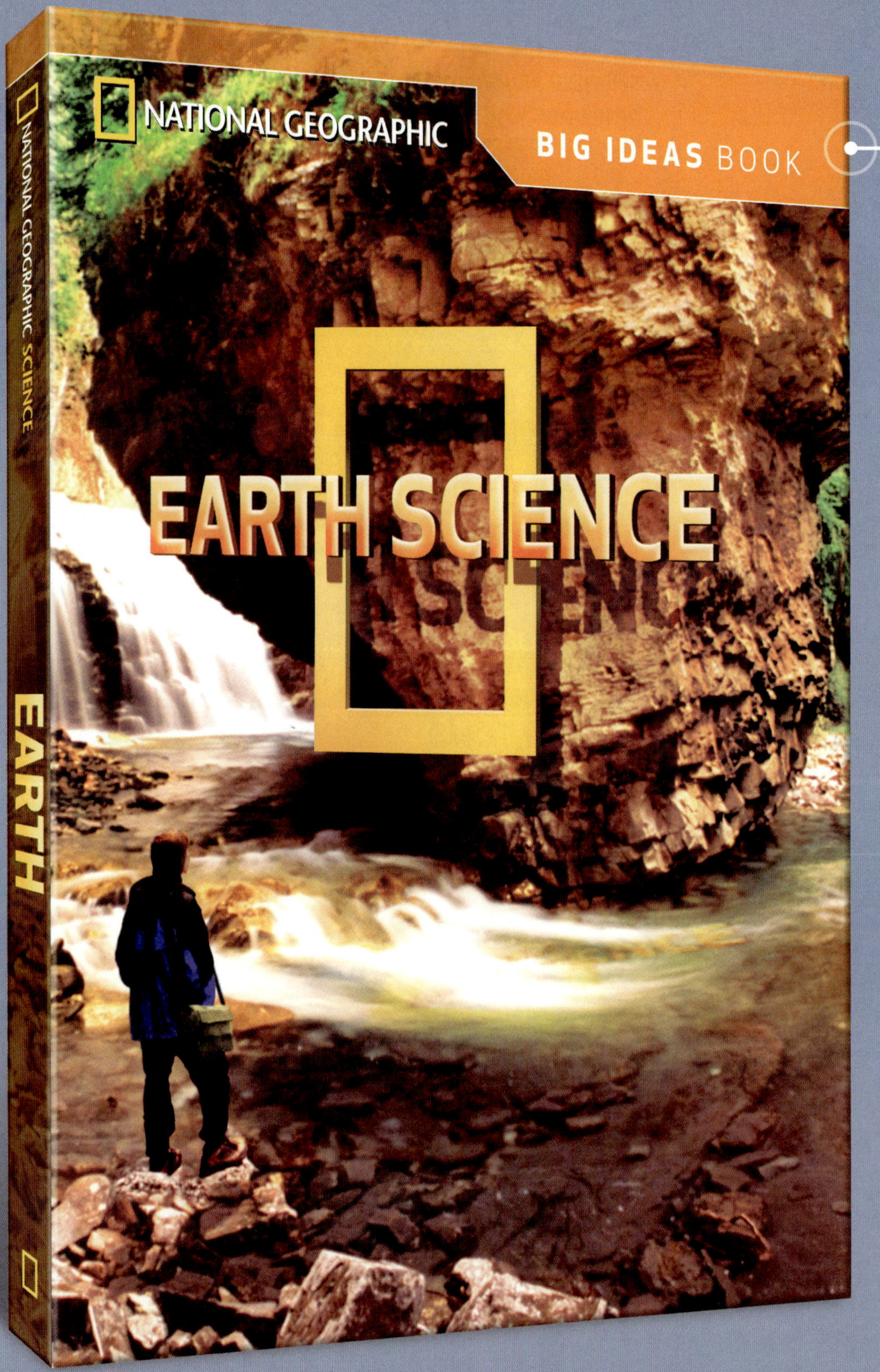

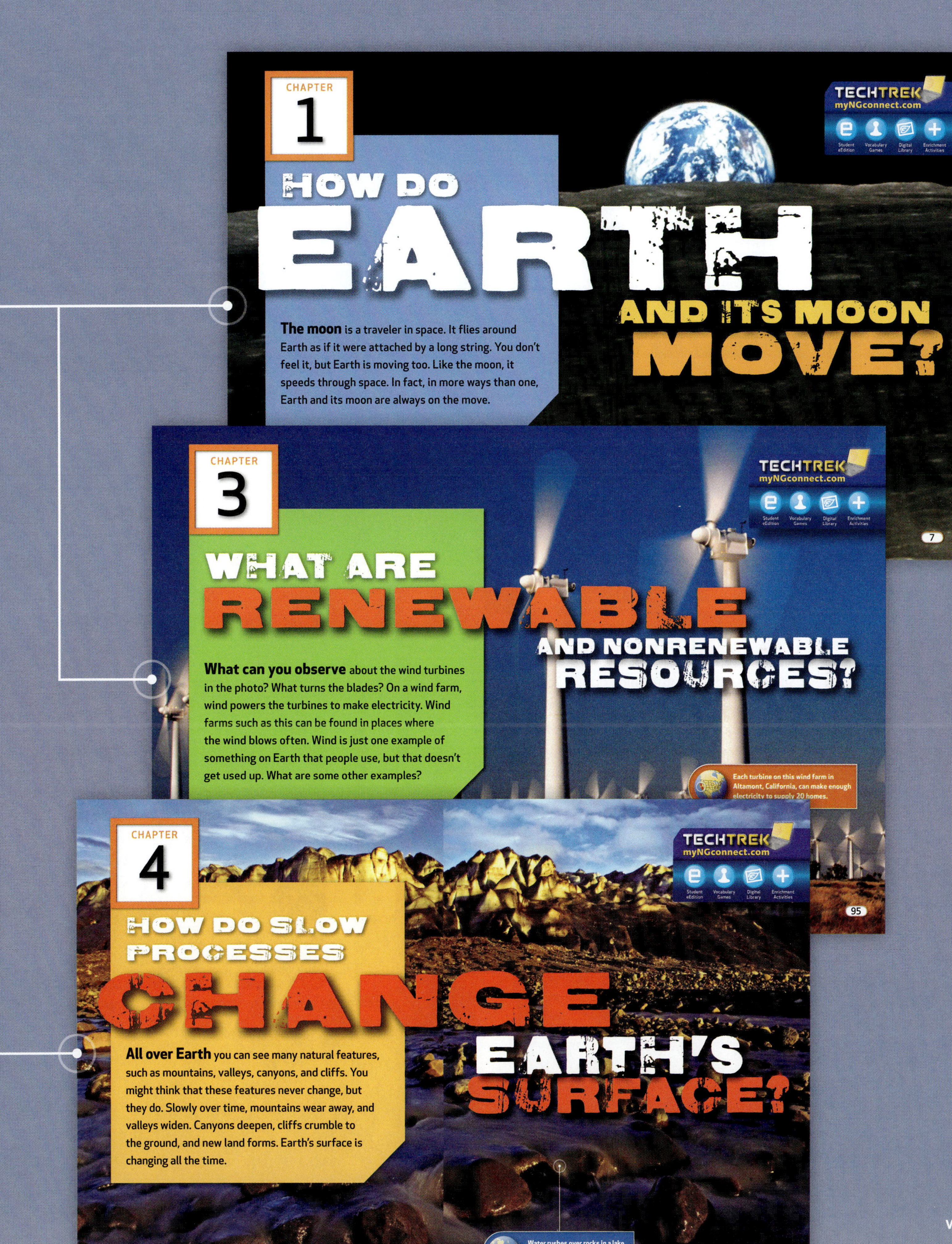

CHAPTER
1

TECHTREK
myNGconnect.com
Student eEdition
Vocabulary Games
Digital Library
Enrichment Activities

HOW DO
EARTH
AND ITS MOON
MOVE?

The moon is a traveler in space. It flies around Earth as if it were attached by a long string. You don't feel it, but Earth is moving too. Like the moon, it speeds through space. In fact, in more ways than one, Earth and its moon are always on the move.

7

CHAPTER
3

TECHTREK
myNGconnect.com
Student eEdition
Vocabulary Games
Digital Library
Enrichment Activities

WHAT ARE
RENEWABLE
AND NONRENEWABLE
RESOURCES?

What can you observe about the wind turbines in the photo? What turns the blades? On a wind farm, wind powers the turbines to make electricity. Wind farms such as this can be found in places where the wind blows often. Wind is just one example of something on Earth that people use, but that doesn't get used up. What are some other examples?

Each turbine on this wind farm in Altamont, California, can make enough electricity to supply 20 homes.

95

CHAPTER
4

TECHTREK
myNGconnect.com
Student eEdition
Vocabulary Games
Digital Library
Enrichment Activities

HOW DO SLOW PROCESSES
CHANGE
EARTH'S
SURFACE?

All over Earth you can see many natural features, such as mountains, valleys, canyons, and cliffs. You might think that these features never change, but they do. Slowly over time, mountains wear away, and valleys widen. Canyons deepen, cliffs crumble to the ground, and new land forms. Earth's surface is changing all the time.

Water rushes over rocks in a lake fed by water from Vatnajokull Glacier in Iceland.

Build Scientific Literacy for Your Future Scientists

Students will build scientific literacy and gain an understanding of the Nature of Science from real-life National Geographic Explorers. Your students will learn that Science is:

- A way of knowing
- Empirically based and consistent with evidence
- Subject to change when new evidence presents itself
- A creative process

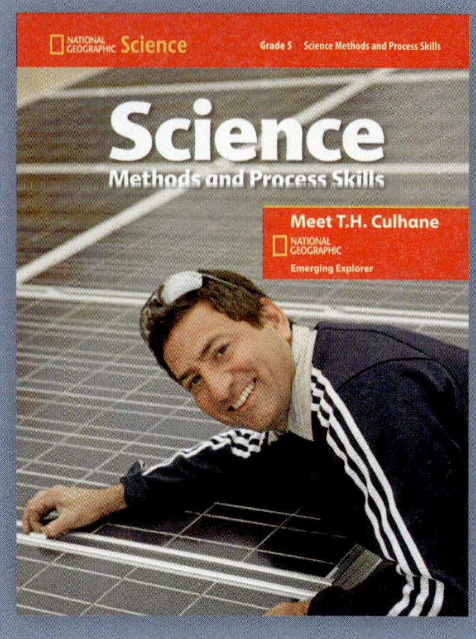

☐ Ask Questions

Scientists ask many kinds of **questions.**

☐ Make a Plan

A scientist needs to make a **plan** to answer

☐ Carry Out a Plan

Plans can take many forms—making

☐ Collect and Record Data

Scientists have a lot to think about. To help them remember, organize, and process all the information, or **data,** they collect from an experiment or investigation, they record the data in some sort of organized way.

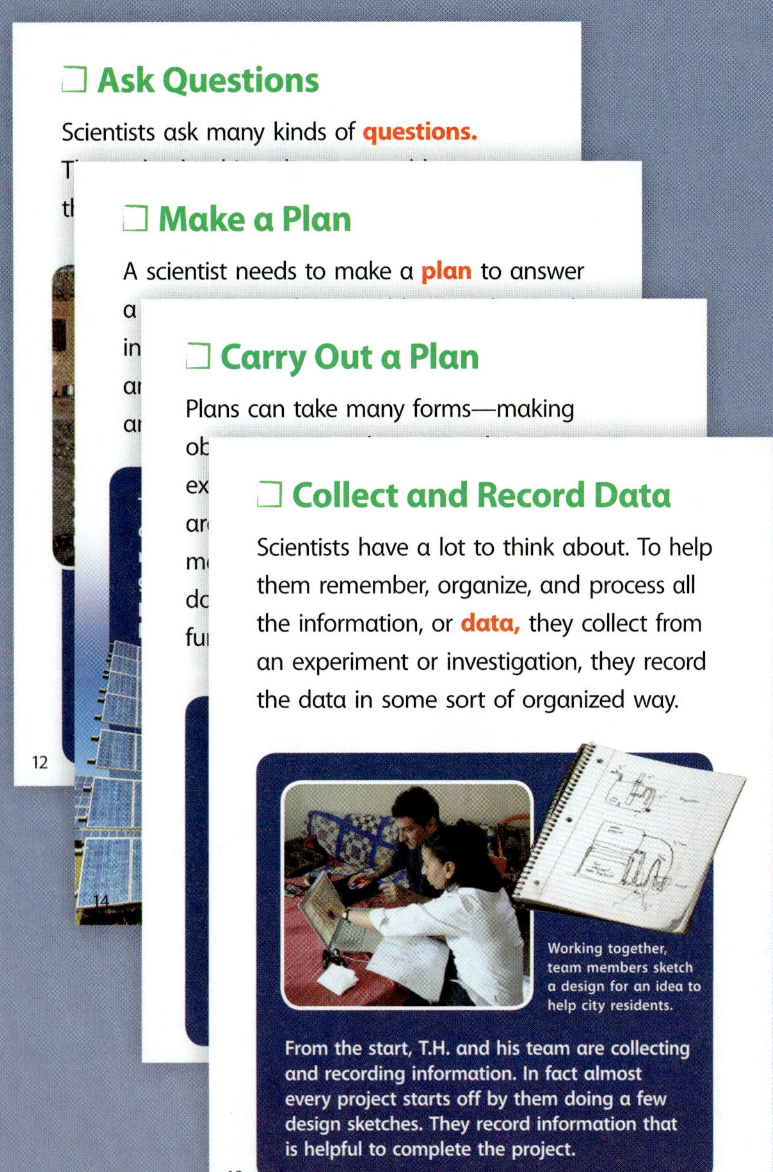

12

14

18

☐ Analyze Data

What do scientists do with the data they

☐ Make Conclusions

After finishing a plan and analyzing data,

☐ Share Results

Scientists **share** their results with other people. They want others to learn what they find out.

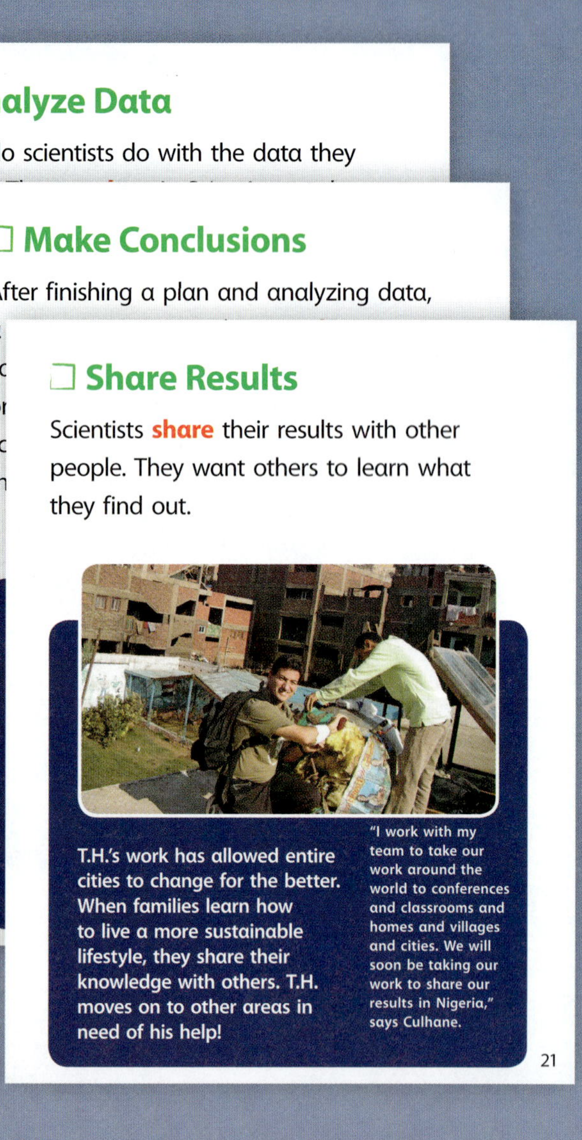

20

21

Access Science in Our World

Become An Expert

Students, through each chapter's Big Idea, learn concepts specifically tied to your state standards.

In the *Become An Expert* section of each chapter, students apply what they learned through concrete examples found throughout our world.

Reading Comprehension Strategies

In the Teacher's Edition, teachers are offered four comprehension strategies to ensure that content learning is deep and lasting.

- Preview and Predict
- Monitor and Fix Up
- Make Inferences
- Sum Up

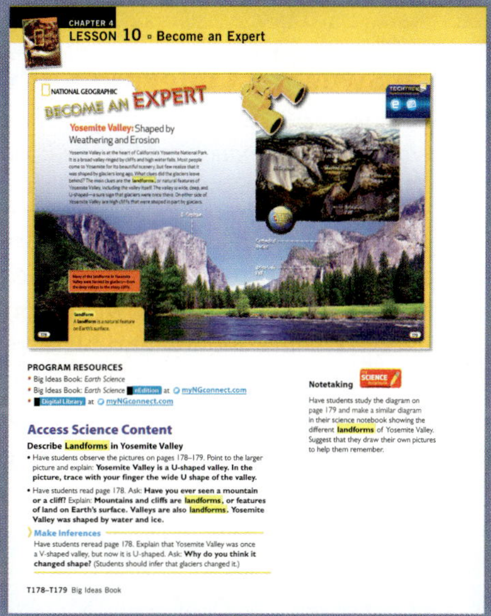
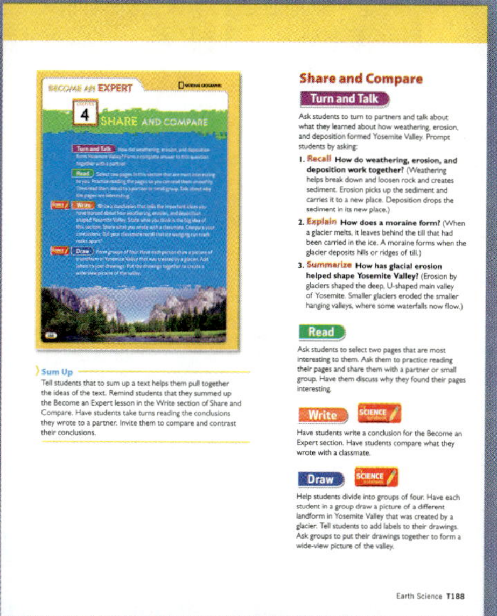

Explore on Your Own Leveled Books

Explore on Your Own books provide an opportunity for group or independent reading. These books are leveled for your low-mid achieving students and for your mid-high achieving students. These books use the same concept at different reading levels to motivate all students to read independently.

Program Authors

Malcolm B. Butler, Ph.D.

Associate Professor of Science Education,
University of South Florida, St. Petersburg, Florida

SCIENCE

Dr. Malcolm B. Butler's teaching and research address multicultural issues in the classroom. With a specialization in physics, he has worked to support typically underserved student populations and has interests in the areas of writing to learn in science, science content for elementary teachers, and coastal and environmental education professional development for teachers. He has written and contributed to several academic journals, including *The Journal of Research in Science Teaching, The Journal of Science Teacher Education, The Journal of Multicultural Education,* and *Science Activities.*

Judith Sweeney Lederman, Ph.D.

Director of Teacher Education, Associate Professor of Science Education,
Department of Mathematics and Science Education,
Illinois Institute of Technology, Chicago, Illinois

SCIENCE

Dr. Judith S. Lederman's teaching experiences include pre-K to 12th grade science, masters and doctoral level science education courses, and post-graduate professional development for in-service teachers. She is known nationally and internationally for her work in the teaching and learning of Scientific Inquiry and Nature of Science in both formal and informal settings. In 2008 she was awarded a Fulbright Fellowship to work with South African university and museum educators and K–12 science teachers. Dr. Lederman served on the Board of Directors of the National Science Teachers Association (NSTA) and as President of the Council for Elementary Science International (CESI).

Randy Bell, Ph.D.

Associate Professor of Science Education,
University of Virginia, Charlottesville, Virginia

SCIENCE

Dr. Randy Bell began his career as a forest researcher in the Pacific Northwest. His interest in sharing science with others led him to pursue a teaching license, and he taught science for six years in rural eastern Oregon. Currently, Dr. Bell teaches pre-service teachers, provides professional development for practicing teachers, and researches and develops curricular materials. He has received numerous teaching awards as well as the Early Career Research Award from the National Association for Research in Science Teaching. Dr. Bell has authored dozens of research articles and books, and his two primary areas of research focus on teaching and learning about the nature of science and assessing the impact of educational technology.

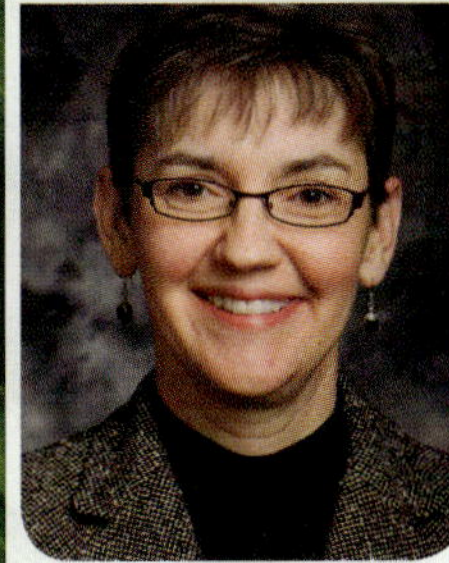

Kathy Cabe Trundle, Ph.D.

Associate Professor of Early Childhood Science Education,
The School of Teaching and Learning,
The Ohio State University, Columbus, Ohio

SCIENCE

Dr. Kathy Cabe Trundle has enjoyed her work in science education programs for more than 25 years, including 10 years as a public school teacher. Her research focuses on children and adults' understandings of Earth and space science concepts, including misconceptions and instructional interventions that promote scientific ideas and understandings. Her publications include research reports, professional articles, and curriculum materials. She is also very active in the development of science teacher education and professional development programs. Dr. Cabe Trundle was the recipient of the 2008 Outstanding Teacher Educator of the Year, presented by the Association for Science Teacher Education, an international organization.

David W. Moore, Ph.D.

Professor of Education, College of Teacher Education and Leadership,
Arizona State University, Tempe, Arizona

LITERACY

Dr. David W. Moore taught reading in Arizona public schools before entering college teaching. He currently is in the Division of Educational Leadership and Innovation at Arizona State University where he researches and teaches courses in literacy across the content areas. He is actively involved with several professional associations. His thirty-year publication record balances research reports, professional articles, book chapters, and books. Noteworthy publications include a history of content area literacy instruction and a quantitative and qualitative review of graphic organizer research. He currently is preparing the sixth edition of *Developing Readers and Writers in the Content Areas: K–12*.

Program Reviewers

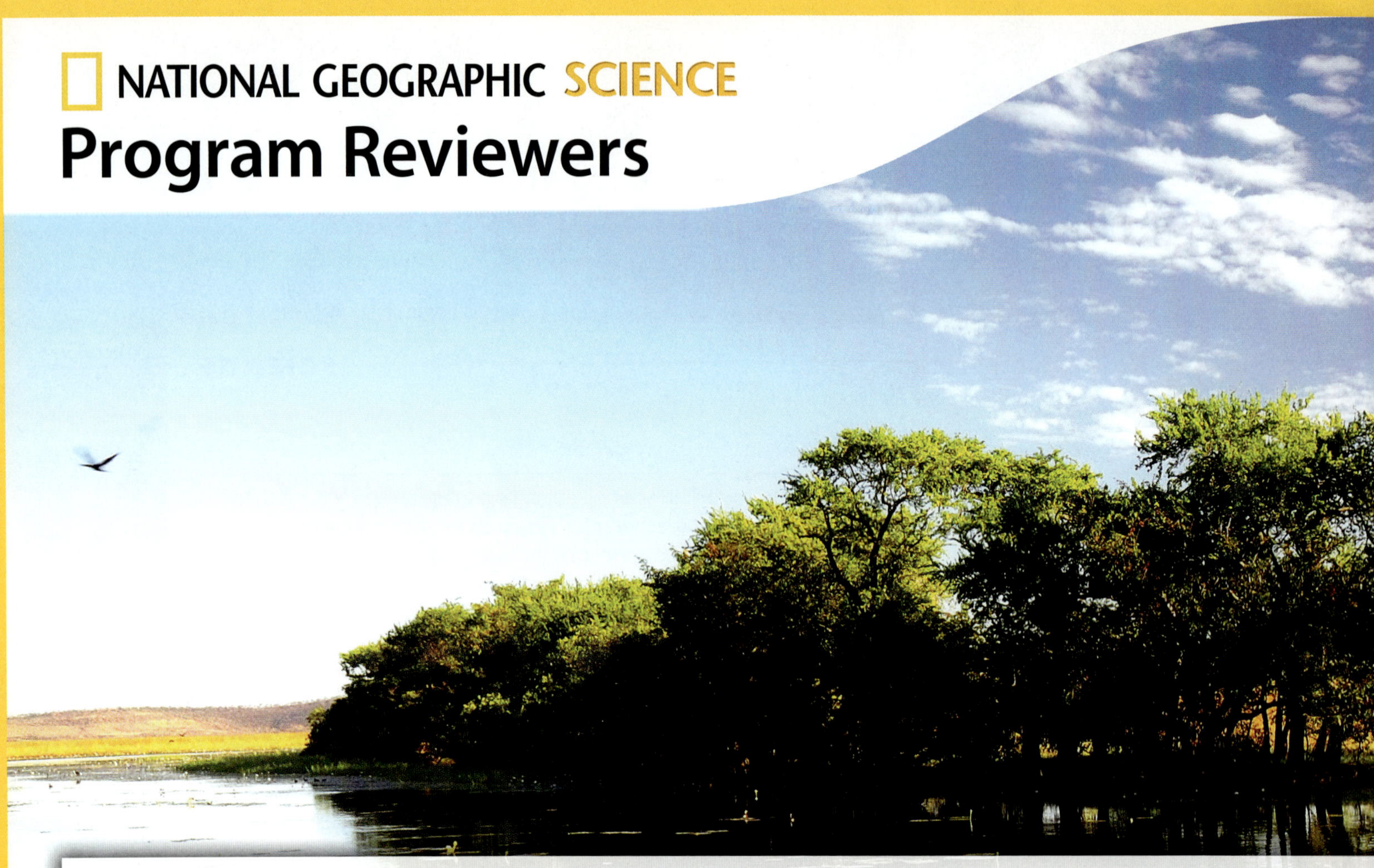

Amani Abuhabsah
Teacher, Dawes Elementary
Chicago, IL

Maria Aida Alanis, Ph.D.
Elementary Science Instructional
Coordinator, Austin Independent
School District
Austin, TX

Jamillah Bakr
Science Mentor Teacher,
Cambridge Public Schools
Cambridge, MA

Gwendolyn Battle-Lavert
Assistant Professor of Education,
Indiana Wesleyan University
Marion, IN

Carmen Beadles
Retired Science Instructional
Coach, Dallas Independent
School District
Dallas, TX

Andrea Blake-Garrett, Ed.D.
Science Educational Consultant,
Newark, NJ

Lori Bowen
Science Specialist, Fayette County
Schools
Lexington, KY

Pamela Breitberg
Lead Science Teacher, Zapata
Academy
Chicago, IL

Carol Brueggeman
K–5 Science/Math Resource
Teacher, District 11
Colorado Springs, CO

Miranda Carpenter
Teacher, MS Academy Leader,
Imagine School
Bradenton, FL

Samuel Carpenter
Teacher, Coonley Elementary
Chicago, IL

Diane E. Comstock
Science Resource Teacher,
Cheyenne Mountain School
District
Colorado Springs, CO

Kelly Culbert
K–5 Science Lab Teacher,
Princeton Elementary
Orange County, FL

Karri Dawes
K–5 Science Instructional Support
Teacher, Garland Independent
School District
Garland, TX

Richard Day
Science Curriculum Specialist,
Union Public Schools
Tulsa, OK

Michele DeMuro
Teacher/Educational Consultant,
Monroe, NY

Richard Ellenburg
Science Lab Teacher, Camelot
Elementary
Orlando, FL

Beth Faulkner
Brevard Public Schools Elementary
Training Cadre, Science Point
of Contact, Teacher, NBCT,
Apollo Elementary
Titusville, FL

Kim Feltre
Science Supervisor, Hillsborough
School District
Newark, NJ

Judy Fisher
Elementary Curriculum
Coordinator, Virginia Beach
Schools
Virginia Beach, VA

Anne Z. Fleming
Teacher, Coonley Elementary
Chicago, IL

Becky Gill, Ed.D.
Principal/Elementary Science
Coordinator, Hough Street
Elementary
Barrington, IL

Rebecca Gorinac
Elementary Curriculum Director,
Port Huron Area Schools
Port Huron, MI

Anne Grall Reichel Ed.D.
Educational Leadership/
Curriculum and Instruction
Consultant,
Barrington, IL

Mary Haskins, Ph.D.
Professor of Biology, Rockhurst
University
Kansas City, MO

Arlene Hayman
Teacher, Paradise Public School
District
Las Vegas, NV

DeLene Hoffner
Science Specialist, Science
Methods Professor, Regis
University, Academy 20 School
District
Colorado Springs, CO

Cindy Holman
*District Science Resource Teacher,
 Jefferson County Public Schools*
Louisville, KY

Sarah E. Jesse
*Instructional Specialist for Hands-
on Science, Rutherford County
 Schools*
Murfreesboro, TN

Dianne Johnson
*Science Curriculum Specialist,
 Buffalo City School District*
Buffalo, NY

Kathleen Jordan
Teacher, Wolf Lake Elementary
Orlando, FL

Renee Kumiega
*Teacher, Frontier Central School
 District*
Hamburg, NY

Edel Maeder
*K–12 Science Curriculum
 Coordinator, Greece Central
 School District*
North Greece, NY

Trish Meegan
Lead Teacher, Coonley Elementary
Chicago, IL

Donna Melpolder
*Science Resource Teacher,
 Chatham County Schools*
Chatham, NC

Melissa Mishovsky
*Science Lab Teacher, Palmetto
 Elementary*
Orlando, FL

Nancy Moore
Educational Consultant,
Port Stanley, Ontario, Canada

Melissa Ray
Teacher, Tyler Run Elementary
Powell, OH

Shelley Reinacher
*Science Coach, Auburndale
 Central Elementary*
Auburndale, FL

Kevin J. Richard
*Science Education Consultant,
 Office of School Improvement,
 Michigan Department of
 Education*
Lansing, MI

Cathe Ritz
Teacher, Louis Agassiz Elementary
Cleveland, OH

Rose Sedely
*Science Teacher, Eustis Heights
 Elementary*
Eustis, FL

Robert Sotak, Ed.D.
*Science Program Director,
 Curriculum and Instruction,
 Everett Public Schools*
Everett, WA

Karen Steele
*Teacher, Salt Lake City School
 District*
Salt Lake City, UT

Deborah S. Teuscher
*Science Coach and Planetarium
 Director, Metropolitan School
 District of Pike Township*
Indianapolis, IN

Michelle Thrift
*Science Instructor, Durrance
 Elementary*
Orlando, FL

Cathy Trent
*Teacher, Ft. Myers Beach
 Elementary*
Ft. Myers Beach, FL

Jennifer Turner
Teacher, PS 146
New York, NY

Flavia Valente
*Teacher, Oak Hammock
 Elementary*
Port St. Lucie, FL

Deborah Vannatter
*District Coach, Science Specialist,
 Evansville Vanderburgh School
 Corporation*
Evansville, IN

Katherine White
*Science Coordinator, Milton
 Hershey School*
Hershey, PA

Sandy Yellenberg
*Science Coordinator, Santa Clara
 County Office of Education*
Santa Clara, CA

Hillary Zeune de Soto
*Science Strategist, Lunt
 Elementary*
Las Vegas, NV

Explorers and Scientists

These explorers and scientists help students understand real-world science, the nature of science, and science inquiry.

CONSTANCE ADAMS

National Geographic Emerging Explorer
Space Architect

"When you have a brand new problem, you need as many tools as you can get. Who knows? An approach from a very different field might give you the insight you need. For example, I'm working to forge communication between advanced engineering and consumer product design to bring more user-centered designs to aerospace."

STEPHON ALEXANDER, Ph.D.

National Geographic Emerging Explorer
Theoretical Physicist

"My childhood was full of surprises. It taught me the idea of embracing the unknown. I cope with unexpected events by making up theories about why they may be happening."

THOMAS TAHA RASSAM "T.H." CULHANE

National Geographic Emerging Explorer
Urban Planner

"During rain forest ecology fieldwork with the Dayak of Borneo and the Maya Itza of Guatemala's jungle villages, I witnessed a culture that used every part of the environment to survive and thrive. This inspired me to rethink urban living along those same ecological principles. Now I want to bring that message to the rest of the world."

LUKE DOLLAR, Ph.D.

National Geographic Emerging Explorer
Conservation Scientist

"When you're in the field, it's muddy, sweaty, stinky, gritty—there's no glamour to it at all. But it's great fun. I wake up every morning knowing I'm one of the luckiest guys on Earth because I'm doing exactly what I want to do, and it's going to make a difference."

MARIANNE DYSON

Science Writer and Former NASA
Flight Controller

"Children who learn to observe, describe, and predict forces in the world will develop the skills to tackle the challenges of a future increasingly dependent on technology."

MARIA FADIMAN, Ph.D.

National Geographic Emerging Explorer
Ethnobotanist

"I was born with a passion for conservation and a fascination with indigenous cultures. Ethnobotany lets me bring it all together. On my first trip to the rain forest, I met a woman who was in terrible pain because no one in her village could remember which plant would cure her. I saw that knowledge was truly being lost, and in that moment, I knew this was what I wanted to do with my life."

BEVERLY GOODMAN, Ph.D.

National Geographic Emerging Explorer
Geo-Archaeologist

"We can learn about our past, present, and future by studying the sea. Coastal regions present a major challenge. I hope the clues I am collecting help avert future catastrophe."

MADHULIKA GUHATHAKURTA, Ph.D.

NASA Astrophysicist

"Earth is inside the atmosphere of the sun. The sun is so big that if you dropped the entire Earth onto the sun's surface, it would barely make a decent sunspot."

ALBERT YU-MIN LIN, Ph.D.

National Geographic Emerging Explorer
Material Scientist, Digital Explorer

"I spent much of my young adulthood dreaming of exploration and have realized that goals can only be reached by many different approaches."

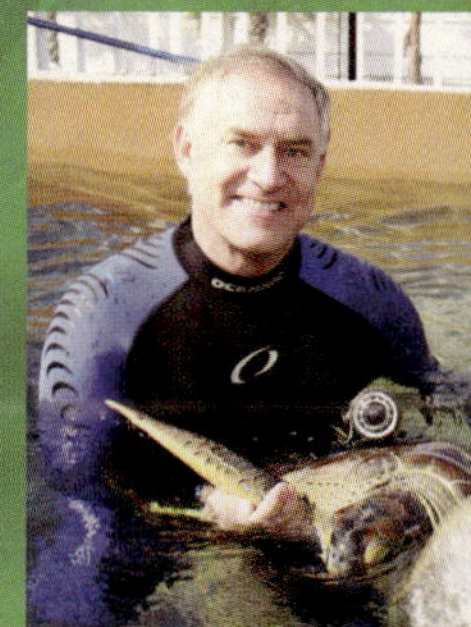

GREG MARSHALL

National Geographic Filmmaker
Marine Biologist, Conservationist, Inventor

"Science is a great adventure in exploration, discovery, and learning new things. We want to expand our knowledge and inspire others to help conserve and protect animals and their habitats."

MIREYA MAYOR, Ph.D.

National Geographic Emerging Explorer
Primatologist, Conservationist

"The more questions I asked, the more it became clear to me that much about our natural world still remained a mystery."

AINISSA RAMIREZ, Ph.D.

Physicist

"In a technologically-driven society, we need citizens who can comprehend science. Knowledge of science is necessary in order to fully appreciate the world's beauty."

TIM SAMARAS

National Geographic Emerging Explorer
Severe-Storms Researcher

"I started to love science when I was a young child. Now I am studying tornadoes to help unlock their secrets for a better understanding of how and why they form."

TIERNEY THYS, Ph.D.

National Geographic Emerging Explorer
Marine Biologist, Filmmaker

"I think what it takes to be an explorer is simply to never lose your curiosity, never lose your desire to learn more, to ask questions, and to keep pushing deeper and deeper into what you're studying. It's a vital time to be an explorer; there's never been a better, more important time."

KATEY WALTER ANTHONY, Ph.D.

National Geographic Emerging Explorer
Aquatic Ecologist, Biogeochemist

"We are researching the greenhouse gas that could have the most powerful effect of all on global warming. It's a worldwide responsibility to reduce our carbon footprint and its effects on the atmosphere."

NATIONAL GEOGRAPHIC SCIENCE

K-5

Take your students from the classroom to the world!

Grade K

Life Science
- Plants
- Animals

Earth Science
- Day and Night
- Weather and Seasons

Physical Science
- Observing Objects
- How Things Move

Grades 1-2

- Living Things
- Plants and Animals
- Sun, Moon, and Stars
- Land and Water
- Pushes and Pulls
- Properties
- Habitats
- Life Cycles
- Weather
- Rocks and Soil
- Solids, Liquids, and Gases
- Forces and Motion

Grades K–5

myNG**connect**.com

Grade 3

Big Ideas Books

Science Inquiry and Writing Book

Explore on Your Own Books

Grade 4

Big Ideas Books

Science Inquiry and Writing Book

Explore on Your Own Books

Grade 5

Big Ideas Books

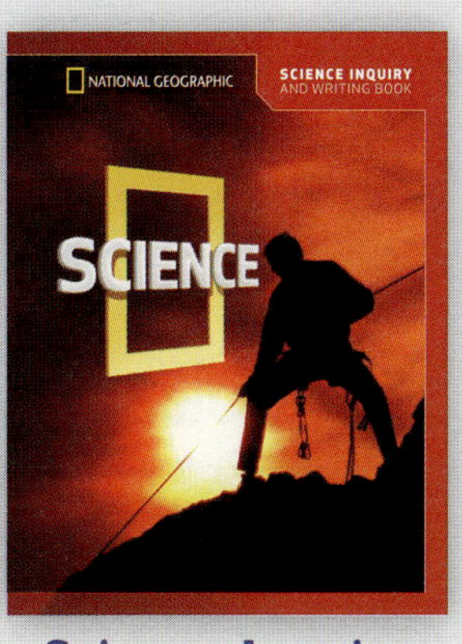

Science Inquiry and Writing Book

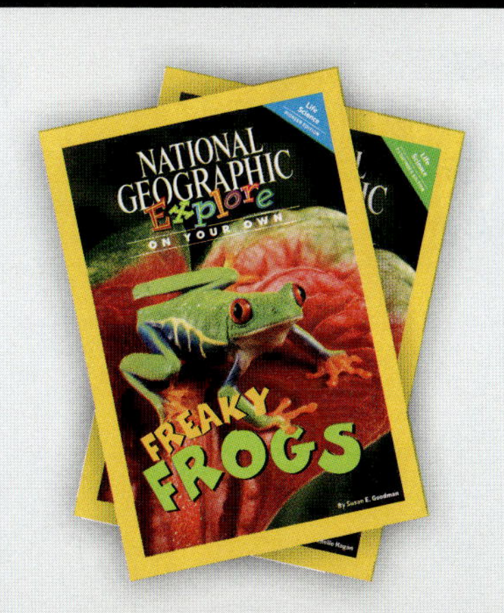

Explore on Your Own Books

STUDENT
Unit Components

Science Inquiry and Writing Book

Big Ideas Books

Unit Launch Videos

eEdition

TECHTREK
myNGconnect.com

Explore on Your Own Books

National Geographic Digital Library

Enrichment Activities

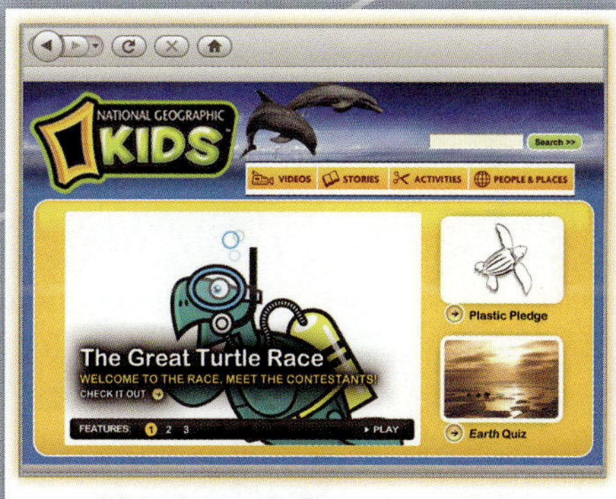

National Geographic Kids Web Site

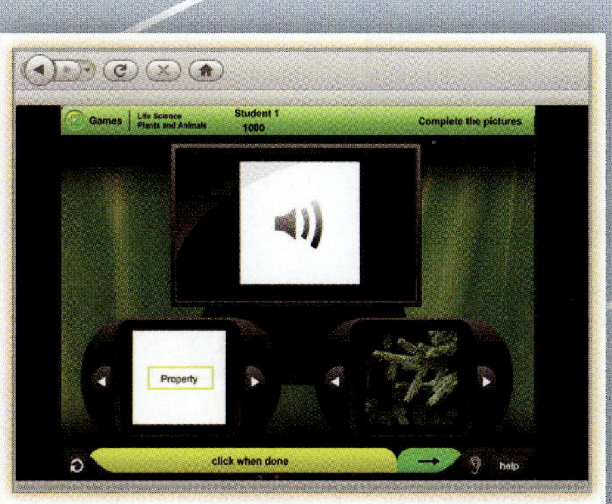

Vocabulary Games

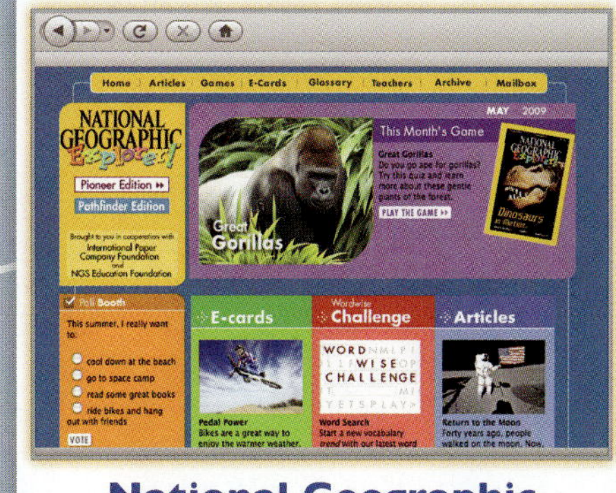

National Geographic Explorer! Web Site

NATIONAL GEOGRAPHIC
SCIENCE

TEACHER
Unit Components

Teacher's Editions

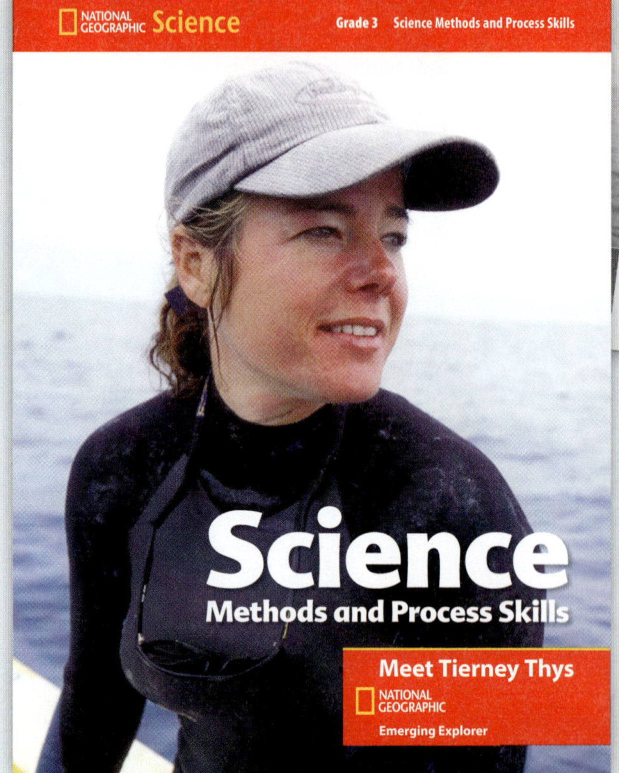

Science Methods and Process Skills Book / Teacher's Guide

www.myNGconnect.com

Integrated Online System

TECHTREK
myNGconnect.com

Assessment Handbook

Science Inquiry Kit

Learning Masters

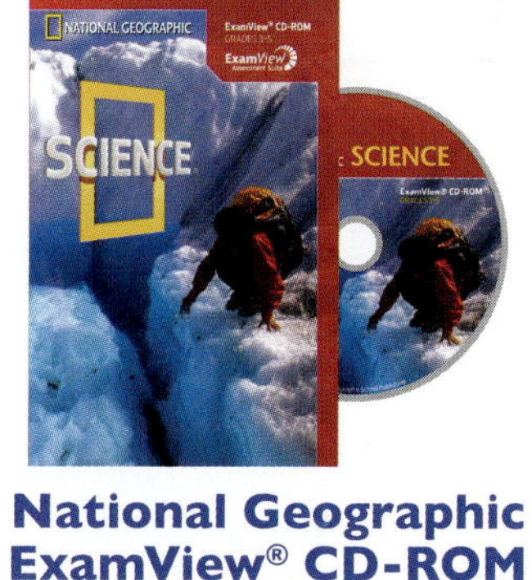

National Geographic ExamView® CD-ROM

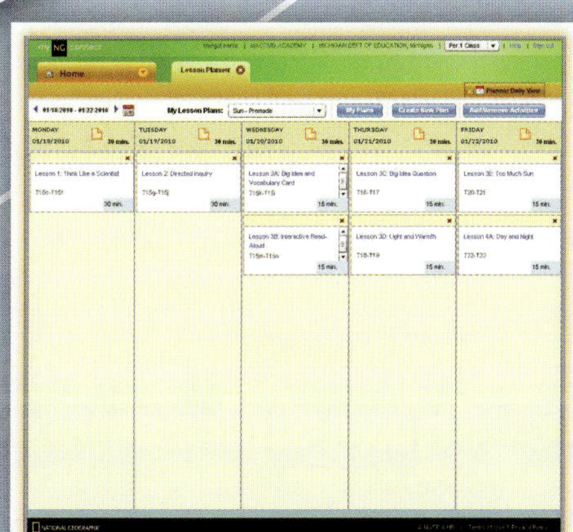

Online Lesson Planner

National Geographic Digital Library

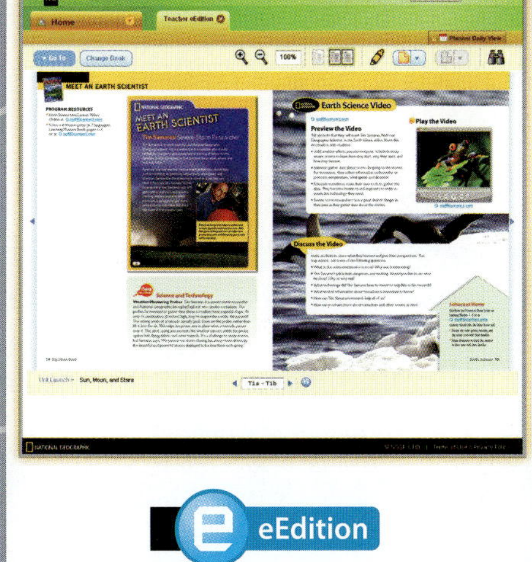

eEdition

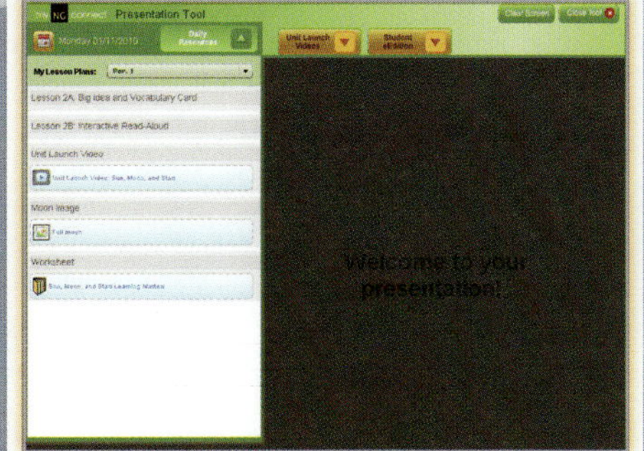

National Geographic Presentation Tool

Assessment

Assessment Design

National Geographic Science assessments have been designed so that frequent, varied assessment informs instruction every step of the way. Chapter and Benchmark Tests provide a window into student thinking about scientific concepts throughout the instructional cycle.

Instruct

Develop an understanding of and provide explicit and systematic instruction in:

- The Big Ideas in Science
- Science Academic Vocabulary
- Scientific Inquiry

Assess to Monitor Progress

Chapter Tests provide opportunities for students to apply their understanding of key concepts. Use **Chapter Tests** for timely information about student progress as you deliver instruction.

Show Success

The **Benchmark Test** measures a student's overall progress in understanding the Life, Earth, and Physical units' Big Ideas.

Assessment Tools

National Geographic Science offers an array of assessments in a variety of formats.

ASSESSMENT TOOLS	Assessment Handbook	PDFs myNGConnect.com	Exam*View*®
CHAPTER TESTS • Chapter Tests provide immediate feedback on students' understanding of standards-based scientific concepts. • The tests are administered at the end of each chapter to provide an early indicator of the students' progress.	✓	✓	✓
CHAPTER SELF-ASSESSMENT • The Chapter Self-Assessment empowers students to rate their own understanding of the chapter's concepts and share their opinions about future interests. • The self-assessment is completed by students before they take the Chapter Test.	✓	✓	
BENCHMARK TEST • The Benchmark Test measures students' progress toward understanding the unit's Big Ideas. • The test is administered at the end of the unit to measure the students' understanding of the science standards taught in the unit.	✓	✓	✓
INQUIRY RUBRICS • Inquiry Rubrics assess students' performance of skills in Inquiry activities. • The rubrics are completed after students have finished each Inquiry activity.	✓	✓	
INQUIRY SELF-REFLECTIONS • Inquiry Self-Reflections engage students in evaluating their own performance of inquiry skills and in sharing their opinions about the activity process. • The self-reflections are completed by students after they have finished each Inquiry activity.	✓	✓	

Unit Overview and Pacing Guide

Nature of Science / Science Notebook

DAY 1	20 minutes	**Setting Up a Science Notebook** • *Big Ideas Book*	*page SN3*

eEdition

Physical Science Big Ideas Book and Inquiry

Unit Launch

DAY 1	30 minutes	**What Is Physical Science?**	*page T2*
		Meet a Scientist / Unit Launch Video • *Big Ideas Book*	*page T4*

Chapter 1

DAY 2	20 minutes	**Explore Activity** *Investigate Physical Properties* • *Science Inquiry and Writing Book, page 166*	*page T5e*
DAYS 3–6	40 minutes each	**How Can You Describe and Measure Matter?** • *Big Ideas Book*	*page T6–T7*
DAY 7	20 minutes	**Directed Inquiry** *Investigate Volume and Mass* • *Science Inquiry and Writing Book, page 170*	*page T23a*
DAY 8	50 minutes	**Conclusion and Review**	*page T24*
		NATIONAL GEOGRAPHIC **PHYSICAL SCIENCE EXPERT** Physical Chemist	*page T26*
		NATIONAL GEOGRAPHIC **BECOME AN EXPERT** *The Properties of Graphene* • *Big Ideas Book*	*page T28–T...*

Chapter 2

DAY 9	15 minutes	**Directed Inquiry** *Investigate Water and Temperature* • *Science Inquiry and Writing Book, page 174*	*page T37e*
DAYS 10–13	40 minutes each	**What Are States of Matter?** • *Big Ideas Book*	*page T38–T3...*
DAY 14	50 minutes	**Think Like a Scientist** Math in Science • *Science Inquiry and Writing Book, page 178*	*page T55a*
		Guided Inquiry *Investigate Melting* • *Science Inquiry and Writing Book, page 182*	*page T55e*
DAY 15	50 minutes	**Conclusion and Review**	*page T56*
		NATIONAL GEOGRAPHIC **PHYSICAL SCIENCE EXPERT** Ice Sculptor	*page T58*
		NATIONAL GEOGRAPHIC **BECOME AN EXPERT** Sweden: Ice Hotel Construction • *Big Ideas Book*	*page T60–T...*
		Explore ON YOUR OWN *Recycling Rules!*	*page T68a*

PIONEER

PATHFINDER

<table>
<tr><td>Chapter 3</td><td>DAY 16</td><td>30 minutes</td><td>Directed Inquiry   Investigate Force and Motion
• Science Inquiry and Writing Book, page 186</td><td>page T69e</td></tr>
<tr><td></td><td>DAYS 17–22</td><td>40 minutes each</td><td>How Does Force Change Motion?
• Big Ideas Book</td><td>page T70–T71</td></tr>
<tr><td></td><td>DAY 23</td><td>30 minutes</td><td>Guided Inquiry   Investigate Motion and Position
• Science Inquiry and Writing Book, page 190</td><td>page T95a</td></tr>
<tr><td></td><td>DAY 24</td><td>50 minutes</td><td>Conclusion and Review
NATIONAL GEOGRAPHIC PHYSICAL SCIENCE EXPERT   Materials Scientist
NATIONAL GEOGRAPHIC BECOME AN EXPERT   High-Speed Trains
• Big Ideas Book</td><td>page T96
page T98
page T100–T101</td></tr>
<tr><td>Chapter 4</td><td>DAY 25</td><td>20 minutes</td><td>Directed Inquiry   Investigate Energy of Motion
• Science Inquiry and Writing Book, page 194</td><td>page T109e</td></tr>
<tr><td></td><td>DAYS 26–31</td><td>40 minutes each</td><td>What Is Energy?
• Big Ideas Book</td><td>page T110–T111</td></tr>
<tr><td></td><td>DAY 32</td><td>30 minutes</td><td>Guided Inquiry   Investigate Vibrations and Sound
• Science Inquiry and Writing Book, page 198</td><td>page T135a</td></tr>
<tr><td></td><td>DAY 33</td><td>50 minutes</td><td>Conclusion and Review
NATIONAL GEOGRAPHIC PHYSICAL SCIENCE EXPERT   Product Designer
NATIONAL GEOGRAPHIC BECOME AN EXPERT   Use Some Energy: Have Some Fun!
• Big Ideas Book</td><td>page T136
page T138
page T140–T141</td></tr>
<tr><td></td><td></td><td></td><td>Explore ON YOUR OWN   The Energy of Water   PIONEER   PATHFINDER</td><td>page T148a</td></tr>
</table>

NATIONAL GEOGRAPHIC

Unit Overview and Pacing Guide, continued

Chapter 5	DAY 34	20 minutes each	**Directed Inquiry** Investigate Light and Heat • Science Inquiry and Writing Book, page 202	page T149e
	DAYS 35–38	40 minutes each	**What Is Light?** • Big Ideas Book	page T150–T151
	DAY 39	30 minutes	**Guided Inquiry** Investigate Light and Objects • Science Inquiry and Writing Book, page 206	page T167a
	DAY 40	50 minutes	**Conclusion and Review**	page 168
			NATIONAL GEOGRAPHIC **PHYSICAL SCIENCE EXPERT** Inventor	page T170
			NATIONAL GEOGRAPHIC **BECOME AN EXPERT** Laser: A Special Kind of Light • Big Ideas Book	page T172–T173
			Explore ON YOUR OWN Scope This Out!	page T180a
Unit Wrap Up	DAY 41	50 minutes	**Open Inquiry** Do Your Own Investigation • Science Inquiry and Writing Book, page 210	page T180j
			Write Like a Scientist Write About an Investigation • Science Inquiry and Writing Book, page 212	page T180l
	DAY 42	30 minutes	**Think Like a Scientist** How Scientists Work • Science Inquiry and Writing Book, page 220	page T180r
	DAY 43	50 minutes (Select one unit project.)	**Create a Web Page About Matter**	page T180v
			Make a Timeline of Water Changes	page T180w
			Demonstrate the Forms of Energy	page T180x
			Make a Movie About Light • Big Ideas Book	page T180y

Contents

Grade 3	Grade 4	Grade 5
CLASSIFY	**PREDICT/HYPOTHESIZE**	**DESIGN EXPERIMENTS**
• There is often no single "right" answer in science.	• Scientific theories provide the base upon which predictions and hypotheses are built.	• There is no single, scientific method that all scientists follow. • There are a number of ways to do science.

Why Use a Science Notebook?

CLASSIFY AND RECORD DATA

A science notebook is a place for students to record data about classifying, much like scientists record their data about classifying. Science notebooks make science more meaningful for students because they are able to classify, record, and reflect on their work. Students will learn through classifying that there is often no single "right" answer in science, an important tenet of the Nature of Science.

DEVELOP INQUIRY SKILLS

Using a science notebook can help students develop inquiry skills and practice science process skills such as observe, sort, infer, predict, collect data, and conclude. Classifying is a key process skill that will help students understand the Nature of Science in *National Geographic Science*.

ENHANCE UNDERSTANDING OF SCIENCE CONTENT

A science notebook can help students understand science content. Writing about what they learn helps students understand science concepts and improves organizational skills. Encourage students to use science notebooks to present findings to share with others.

DIFFERENTIATE INSTRUCTION

A science notebook provides a way to support differentiated instruction. You can design science notebook use in your classroom so that students move at their own pace. This may include flexible time frames for completing work or small groups to support student learning.

EVALUATE

You can use the science notebooks for evaluation. Review the notebooks and see what understandings students have reached and what material you need to teach again. You may have students use their notebooks for self-reflection by marking places where they think they have successfully thought about and accomplished a section.

Setting Up a Science Notebook

Guide your students to set up a structure that will support their learning and your evaluation of science understanding.

1 Introduce the science notebook to students.

Say: **Like scientists, you will use this science notebook for your science work. It's your own notebook to use every time you do science. Like a real scientist, you will write notes in it, draw pictures, write about experiments, and write vocabulary words.**

2 Have students choose an organization style.

DAILY RECORD: Students can make entries, in order, as they move through the unit. They can take notes about the Nature of Science, content lessons, and inquiry activities.

SEPARATE SECTIONS: Students can create separate sections for classifying, inquiry activities, vocabulary, notetaking, questions, reflections, and so on.

3 Have students choose the notebook.

Some students may want to put looseleaf paper in a binder.

Other students may choose a spiral notebook, a two-pocket folder, or a composition book. Still other students might choose a combination of these.

4 Have students make and/or decorate the cover.

Instruct students to include their names and grade on the cover.

5 Have students write "Contents" on the first right-hand page.

Suggest that students save several pages for the Table of Contents and fill in the information as they continue through the unit.

6 Tell students that they should number the pages as they go along.

Tell students to add the appropriate page numbers to the Table of Contents as they proceed through the unit.

TEACHING TIPS

- Have students use the science notebook daily or on a regular basis. Assign a student to pass them out at the beginning of each science period. Collecting the notebooks each day and keeping them in a central place may help them last longer.

- Students may want to tape or staple a ribbon into the notebook and use it to mark their place.

- Tell students to use labels and/or captions for all drawings.

- Students can include graphic organizers to help them understand and organize information. A variety of graphic organizer options are found in the Learning Masters Book.

- You may want to use sticky notes to indicate corrections or additions that students need to make in their science notebook.

What's In a Science Notebook?

Science notebook references are included throughout the Teacher's Edition. A science notebook is a place to record data about classifying. Students can record their data about classifying in a variety of ways:

STUDENT DRAWINGS

- Have students draw pictures to illustrate the Nature of Science (science is based on classifying) and their understanding of science concepts.

TABLES, CHARTS, AND GRAPHS

- Have students draw tables, charts, and graphs to record information or data.

NOTES

- Encourage students to jot down notes from each lesson in their science notebook. They can include graphic organizers, charts, lists, questions, and sketches. Suggestions for notetaking appear in the Teacher's Edition at point of use.

States of Matter

Solid	Liquid	Gas
ice	water	water vapor

States of Matter

This icicle is a solid.

How long will it take for this solid icicle to become liquid if I break it off and bring it inside?

COLLECTED OBJECTS

- photos
- magazine and newspaper clippings
- examples of matter

REFLECTIVE AND ANALYTICAL ENTRIES

- You might want to give students prompts or frames to guide them as they write in their science notebook. For example:

  I want to find out __________.

  If __________, then __________.

  What would happen if I changed __________?

  I think __________ because __________.

  The most important thing I learned in this chapter was __________.

  I was surprised to learn __________.

OTHER QUESTIONS STUDENTS HAVE

- Students may have a variety of questions. Have them record questions and help them research the answers.

Integrated Technology

- **Digital Camera** Suggest that students use digital cameras to take photos. The photos can be included in their science notebook.
- **Computer Presentation** Encourage students to share their ideas. They can share their notebook with each other, present their ideas to the class, or talk about their ideas in small groups. They can also make computer presentations as appropriate.

What's In a Science Notebook? continued

STUDENT REPORTS

- Students can answer one or two reflective questions at the end of each chapter. Or you can assign special projects or reports for them to write in their science notebook.

Name ______________________________ Date ______________ 

Chapter 2 Science Vocabulary

Write the word that fits each description. Use the words in the box.
You will use some words more than once.

states of matter solid liquid gas evaporation condensation

1. Always keeps its shape: _solid_

2. Gas changes to a liquid: _condensation_

3. Solid, liquid, gas: _states of matter_

4. Takes the shape of its container: _liquid_

5. Liquid changes to a gas: _evaporation_

6. Spreads out to fill a space: _gas_

7. Water vapor turns to drops of water: _condensation_

8. Liquid water turns to water vapor: _evaporation_

Complete the chart with vocabulary words.

States of Matter	Ways Matter Changes State
solid	evaporation
liquid	condensation
gas	

Learning Master   **195**   Physical Science
© NGSP & HB

SCIENCE ACADEMIC VOCABULARY

- You can have students include the Vocabulary Learning Masters in their science notebook. See page SN11 for other suggestions for using the science notebook with Science Academic Vocabulary.

Name ______________________________ Date ______________

Directed Inquiry

Investigate Water and Temperature

What happens to water as the temperature changes?

Predict

What will happen to the water in the bags when you put the bags in the freezer in step 2?

Predictions will vary. Students may predict that the water in both bags will freeze, and its shape will no longer be able to change.

What will happen to the water in the bags when you put the bags in sunlight in step 5?

Predictions will vary. Students may predict that the water in the open Bag 1 will evaporate, but the water in the closed Bag 2 will evaporate and condense inside the closed bag.

Learning Master   **191**   Physical Science
© NGSP & HB

INQUIRY ACTIVITIES

- Use Learning Masters or have students write notes about the activities in their notebook. See pages SN8–SN10 for suggestions for using a science notebook with inquiry activities.

Using a Science Notebook for Inquiry Activities

The inquiry activities in *National Geographic Science* provide an opportunity for students to ask questions and do investigations much like scientists do. Writing what they learn will help students understand why they are doing the activity and what it teaches them.

ASK A QUESTION

• Every inquiry activity begins with a question that shows the purpose of the activity. Have students write the question in their science notebook.

BUILD VOCABULARY

• Have students write the Science Process Vocabulary words and their definitions in their science notebook.

MAKE A PREDICTION

• Have students write a statement predicting what will happen in the activity. Encourage students to use their prior knowledge and experience to make the prediction.

What happens to water as the temperature changes?

observe:
When you observe, you use your senses to learn about objects or events.
compare:
When you compare, you tell how objects are alike and different.

I predict that the water will change state.

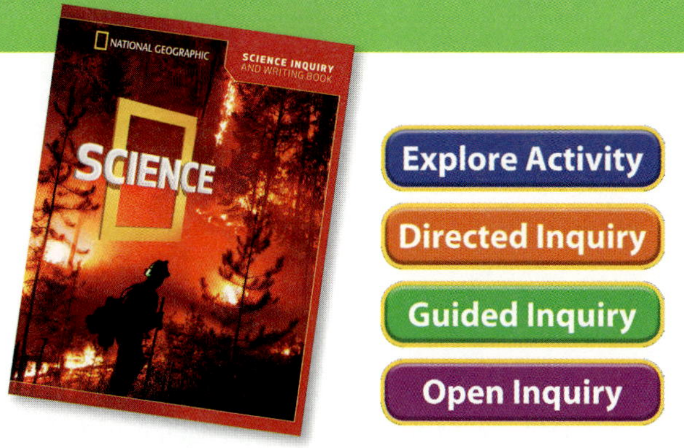

1. Use the graduated cylinder to measure 100 mL of water.

The water is frozen. Ice is a solid.

WHAT TO DO

- Have students write or draw the steps of the activity in their notebook.

CLASSIFY AND RECORD

- Have students record their data on classifying in the table on the Learning Master or have them draw and fill in their own table or graph. Students can write or make drawings to record their data on classifying.

Using a Science Notebook for Inquiry Activities continued

EXPLAIN AND CONCLUDE

- Have students examine data, or evidence, and use this evidence to classify, develop explanations, and draw conclusions about their results. New conclusions can then be made based on further observations.

- In doing these activities, students are ready to learn that scientific conclusions are based on evidence and can change with new evidence.

I can classify matter by what state it is in. The water that was in Bag 1 is a gas. The water in Bag 2 is liquid and some is gas.

THINK OF ANOTHER QUESTION

- Have students reflect about what they have learned. Then have them use their observations to think of other questions that they could study through an investigation. Have them write their questions in their science notebook.

How can I classify other kinds of matter?

Science Academic Vocabulary

Students can use their science notebook as their own vocabulary resource.

Have students designate a special section of the science notebook for Science Academic Vocabulary.

- Write the vocabulary words and their definitions on a Science Word Wall for students to copy.

- Encourage students to use the graphic organizers from the Learning Masters Book or draw their own for recording vocabulary.

  - Students can write the word, draw a picture, write a definition in their own words, and write a sentence using the word.
  - Students can fill in a Word Web.
  - Students can make their own vocabulary cards or use the Vocabulary Cards Graphic Organizer, writing the word on one side and a picture and/or a definition on the other side.

- Encourage students to write ideas about vocabulary in their science notebook.

Using a Science Notebook for Unit Activities

Students can summarize and synthesize the unit's content with culminating unit entries.

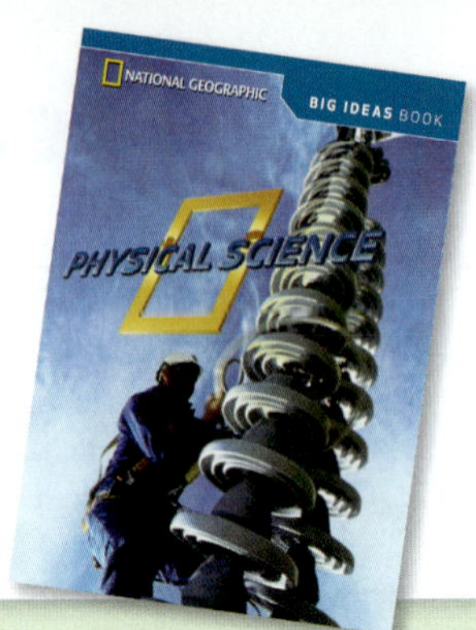

STUDENT REFLECTIONS

At the end of each chapter, help students think about what they learned. Writing will help them understand the Nature of Science, science content, and science inquiry. This helps students relate the three parts of science to their daily lives. Ask them to write answers to questions such as these in their science notebook:

- What is the main idea of this chapter?
- What is the most surprising thing I learned?
- What is the most important thing I learned?
- What are some examples of matter changing state?
- How do I use different states of matter in my life?
- What states of matter do I observe around me?

USING THE SCIENCE NOTEBOOK WITH THE UNIT WRAP UP

- Use the science notebook as a place to respond to the Unit Wrap Up projects found on pages T180v–T180y.

1

CONTENTS

❶ Introduce

Tap Prior Knowledge

- Have students share what they know about the word *physical*.

- Ask: **What do you think physical scientists study?** (Sample answers: things that move, things you can touch, feel, and see)

Set a Purpose and Read

- Read the heading. Tell students that this is a preview of what they will learn in the *Physical Science* unit.

- Have students read pages 2–3.

❷ Teach

How Can You Describe and Measure Matter?

- Read aloud the properties of matter listed on the page: size, shape, color, texture, hardness, mass, and volume. Ask: **What colors do you see on the hot air balloons?** (yellow, red, orange, green, blue, purple) **What shape are the balloons?** (round)

- Say: **The hot air balloons are made of matter. Describe the walls in this room by using two properties of matter.** (Sample answers: white, rough, square, tall, hard)

What Are States of Matter?

- Ask: **How have you used water today?** (drinking, bathing, swimming) **If the water looked like the droplets in the photo, it was liquid.**

- Ask: **What is an ice cube or a frozen lake?** (solid water) Say: **Solid is another state of water.**

- Ask: **What type of weather changes rainwater to snow?** (cold) Explain: **When temperatures go below freezing, water changes from liquid to solid.**

What Is Physical Science?

Physical science is the study of the physical world around you. This type of science investigates the properties of different objects, as well as how those objects interact with each other. Physical science includes the study of matter, motion and forces, and many kinds of energy, including light and electricity. People who study how all of these things work together are called physical scientists.

You will learn about these aspects of physical science in this unit:

HOW CAN YOU DESCRIBE AND MEASURE MATTER?

Matter is anything that has mass and takes up space. Physical scientists study all of the different properties of matter. These include size, shape, color, texture, and hardness, as well as mass and volume.

WHAT ARE STATES OF MATTER?

Matter can be found in many different states. Each of these states has its own special properties. Physical scientists study these different states of matter. They also study how temperature changes can cause matter to change from one state of matter to another.

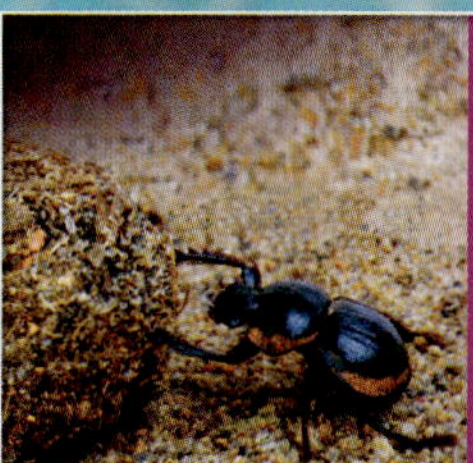

HOW DOES FORCE CHANGE MOTION?

Physical scientists study how objects move. They also study the forces that act on objects, and how those objects respond to different kinds of forces. Physical scientists also observe and measure the changes that forces cause.

2

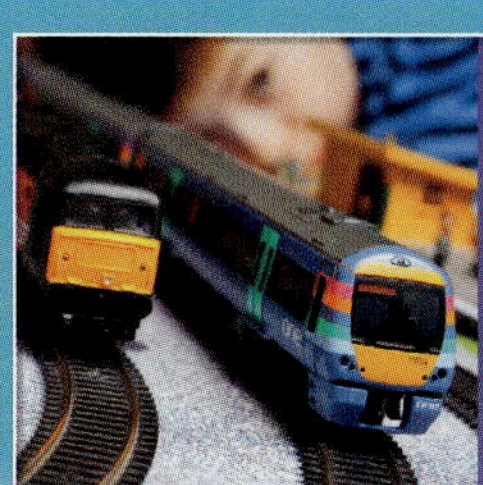

WHAT IS ENERGY?

Physical scientists study energy in all of its forms. They also learn about how energy can cause changes in the physical world. The different kinds of energy include light, sound, electrical, and mechanical.

WHAT IS LIGHT?

Light is a kind of energy that you can see. Physical scientists study the properties of light. They also learn about how light can change direction and even cause objects to heat up.

③

How Does Force Change Motion?

- Ask: **Think of a sport you have played. What things were moving during the sport?** (Sample answers: people running, balls rolling) **Did the moving objects keep moving? What stopped them?** (Sample answers: another player, a wall)

- Say: **Take a look at the photo.** Ask: **What is the bug doing?** (It is pushing something.) **What is the opposite of a push?** (a pull)

What Is Energy?

- Say: **Energy is very important. Mechanical energy helped each one of us get to school today.**

- Ask: **How do you think electrical energy helped you today?** (Sample answers: It powered my TV. It warmed my breakfast.)

- Say: **Name one type of energy that is important to you at this moment and explain why.** (Sample answers: light energy so I can see; sound energy so I can hear)

What Is Light?

- Say: **Light reflects off objects.** Ask: **What happens when the sun hits a shiny object?** (The object reflects the light.)

- Say: **Take a look at the photo.** Ask: **What do you think causes the shadows?** (The people are blocking that shape of light from hitting the ground.)

- Ask: **What do you think is the light source in the photo?** (the sun) **How do you know the sun causes objects to heat up?** (Sample answers: I feel the sun warm my skin. I've seen the sun melt ice.)

❸ Wrap Up

- Ask: **In summary, what are the things that physical scientists study?** (matter, including its properties and different states; motion and forces; many kinds of energy, including light and electricity)

- Ask: **Which of the physical science topics do you want to read more about? Why?** (Answers will depend on individuals.)

Differentiated Instruction

ELL **Language Support for Describing Parts of Physical Science**

BEGINNING	INTERMEDIATE	ADVANCED
Help students use language to describe parts of physical science. Ask yes/no questions: **Are ice and rain both water? Can you see light energy?**	Help students use Academic Language Frames to describe physical science. • _______ *is a plentiful substance on Earth.* • *Energy includes light, sound, electrical, and* ____.	Have students use Academic Language Stems to describe parts of physical science. • *You can describe matter using …* • *The physical world …*

PROGRAM RESOURCES

- *Physical Science* Unit Launch Video:
  Online at **myNGconnect.com**
- Science at Home Letter (in 7 languages):
  Learning Masters Book, pages 173–179,
  or at **myNGconnect.com**

Constance Adams: Space Architect

Constance Adams is a space architect with NASA and a National Geographic Emerging Explorer. One of her first projects with NASA was TransHab, designed to be a transit habitat for the first human mission to Mars.

Constance and her team of experts bring together innovations from diverse disciplines such as architecture, engineering, industrial design, and sociology with the hopes of solving complex design issues. "When you have a brand-new problem, you need as many tools as you can get. Who knows—an approach from a very different field might give you the insight you need. For example, I'm working to forge communication between advanced engineering and consumer-product design to bring more user-centered designs to aerospace," says Constance.

As a space architect, Constance works with space agencies, including NASA, to design and build components for spacecraft and the International Space Station.

Science and Technology

TransHab, Transit Habitat to Mars Constance Adams is a space architect and National Geographic Emerging Explorer that works for NASA. Adams set out to design a habitat that could start out small for an easier launch and then expand once it was in space. Adams's multidisciplinary background has helped her come up with many solutions while working at NASA. She has a bachelor's degree in social studies and a master's degree in architecture. Adams likes to bring together ideas from many different fields to solve design problems on aerospace projects, such as TransHab.

Physical Science Video

myNGconnect.com

Preview the Video

Tell students that they will meet Constance Adams, National Geographic Emerging Explorer, in the *Physical Science* video. Share this information with students:

- Habitation of a spacecraft going to Mars presents many challenges. Vehicles and other equipment need to be designed to withstand being constantly showered by very small particles.

- Scientists also need to find a way to protect the crew from radiation. They need to make the bulky space suits flexible so that the crew can move around freely.

- Solving problems for something as complicated as space travel requires many different sciences. Other fields of study, such as architecture and social studies, give scientists new insight to see a problem differently and solve it.

Play the Video

myNGconnect.com

Discuss the Video

Invite students to share what they learned and give their perspectives. You may want to ask some of the following questions.

- What in the video would you like to learn more about? Why was it interesting to you?

- What other fields of study did Constance Adams use to solve her design problems?

- Constance Adams's job is both interesting and challenging. Would you like to do what she does? Why or why not?

- What were some of the design concerns for the TransHab project?

- TransHab needed to expand three times to house a six-person crew. Can you think of a common item that expands before use?

Science at Home

Distribute the Science at Home Letter on Learning Masters 173–179, or at

myNGconnect.com

Students should take the letter home and

- Discuss matter, water, energy, and light with their families.

- Follow the directions to make a list of objects and their properties with their families.

Science Through Literacy

Teaching students reading comprehension strategies can help them access science content and support their scientific inquiries. National Geographic Science focuses on four key reading comprehension strategies.

Each chapter throughout the unit includes notes to support your teaching of the four key reading comprehension strategies.

Reading Comprehension Strategies

Preview and Predict

- Look over the text.
- Form ideas about how the text is organized and what it says.
- Confirm ideas about how the text is organized and what it says.

Monitor and Fix Up

- Think about whether the text is making sense and how it relates to what you know.
- Identify comprehension problems and clear up the problems.

Make Inferences

- Use what you know to figure out what is not said or shown directly.

Sum Up

- Pull together the text's big ideas.

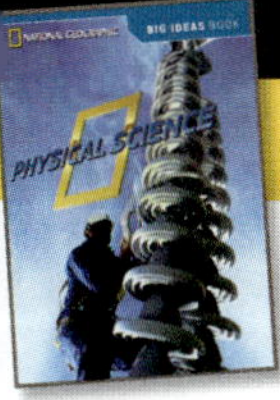

How Can You Describe and Measure Matter?

PHYSICAL SCIENCE

After reading Chapter 1, you will be able to:

- Recognize that all objects and substances in the world are made of matter. **PROPERTIES OF MATTER**

- Recognize that matter takes up space and has mass. **PROPERTIES OF MATTER**

- Identify the properties of objects and substances, such as size, shape, color, texture, and hardness. **PROPERTIES OF MATTER**

- Show that the property of color may be dependent on the surroundings in which the object exists. **PROPERTIES OF MATTER**

- Recognize that materials may be composed of parts too small to be seen without magnification. **PROPERTIES OF MATTER**

- Explain how to measure and compare the mass and volume of solids and liquids. **MEASURING MASS, MEASURING VOLUME**

- **Science in a Snap!** Show that the property of color may be dependent on the surroundings in which the object exists. **PROPERTIES OF MATTER**

5

❯ Preview and Predict

Tell students that they can preview a text by looking over it to get an idea of what the text is about and how it is organized. Tell students that predicting is forming ideas about what they will read. Students should confirm their ideas with others.

Have students preview and make predictions about the chapter. Remind them to use section heads, vocabulary words, pictures, and their background knowledge to make predictions.

CHAPTER 1 ▫ How Can You Describe and Measure Matter?

LESSON	PACING	OBJECTIVES
Explore Activity **1** *Investigate Physical Properties* pages T5e–T5h **Science Inquiry and Writing Book** pages 166–169	**20** minutes	Investigate and Explore (answer a question; make and compare observations; collect and record data and observations; share findings). Investigate that objects can be described in terms of the materials they are made of and their physical properties. Investigate that physical properties of a material remain the same as the material is reduced in size. Explain how technology is used and how it can extend human abilities. Observe and discuss objects and events and record observations.
2 **Big Idea Question and Vocabulary** pages T6-T7–T8-T9	**25** minutes	Recognize that all objects and substances in the world are made of matter.
3 **Properties of Matter** pages T10–T15	**30** minutes	Recognize that all objects and substances in the world are made of matter. Recognize that matter takes up space and has mass. Identify the properties of objects and substances, such as size, shape, color, texture, and hardness. Show that the property of color may be dependent on the surroundings in which the object exists. Recognize that materials may be composed of parts too small to be seen without magnification.
4 **Measuring Mass** pages T16–T17 **Measuring Volume** pages T18–T21	**30** minutes	Explain how to measure and compare the mass and volume of solids and liquids.
5 NATIONAL GEOGRAPHIC **Giants of the Pumpkin Patch** pages T22–T23	**20** minutes	Compare objects according to properties such as size, shape, color, texture, and mass.

VOCABULARY	RESOURCES	ASSESSMENT
observe predict	Science Inquiry and Writing Book: *Physical Science* Science Inquiry Kit: *Physical Science* Explore Activity: Learning Masters 180–183	Inquiry Rubric: Assessment Handbook, page 211 Inquiry Self-Reflection: Assessment Handbook, page 217 Reflect and Assess, page T5h
	Vocabulary: Learning Master 184 Big Ideas Book: *Physical Science*	
matter texture		Assess, page T15
mass volume	Extend Learning: Learning Master 185 Science Inquiry and Writing Book: *Physical Science*	Before You Move On, page T17 Assess, page T21
	Share and Compare: Learning Master 186	Assess, page T23

TECHNOLOGY RESOURCES

STUDENT RESOURCES

myNGconnect.com

Student eEdition

Big Ideas Book

Science Inquiry and Writing Book

Read with Me

Vocabulary Games

Digital Library

Enrichment Activities

National Geographic Kids

National Geographic Explorer!

TEACHER RESOURCES

myNGconnect.com

Teacher eEdition

Teacher's Edition

Science Inquiry and Writing Book

Online Lesson Planner

National Geographic Unit Launch Videos

Assessment Handbook

Presentation Tool

Digital Library

NGSP ExamView CD-ROM

IF TIME IS SHORT...
FAST FORWARD.

CHAPTER 1 □ How Can You Describe and Measure Matter?

LESSON	PACING	OBJECTIVES
Directed Inquiry **6** *Investigate Volume and Mass* pages T23a–T23d **Science Inquiry and Writing Book** pages 170–173	**20** minutes	Investigate through Directed Inquiry (answer a question; make and compare observations; collect and record data and observations; generate explanations and conclusions based on evidence; share findings; ask questions based on observations to increase understanding). Recognize that matter takes up space and has mass, and that two objects cannot occupy the same place at the same time. Use a balance and a graduated cylinder to gather information and solve problems. Compare the mass and volume of solids and liquids using appropriate mathematical signs. Judge whether measurements and computation of quantities are reasonable.
7 **Conclusion and Review** pages T24–T25	**15** minutes	
8 NATIONAL GEOGRAPHIC **PHYSICAL SCIENCE EXPERT** *Physical Chemist* pages T26–T27 NATIONAL GEOGRAPHIC **BECOME AN EXPERT** *The Properties of Graphene* pages T28-T29–T36	**35** minutes	Describe the biographies of and contributions made by various scientists and inventors from diverse gender and ethnic backgrounds. Identify the properties of objects and substances, such as size, shape, color, texture, and hardness.

FAST FORWARD ▶▶▶
ACCELERATED PACING GUIDE

DAY 1  **20** minutes

Explore Activity

Investigate Physical Properties, page T5e

DAY 2 🕐 **35** minutes

NATIONAL GEOGRAPHIC **PHYSICAL SCIENCE EXPERT** *Physical Chemist,* page T26

NATIONAL GEOGRAPHIC **BECOME AN EXPERT** *The Properties of Graphene,* page T28–T29

VOCABULARY	RESOURCES	ASSESSMENT
measure **predict**	Science Inquiry and Writing Book: *Physical Science* Science Inquiry Kit: *Physical Science* Directed Inquiry: Learning Masters 187–190	Inquiry Rubric: Assessment Handbook, page 211 Inquiry Self-Reflection: Assessment Handbook, page 218 Reflect and Assess, page T23d
		Assess the Big Idea, page T25 Chapter 1 Test: Assessment Handbook, pages 88–90 NGSP ExamView CD-ROM
		Assess, pages T27, T30–T31, T34–T35

TECHNOLOGY RESOURCES

STUDENT RESOURCES

myNGconnect.com

- Student eEdition
- Big Ideas Book
- Science Inquiry and Writing Book
- Read with Me
- Vocabulary Games
- Digital Library
- Enrichment Activities

National Geographic Kids

National Geographic Explorer!

TEACHER RESOURCES

myNGconnect.com

- Teacher eEdition
- Teacher's Edition
- Science Inquiry and Writing Book
- Online Lesson Planner
- National Geographic Unit Launch Videos
- Assessment Handbook
- Presentation Tool
- Digital Library

NGSP ExamView CD-ROM

DAY 3 **20** minutes

Directed Inquiry

Investigate Volume and Mass, page T23a

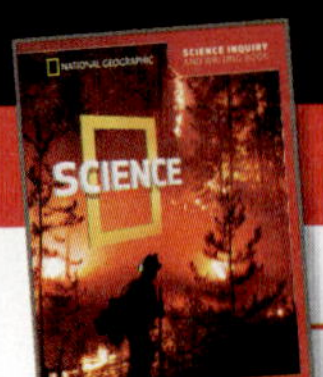

Objectives

Students will be able to:

- Investigate and Explore (answer a question; make and compare observations; collect and record data and observations; share findings).
- Investigate that objects can be described in terms of the materials they are made of and their physical properties.
- Investigate that physical properties of a material remain the same as the material is reduced in size.
- Explain how technology is used and how it can extend human abilities.
- Observe and discuss objects and events and record observations.

Science Process Vocabulary

observe, predict

PROGRAM RESOURCES

- Science Inquiry and Writing Book: *Physical Science*
- Science Inquiry and Writing Book **eEdition** at ⊘ **myNGconnect.com**
- **Inquiry eHelp** at ⊘ **myNGconnect.com**
- Science Inquiry Kit: *Physical Science*
- Learning Masters Book, pages 180–183, or at ⊘ **myNGconnect.com**
- Inquiry Rubric: Assessment Handbook, page 211, or at ⊘ **myNGconnect.com**
- Inquiry Self-Reflection: Assessment Handbook, page 217, or at ⊘ **myNGconnect.com**

MATERIALS

Kit materials are listed in italics.

burlap; hand lens; microscope; battery; scissors; *wool; sandpaper;* paper towel; *silk*

❶ Introduce

Tap Prior Knowledge

- Lead a discussion with students about everyday tools and their uses. Ask: **What are some tools you use? What do you use them for?** Make a table that lists each tool and one or two uses. Include technology, such as computers and electrical appliances, and discuss which ones extend human senses.

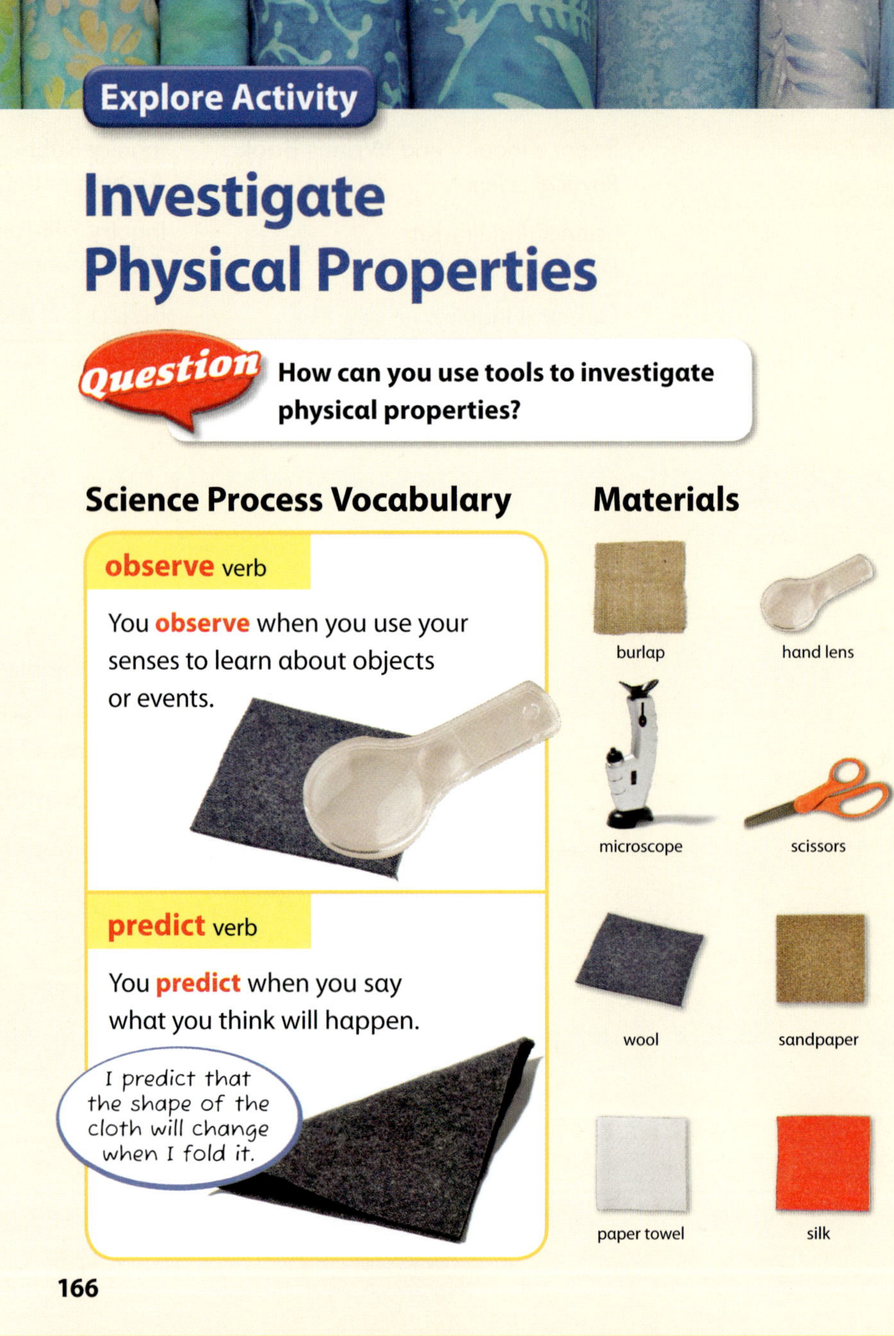

MANAGING THE INVESTIGATION

Time

20 minutes

Groups

Small groups of 4

Advance Preparation

- Cut a 10 x 10 cm square of burlap, wool, silk, sandpaper, and paper towel for each group.
- Model how to use and focus a microscope for students.

What to Do

1 **Observe** the physical properties of the burlap with just your eyes. Look at properties such as its color and whether it looks rough or smooth. Record your observations in your science notebook.

2 Observe the properties of the burlap with the hand lens. Record your observations.

3 **Predict** how your observations will change if you observe the burlap with a microscope. Record your prediction. Then observe the burlap with the microscope. Record your observations.

Turn this knob to focus.

167

Teaching Tips

• Tell students to observe each material along the cut edge as well as the top and bottom surface.

• Make sure the microscope light is on as students observe the materials. Guide them to turn the knobs slowly and carefully until the image comes into focus.

What to Expect

Students will observe that the physical properties of burlap, wool, sandpaper, paper towels, and silk stay the same after the materials have been cut into small pieces.

Introduce, continued

Connect to the Big Idea

• Review the Big Idea Question, *How can you describe and measure matter?* Explain to students that in this activity they will use their senses, a hand lens, and a microscope to investigate matter.

• Have students open their Science Inquiry and Writing Books to page 166. Read the Question and invite students to share ideas about the physical properties of matter.

❷ Build Vocabulary

Science Process Words: observe , predict

Use this routine to introduce the words.

1. **Pronounce the Word** Say **predict**. Have students repeat it in syllables.

2. **Explain Its Meaning** Choral read the sentence. Ask students for other words or phrases that mean the same as **predict**. (tell what you think will happen)

3. **Encourage Elaboration** Ask: **What do you predict sandpaper will look like under a microscope?** (Possible answer: I predict sandpaper will look bumpy under the microscope.)

ELL Use Cognates

Spanish speakers may know the word **predicir** in their home language. Use it to help them access the English word **predict**.

Repeat for the word **observe**. To encourage elaboration, ask: **What sense do you use to observe color?** (sight)

❸ Guide the Investigation

• Distribute materials. Have students match each pictured material with the real material as they say its name.

• Read the steps on pages 167–168. Move from group to group and clarify steps if necessary.

• Ask students to compare the textures, strengths, and colors of the fabric. They may pull the ends of the fabric or scratch the surfaces.

Guide the Investigation, continued

- Remind students to record observations in their science notebooks. Tell them that their observations should be as detailed as possible.

- In step 4, have students discuss their predictions with each other. Tell them to explain why they made their predictions.

⚠ SafetyFirst

Remind students to handle scissors carefully.

❹ Explain and Conclude

- Students should record their predictions and observations in tables similar to the ones on page 169. They should then compare the physical properties of the materials.

- Have each group compare predictions and results and then share ideas with the class. Ask students to compare what they observed when using their eyes, the hand lens, and the microscope.

- Help students understand that the hand lens and microscope extend the human sense of sight. Ask: **What properties were you able to observe when the objects were magnified?** (the stitches on the silk; the grains and coating in the sandpaper; the patterns in the paper towel)

- Ask: **How else might technology be used in science?** Guide students to discuss how technology can be used to make measurements; collect and treat samples; store, collect, and retrieve data; and communicate information.

- Ask: **What properties remained the same when the material is cut up?** (color, texture, and what the material is made of) Students should actively listen for other interpretations and ideas.

- Invite students to describe each material in terms of what it was made of. Explain that silk is a fiber produced by silkworms, burlap is made from a plant, paper towels come from trees, and sandpaper contains paper, glue, and sand or other rough materials.

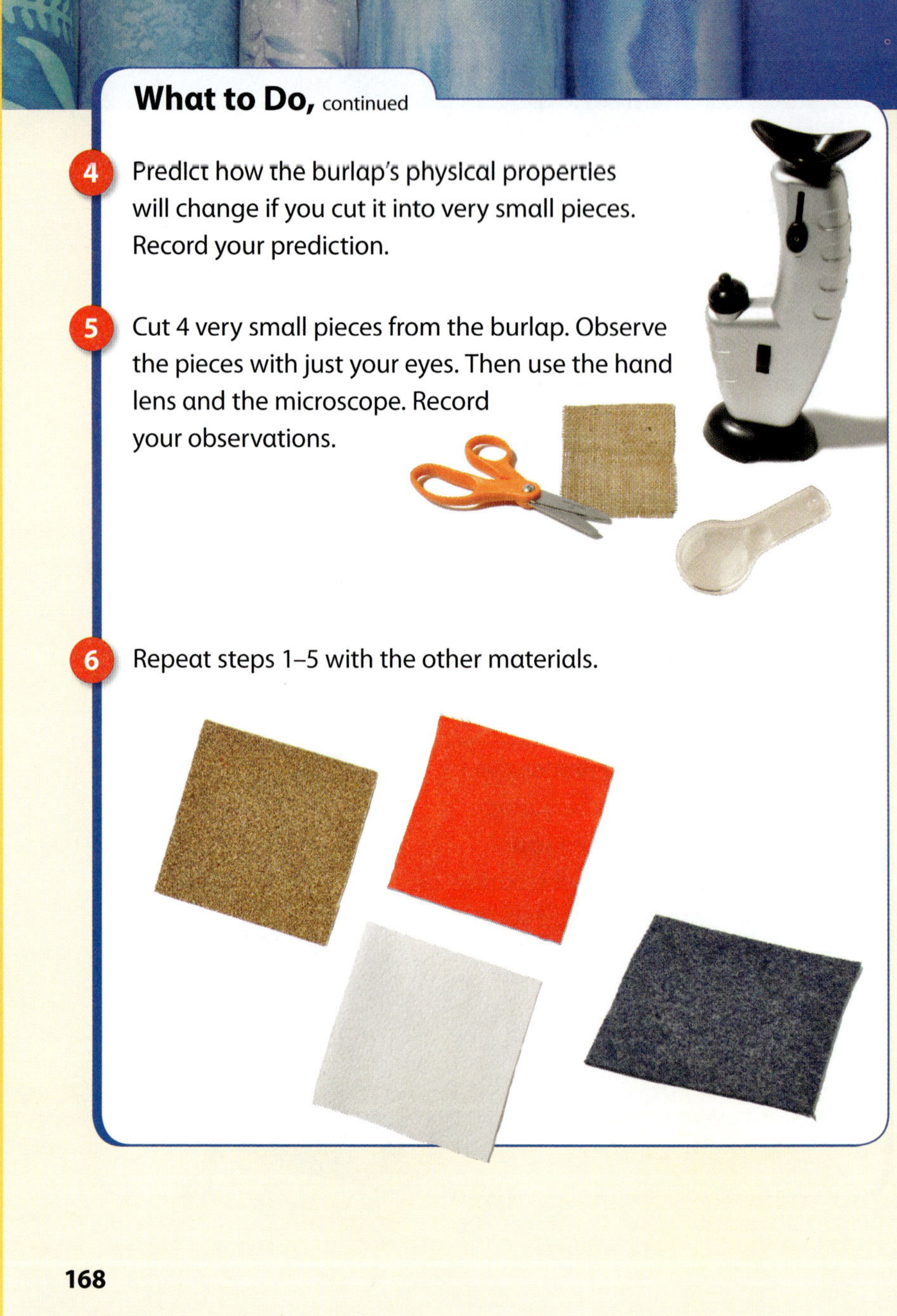

Differentiated Instruction

ELL **Sharing Results**

BEGINNING	INTERMEDIATE	ADVANCED
• When observing the wool say: **I observe that the wool is gray with white specks.** Have students echo your words. • Repeat for each of the other materials, completing the sentence with an appropriate observation.	When observing the wool, have students complete the following sentence: • *I observe that the wool is _____.*	• When observing the wool, prompt students to describe the color of the wool as an observation that begins: • *I observe that _____.*

Record

Write in your science notebook.
Use tables like these.

Predictions		
Material	How will my observations change when I use the microscope?	How will the material's physical properties change when I cut it?
Burlap		

Observations of Physical Properties					
Material	Before Cutting			After Cutting		
	Just Eyes	Hand Lens	Microscope	Just Eyes	Hand Lens	Microscope
Burlap						

Explain and Conclude

1. Did your results support your **predictions?** Explain.
2. How did tools help you to better **observe** the objects?
3. **Compare** the physical properties of the large pieces of the materials with the physical properties of the very small pieces.

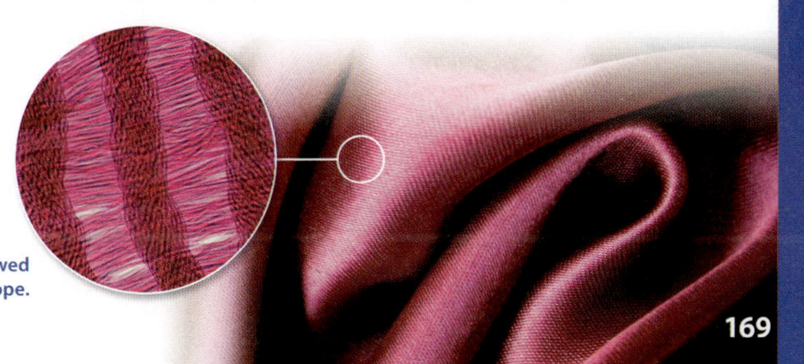

This cloth looks different when viewed through a microscope.

169

Inquiry Rubric at **myNGconnect.com**	Scale			
The student **observed** materials using the senses and tools such as a hand lens and a microscope.	4	3	2	1
The student described details that could be seen with and without tools.	4	3	2	1
The student observed and **compared** the physical properties of materials after they were cut into small pieces.	4	3	2	1
The student drew **conclusions** about the physical properties of materials before and after they were cut into very small pieces.	4	3	2	1
The student **shared** observations and conclusions with others.	4	3	2	1
Overall Score	4	3	2	1

Explain and Conclude, continued

Answers

1. Answers will vary depending on students' predictions.
2. Possible answer: I was better able to see that the objects were made tiny parts.
3. Possible answer: The size of the pieces changed, but the physical properties of the material did not change.

❺ Reflect and Assess

- To assess student work with the Inquiry Rubric shown below, see Assessment Handbook, page 211, or go online at **myNGconnect.com**
- Score each item separately and then decide on one overall score.
- Use the Inquiry Self-Reflection on Assessment Handbook, page 217, or at **myNGconnect.com**

❻ Extend

- Ask students for suggestions of other materials they could observe, such as salt, feathers, and newspaper. Choose new materials that can be cut or broken into small pieces and repeat the activity.

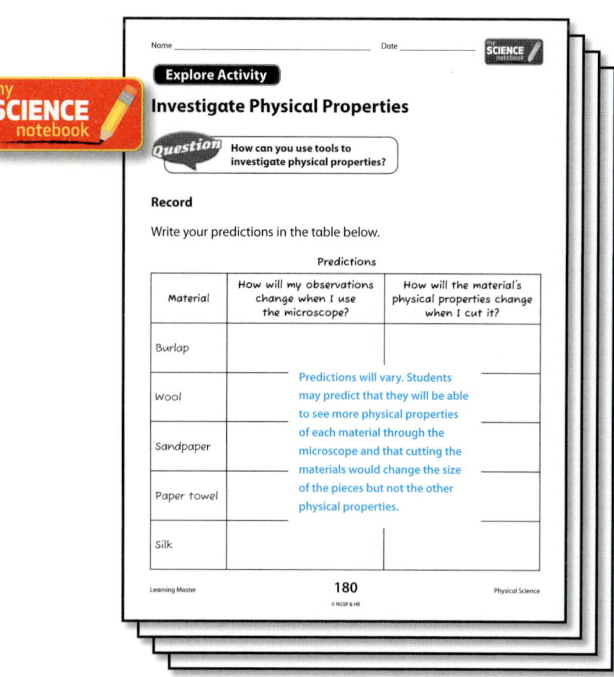

**Learning Masters 180–183
or at myNGconnect.com**

Objectives

Students will be able to:

- Recognize that all objects and substances in the world are made of matter.

Science Academic Vocabulary

matter, texture, mass, volume

PROGRAM RESOURCES

- Big Ideas Book: *Physical Science*
- Big Ideas Book: *Physical Science* ■ eEdition
  at ◉ **myNGconnect.com**
- ■ Vocabulary Games at ◉ **myNGconnect.com**
- ■ Digital Library at ◉ **myNGconnect.com**
- ■ Enrichment Activities at ◉ **myNGconnect.com**
- ■ Read with Me at ◉ **myNGconnect.com**
- Learning Masters Book, page 184,
  or at ◉ **myNGconnect.com**

❶ Introduce

Tap Prior Knowledge

- Have students name the five senses. (hearing, smell, taste, touch, and sight) Then ask them to brainstorm a list of words that describe each sense. For example, *sour* can describe *taste*.

❷ Focus on the Big Idea

Big Idea Question

- Read the Big Idea Question aloud, and have students echo it.
- Preview pages 10–15, 16–17, and 18–21, linking the headings with the Big Idea Question.

Differentiated Instruction

ELL Vocabulary Support

BEGINNING	INTERMEDIATE	ADVANCED
From the classroom, select pairs of objects with different surfaces, and have students point to the one that has the smoother texture. Have students repeat the words *smooth texture* and *rough texture*.	Point to various objects in the classroom and have students use these Academic Language Frames to identify the texture: • *The _____ has a smooth texture.* • *The _____ has a rough texture.*	Have students list the names of objects that have a smooth texture and objects that have a rough texture. Have them describe the objects using this Academic Language Frame: *The _____ has a _____ texture.*

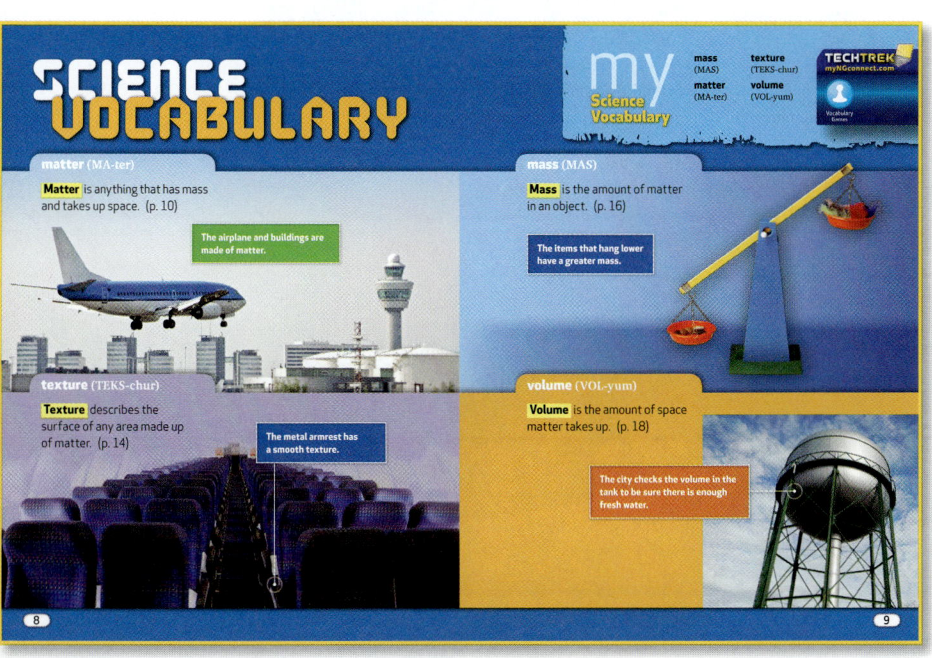

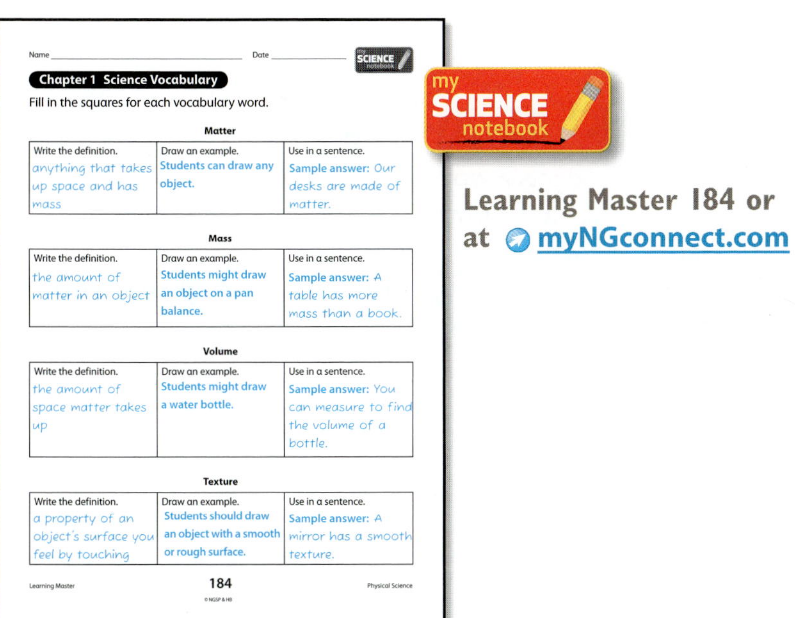

my SCIENCE notebook

Learning Master 184 or at ⊘ myNGconnect.com

Focus on the Big Idea, continued

- With student input, post a chart that displays the chapter headings. Have students orally share what they expect to find in each section. Then read page 6 aloud.

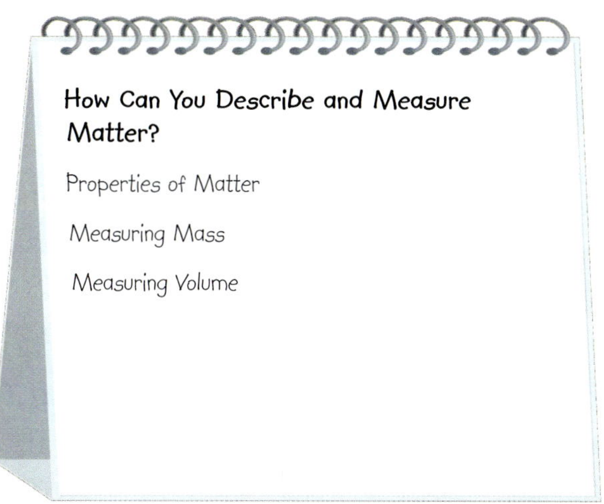

❸ Teach Vocabulary

Have students look at pages 8–9, and use this routine to teach each word. For example:

1. **Pronounce the Word** Say **matter** and have students repeat it.

2. **Explain Its Meaning** Read the word, definition, and sample sentence aloud. Have students point to and name everything in the photo. Say: **Everything in the photo is made of matter.**

3. **Encourage Elaboration** Ask: **What might be inside the plane and buildings?** (people) **How can you use the definition to find out if people are made of matter?** (Ask if people take up space and have mass.)

Repeat for the words **texture**, **mass**, and **volume** using the following Elaboration Prompts:

- **What does texture describe?** (the surface of any object made of matter)

- **What describes the amount of matter?** (mass)

- **Volume is a property of what?** (matter)

LESSON **3** ▫ Properties of Matter

Objectives

Students will be able to:

- Recognize that all objects and substances in the world are made of matter.
- Recognize that matter takes up space and has mass.
- Identify the properties of objects and substances, such as size.

Science Academic Vocabulary

matter

❶ Introduce

Tap Prior Knowledge

- Have students look around their classroom, name objects, and then describe them. Ask: **How can you tell different objects apart?** (by using your senses to observe their properties)

Set a Purpose

- Read aloud the heading. Tell students they will read about ways **matter** can be observed.

❷ Teach

Academic Vocabulary: *matter*

- Point out the highlighting around **matter**. Say: **The highlighting around this word tells you that it is important.**

- Read the sentence that defines **matter**.

Recognize That All Objects and Substances Are Made of **Matter**

- Have students read pages 10–11.

- Randomly point to an object in the room. Ask: **Is this matter?** (yes) **How can you tell?** (You can observe it with your senses; it takes up space and has mass.) Explain that properties can be observed with your senses. Properties of matter can also include odor, sound, and taste.

- Say: **All objects and substances are made of matter. Since matter takes up space and has mass, no two objects can be in exactly the same place at exactly the same time.**

Properties of Matter

What do you see in this photo? Everything you see is made of matter . Matter takes up space and has mass. Each piece of matter takes up its own space. Two airplanes, for example, could not be in the same place at the same time!

You can describe matter by telling about its properties. You observe properties with your senses. This airplane is big, blue, and white.

Social Studies in Science

Use a Map Show students a world map that contains political boundaries. Ask: **What are some large countries?** (Russia, Canada) **What are some small countries?** (Belgium, Monaco, Luxembourg) **What are the properties of this map that you use to read it?** (color and shape)

Size Size is a property of matter. How long do you think this airplane is? You could measure it to find out.

Measuring the size of matter can be important. The runway in the photo needs to be long enough for the airplane to safely land. You could measure both the airplane and runway to find out how long they are.

Compare Matter by Size

- Have students look at the photo and say: **Name all the objects you can find that are larger than you.** (Possible answers: buildings, airplanes, towers)

- Ask: **Name an object that is much smaller than you.** (Possible answers: a blade of grass or a pencil) **How do you think you could find out how much smaller it is?** (I could use a ruler to measure the object and compare its size to mine.)

- Say: **Size is a property of matter. You can measure something's size.**

Making and Recording Observations

Have students create a chart and list the objects in the order of their size, from largest to smallest. Encourage them to measure objects with a ruler to confirm their predictions about size.

Differentiated Instruction

ELL **Language Support for Comparing Matter by Size**

BEGINNING	INTERMEDIATE	ADVANCED
Have students compare size using either/or questions. For example: **Is the chair larger or smaller than the desk? Which is bigger, the pencil or the book?**	Prompt students to select objects and organize them based on their properties. • *Which is the smallest?* • *Which is the largest?*	Have students focus on size and complete the following Academic Language Stems: • *The chair is …* • *The desk is …* • *The book is …*

LESSON 3 ▫ Properties of Matter

Objectives

Students will be able to:

- Identify the properties of objects and substances, such as shape and color.
- Show that the property of color may be dependent on the surroundings in which the object exists.

PROGRAM RESOURCES

-  at ⊘ myNGconnect.com

Teach, continued

Identify and Compare Colors

- Have students read page 12 and the captions on pages 12–13.
- Ask: **What are the different colors of luggage?** (black, blue, green, purple) **Which of your senses do you use to observe color?** (sight)
- Say: **Some pieces are similar in size and shape. How else can travelers spot their luggage?** (by color)

⊘ myNGconnect.com

Have students use the ■ Digital Library to find photos of collections of objects of different colors.

🖥 **Integrated Technology**

Digital Booklet Have students use the photos and information to make a digital booklet with captions that refer to groupings of objects by color.

Color and Shape When you leave an airplane, you have to find your suitcase. How do you find it? You probably observe the properties of color and shape. These suitcases are different shapes and colors. Which do you think would be easiest to find?

Changing Colors 

Materials flashlights, white paper, different colors of cellophane

Have students work in small groups. Distribute one flashlight to each group of students. Have students read and follow the instructions on page 13 of the Big Ideas Book. If you wish, students can use a rubber band to hold the cellophane in place. Students should record their observations and results in their science notebook. Instruct students to exchange flashlights with another group so that each group can view each color.

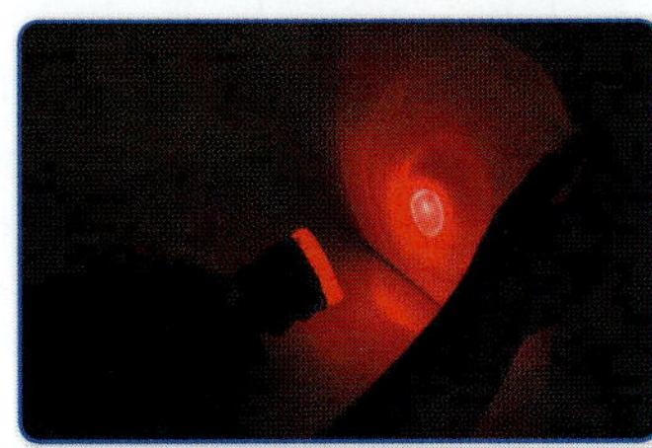

Shine a flashlight onto a white piece of paper. Then cover the flashlight with a piece of colored cellophane.

Shine the flashlight on a white piece of paper.

What happens to the color of the light?

13

Teach, continued

Identify Shapes

- Ask: **Can you think of any shapes that are found in nature?** (round, oval, and irregular shapes, such as the shapes of rocks, trees, etc.)

- Have students look at luggage in the photos on pages 12–13 . Ask: **What two properties of matter might make this luggage hard to fit on a plane?** (size and shape)

- Say: **Name other objects that need a specific shape to fit into something. For example, a key needs a very special shape to fit into a lock.** (Possible answer: Toy building blocks must have matching shapes so that they fit together.)

Compare Shapes

- Ask: **Besides a square, what are some other common shapes of matter?** (Possible answer: circle, triangle) Say: **The shape of a wheel makes it useful for certain activities. Name all the ways you have used a wheel today.** (car, bicycle, skateboard)

- Say: **Square buildings are useful, but square wheels are not.** Encourage students to compare the advantages and disadvantages of other shapes.

What to Expect Students will see different colors according to what color cellophane is covering the flashlight.

Quick Questions Ask students the following questions:

- **Describe what happened.** (When the flashlight was covered with blue cellophane, the light shining on the white paper looked blue.)

- **What does this tell you about the color of an object?** (Color depends on the light that shines on it.)

Physical Science **T13**

Objectives

Students will be able to:

- Identify the properties of objects and substances, such as texture and hardness.
- Recognize that materials may be composed of parts too small to be seen without magnification.

Science Academic Vocabulary

texture

Teach, continued

Academic Vocabulary: *texture*

- Have students read pages 14–15.
- Display the word **texture**. Say it with students and read the definition. Have students reread the text around **texture**. Say: **The words** *fuzzy* **and** *rough* **help explain the meaning of** **texture**.
- Ask: **What is the** **texture** **of your shirt?** (Possible answer: soft, fuzzy)

Identify and Compare Textures

- Pass around a box containing different objects. Say: **Use only your sense of touch to identify each object's** **texture**. (Sample responses: smooth, scratchy, slippery)
- Have students answer the question in the caption on page 14 by describing all of the **textures** they can see inside the airplane.

Texture and Hardness Airplane passengers want comfortable seats. That's why the chairs are covered with smooth fabric like cotton instead of scratchy fabric like wool. *Smooth* and *scratchy* are words that describe **texture**. Texture is a property of an object's surface you feel by touching.

If you look at cotton and wool from a distance, they look very similar. If you look at them up close with a magnifying glass, though, the wool looks bumpier than the cotton.

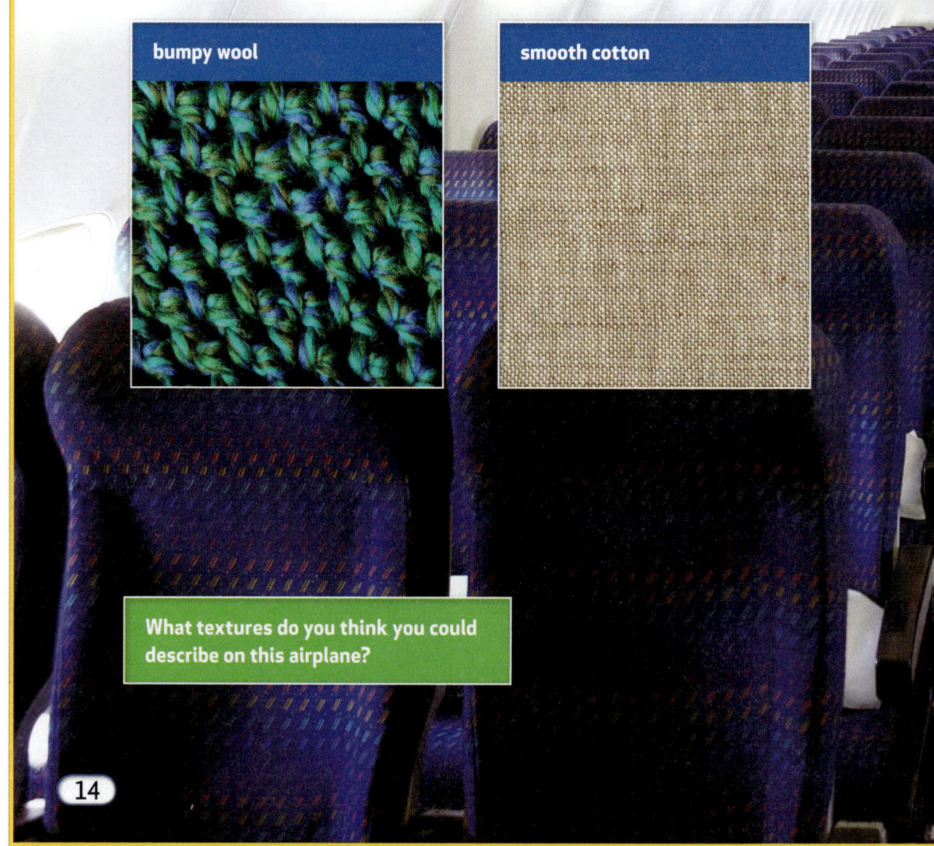

Extend Learning

Compare and Contrast Texture of Materials

Find Out	Think and Do	Describe and Compare
Have students work in small groups. Ask groups to find objects showing examples of different types of textures. Tell students that the surface of each object has different ridges and bumps to give it its distinct texture.	Have students make a chart describing the object, its texture, why it has that texture, and what possible benefit, if any, that texture gives that object. Have students look at the object with a magnifying glass or find magnified photos of similar objects. Ask them to explain why it feels the way it does.	Have groups take turns sharing their findings with the class, including: • the identity and description of the object. • why the object's surface has its texture. • what possible purpose the texture could serve. Ask students to discuss the ways in which the objects are alike and different.

If you pushed your finger against an airplane seat, your finger would make a dent. If you pushed your finger against the outside of an airplane, your finger could not press into the surface. The outside of an airplane is harder than an airplane seat. Hardness is another property that describes matter.

Before You Move On

1. Name properties that you can use to describe matter.
2. Choose an object in the classroom and describe its properties.
3. **Compare** How is texture different from hardness? How can you observe both properties?

15

NATIONAL GEOGRAPHIC **Raise Your SciQ!**

Diamond Nanorods It is traditionally thought that diamonds are the hardest substance known. This is not true. A man-made substance made of aggregate diamond nanorods is actually harder than diamonds. This material is created by applying tremendous amounts of heat and pressure to a substance called fullerite powder. The end result is a material that can scratch a diamond.

Teach, continued

- Have students study the wool and cotton on page 14. Explain that the **texture** of some materials is easier to see when they are magnified. Say: **These photos show the texture of wool and cotton through a magnifying glass.** Ask: **Which has a bumpier texture?** (the wool) **Do smooth surfaces also have bumps and ridges?** (yes) **What gives them a smoother texture?** (Smooth surfaces have fewer and smaller bumps and ridges.) **Would you be able to see these differences if they weren't magnified?** (no)

Identify and Compare Hardness

- Have students compare the hardness of objects in the classroom. Point to objects, such as books, desks, tissues, and papers, and have students identify them as *hard* or *soft*.

- Say: **Hardness describes the ability of a material to resist under pressure.** Ask: **What are some things you use that you want to have a hard surface?** (Possible responses: baseball bats, doors) **What things do you want to have a soft surface?** (Possible responses: seats, shoes, beds, bicycle tires)

> **Monitor and Fix Up**

Some students might struggle with the idea that **texture** is a property of matter. Discuss possible fix-up strategies. For example, ask students to compare two objects that have very different **textures**. Have them look at and feel each object.

❸ Assess

» Before You Move On

1. **Recall** Name properties that you can use to describe matter. (size, shape, color, texture, and hardness)

2. **Explain** Choose an object in the classroom and describe its properties. (Answers will vary.)

3. **Compare** How is texture different from hardness? How can you observe both properties? (Texture describes the surface of an object. Hardness describes the ability of a material to resist under pressure. Texture and hardness can be observed by feeling the objects. Texture can also be observed by looking at the object.)

LESSON 4 □ Measuring Mass

Objectives

Students will be able to:

- Explain how to measure and compare the mass of solids and liquids.

Science Academic Vocabulary

mass

PROGRAM RESOURCES

- **Enrichment Activities** at ⊕ **myNGconnect.com**

❶ Introduce

Tap Prior Knowledge

- Have students recall a visit to the doctor's office or school nurse. Ask: **Did the doctor or nurse weight you?** Have students describe how they were weighed and what tool was used.

Set a Purpose and Read

- Read the heading. Say: **Mass is another property of matter that you can observe and measure. Now you will find out how mass is measured.**

- Have students read pages 16–17.

❷ Teach

Academic Vocabulary: *mass*

- Point out the balance, and read aloud the caption. Say: **We know that the stones and the feathers are both made of matter. To find out how much matter they each have, we can measure their mass.**

- Ask: **What is mass?** (Mass is the amount of matter in an object.) **Which depends on the amount of gravity—mass or weight?** (weight)

Measure and Compare Mass of Solids

- Say: **Look around for an object that has a greater mass than the stones and could replace the feathers.** (textbook, rock, radio)

- Have pairs take turns measuring two objects on a balance.

Measuring Mass

You can compare sizes of objects by measuring their **mass**. Mass is the amount of matter in an object. Mass can be measured in grams or ounces.

A balance can help you compare the masses of different solids. The objects with the greater mass make one tray hang lower. However, a balance cannot tell you the weight of the objects. This is because mass and weight are not the same thing. An object weighs more when there is more gravity. Earth has more gravity than the moon, so you would weigh more on Earth. You would have the same amount of matter, though, in both places. Your mass on the moon would be the same as your mass on Earth.

16

Differentiated Instruction

Extra Support

Help students draw sequence graphic organizers by modeling one on the board. Ask them to fill in the steps on how to measure a liquid. You may want to have students work with partners to come up with the sequence.

Challenge

Tell students that 1,000 grams equals one kilogram. Have them create a chart that shows how many grams are in 1, 5, 10, and 100 kilograms.

How do you measure the mass of a liquid? First, measure the mass of an empty container. Add the liquid to the container. Then, measure the mass of the container and the liquid. Subtract the first measurement from the second measurement. The remainder is the mass of the liquid.

Before You Move On

1. Identify the units for measuring the mass of an object.
2. Describe how to measure the mass of a liquid.
3. **Compare** How does your mass on Earth compare to your mass on the moon? How does your weight on Earth compare with your weight on the moon?

17

Science **Misconceptions**

Size Versus Mass Some students may think that any larger object has more mass than a smaller object. Explain that mass measures more matter than we see on the surface of an object. For example, a large beach ball filled with air has less mass than a small baseball filled with rubber, cork, and other materials.

Teach, continued

Measure and Compare Mass of Liquids

- Point out the process diagram on page 17. Ask: **Why is just the container on the scale in the first step?** (To find the mass of only the liquid, you need the mass of the empty container.) **What is happening in the second step?** (The container and the liquid are being measured together.) Ask: **What is the third step?** (Subtract the mass of the container, or 222, from the mass of the liquid, or 548.) **What is the liquid's mass?** (326 grams)

- Ask: **How could you measure and compare the masses of two different liquids?** (Possible answer: Measure the mass of an empty container. Add one liquid and measure the mass of the container and liquid. Empty and clean the container. Add the same amount of the other liquid and measure the mass. Determine which one had a greater mass.)

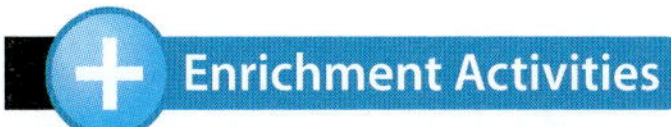

myNGconnect.com

Have students use the Enrichment Activities to compare the masses of different objects.

Digital Slideshow Students can use images they found in the Enrichment Activities to make a slideshow about the mass of various objects.

» Before You Move On

1. **Identify** Identify the units for measuring the mass of an object. (grams or ounces)

2. **Describe** Describe how to measure the mass of a liquid. (Measure the container first. Then take the mass of the container and liquid. Subtract the mass of the container from the mass of the container and liquid.)

3. **Compare** How does your mass on Earth compare to your mass on the moon? How does your weight on Earth compare with your weight on the moon? (Your mass on Earth is the same as your mass on the moon. Your weight would be less on the moon.)

LESSON 4 □ Measuring Volume

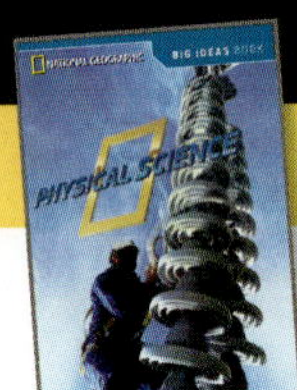

Objectives

Students will be able to:

- Explain how to measure and compare the volume of liquids.

Science Academic Vocabulary

volume

PROGRAM RESOURCES

- Science Inquiry and Writing Book: *Physical Science*
- Science Inquiry and Writing Book ■ eEdition at ◉ **myNGconnect.com**
- ■ Digital Library at ◉ **myNGconnect.com**

❶ Introduce

Tap Prior Knowledge

- Have students list things that can hold liquids, such as a glass. Ask: **How do you know how much a container holds?** (Possible response: Some containers, such as a water bottle, may have the measure printed on the container.)

Set a Purpose

- Read the heading. Say: **You learned how to measure mass. You can measure volume, too. You will read to find out how.**

❷ Teach

Academic Vocabulary: *volume*

- Pronounce **volume** and write it on the board. Explain that the word has many different meanings. Say: **In the scientific measuring system, volume is the amount of space occupied by matter.**

Explain How to Compare and Measure the Volume of Liquids

- Have students read pages 18–19.
- Direct students to the photos on page 18. Ask: **What space does water take up?** (The spaces inside the water bottle and the water tower.) **Which contains a greater volume?** (the water tower)

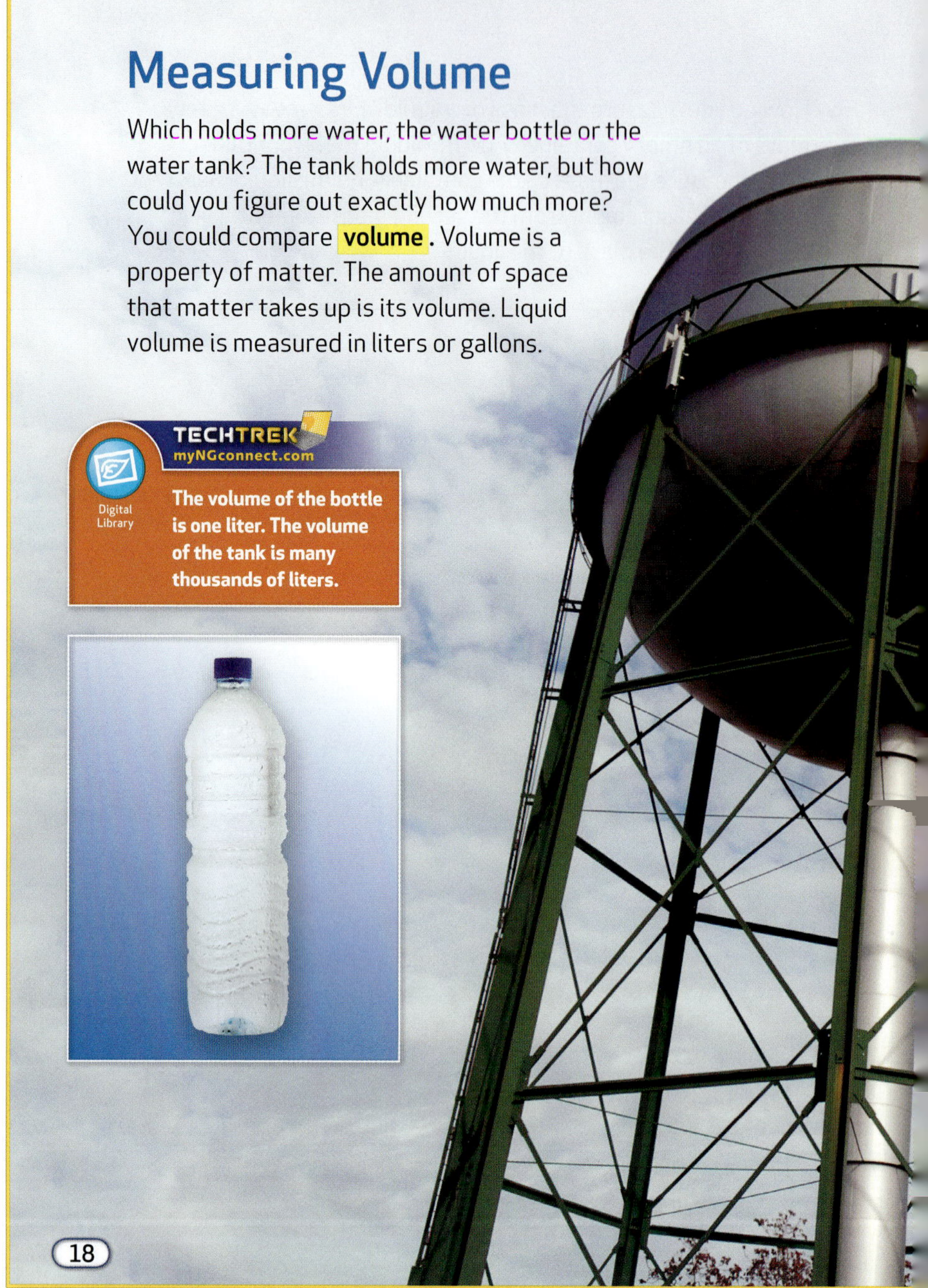

Measuring Volume

Which holds more water, the water bottle or the water tank? The tank holds more water, but how could you figure out exactly how much more? You could compare volume . Volume is a property of matter. The amount of space that matter takes up is its volume. Liquid volume is measured in liters or gallons.

18

NATIONAL GEOGRAPHIC **Raise Your SciQ!**

States of Matter Solids, liquids, and gases are all types of matter that can be measured. Solids have a definite volume and shape; liquids have a definite volume; and gases have neither. A gas, such as helium, spreads out to fill the balloon it is in. A gas can spread out indefinitely.

Look at the two containers below. Do you think they are holding the same volume of liquid? The volume of liquid in one container looks greater than the volume in the other container. Both containers hold the same amount of liquid, though. The shape of a container can fool our eyes into thinking there is more liquid in one container than in another.

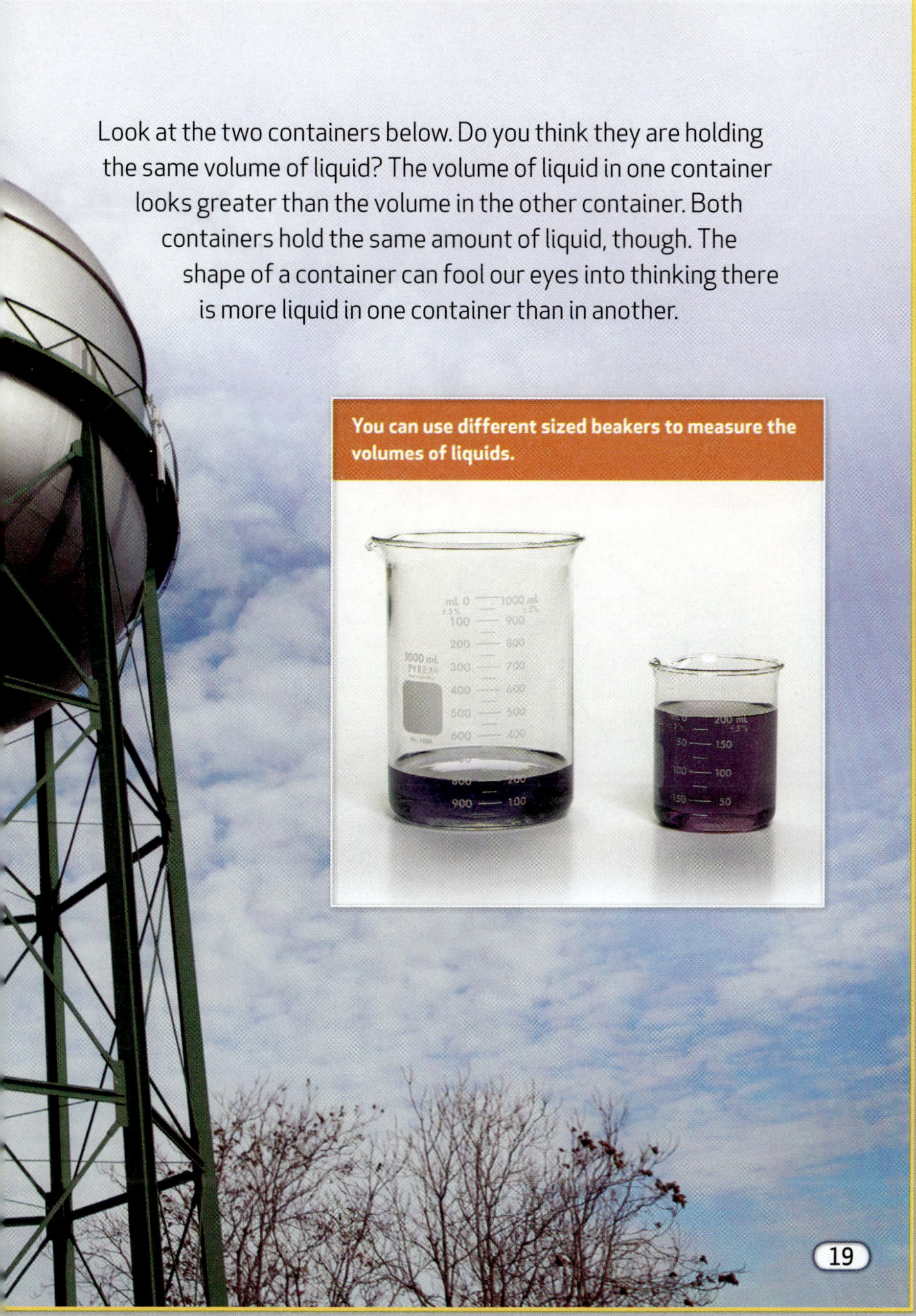

19

Think Like a Scientist — Math in Science

Units of Measurement Ask students to create a chart with the title "Units of Measurement." Have them use the following column heads: Mass, Volume; and row labels: Solids, Liquids. Instruct them to fill in the appropriate cells with gram (g), liter (L) or cubic meters (m^3). Explain that a solid measured in meters has a volume listed in cubic meters.

Teach, continued

- Direct students to the beakers pictured on page 19. Ask: **What is the volume of liquid in the large beaker on the left?** (200 mL) **How does this compare with the volume in the small beaker on the right?** (It is the same.) **How can you know that both hold the same amount of liquid?** (You can match the liquid levels to the measure printed on each beaker.)

- Draw three containers on the board of different shape and size. Say: **Imagine you have these three beakers. One is empty. The other two both hold liquids. How do you figure out which of these two liquids has a greater volume?** (You can pour one into the empty beaker, measure how full the beaker gets, and then repeat with the other liquid. The one that fills up the empty beaker more has a greater volume.)

myNGconnect.com

Have students use the  to find photos of liquids in various containers.

Computer Presentation Students can use the photos to create a computer presentation about the volume of liquids.

Water Predictions

Science Inquiry and Writing Book: *Physical Science,* page 162

Materials 2 translucent containers (1 tall and narrow, 1 short and wide), 200 mL water, measuring cup

Have students work in small groups. Distribute two containers of water and a measuring cup to each group. Make sure both containers hold the same amount of water. Then read and follow the instructions together. Students should compare results and record them in their science notebook.

What to Expect Students might think that the tall container holds more water than the short container. They will find that each container holds the same amount of water.

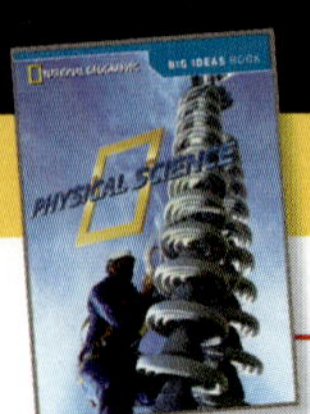

LESSON 4 □ Measuring Volume

Objectives

Students will be able to:
- Explain how to measure and compare the volume of solids.

PROGRAM RESOURCES

-  **Digital Library** at 🌐 **myNGconnect.com**
- Learning Masters Book, page 185, or at 🌐 **myNGconnect.com**

Teach, continued

Explain How to Measure and Compare the Volume of Solids

- Have students read page 20. Write the equation for volume on the board. Using a measuring stick, demonstrate how to measure and calculate the volume of a dictionary. Then give students the approximate length, width, and height of the classroom. Have them calculate its volume.

- Ask: **What is the volume of the classroom compared to the dictionary?** (greater) **How can you figure out how many dictionaries will fit in this room?** (by dividing the volume of the room by that of the dictionary)

 **Digital Library**

🌐 **myNGconnect.com**

Have students use the **Digital Library** to find photos of cubes and rectangles.

You can use math to figure out the volume of a solid. First, measure the solid's length (l). Length means how long something is. Measure its width (w). Width means how wide something is. Measure its height (h). Height means how tall something is. Then, multiply length by width by height. The result is the volume of the solid.

Extend Learning

MANAGING THE INVESTIGATION

Time

 5 minutes for setup
15 minutes for measuring

Groups

four students each

PROGRAM RESOURCES

- Learning Master 185

MATERIALS

- meter sticks
- large cardboard boxes
- several hardcover books

Investigate Volume

Preview	What To Do
 **Question** **How much space inside the box will the books occupy?** Students will investigate the amount of space inside a box. They will measure the volume of the box and several books in order to know how many books can fit inside the box.	1. Measure the inside of the box to find its volume. Remember to multiply length times width times height. 2. Find the volume of each book. 3. Record volume in cubic centimeters. 4. Decide how many books can fit inside the box. 5. Pack the box and see if you are correct.

If you want to find the volume of a solid block, you can measure the length, width, and height. What if you want to find the volume of an object with an odd shape, like a rock? Put water in a beaker. Measure its volume. Then, put the object in the beaker. Measure the volume again to see how far the water rose. The difference in the water's volume is the volume of the object.

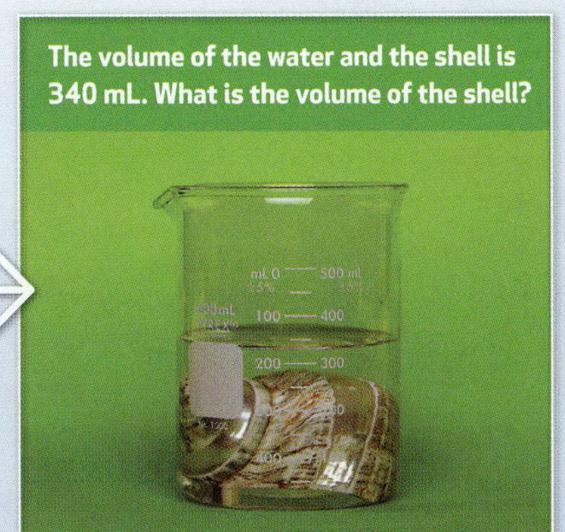

Before You Move On

1. Define volume.
2. Summarize how to measure the volume of a liquid.
3. **Describe** Suppose you have a solid marble. You want to measure its volume. Describe what you would do.

21

Teach, continued

- Have students read page 21 and observe the images carefully. Ask: **Why is it difficult to measure the sides of an oddly-shaped object, such as a shell, with a ruler?** (Possible answers: it doesn't have sides or flat even surfaces) Say: **This is why you measure the volume of objects with odd shapes in a different way.**

- Ask: **What does the beaker measure read in the first photo?** (300 mL) **What is the volume in the second photo?** (340 mL)

- Say: **The second photo does not show more liquid being added.** Ask: **Why does the beaker now read 340 mL?** (because the shell was added)

- Ask: **What is the volume of the shell?** (40 mL)

❸ Assess

❯❯ Before You Move On

1. **Define** Define volume. (Volume is the amount of space an object occupies.)

2. **Summarize** **Summarize how to measure the volume of a liquid.** (Pour the liquid into a graduated cylinder or a beaker. Then read the volume from the marks on the container.)

3. **Describe** **Suppose you have a solid marble. You want to measure its volume. Describe what you would do.** (Put water in a beaker. Measure its volume. Put the marble in the beaker. Measure the volume again. Subtract the volume of the water-filled beaker from the volume of the beaker with the marble in it. The difference is the volume of the marble.)

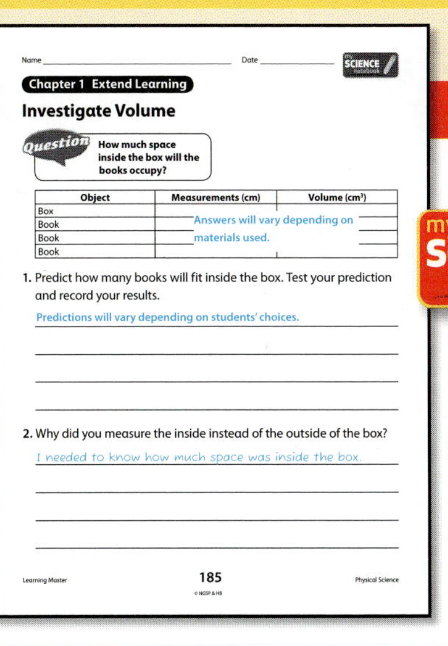

**Learning Master 185
or at**
myNGconnect.com

Explain Results

Students should explain that:

- It is important to measure a container's volume to learn how much it can hold.

- Knowing the volume of each book helps to estimate how many can fit inside the box.

- Measuring the volume of solids can help people pack.

Objectives

Students will be able to:

- Compare objects according to properties, such as size, shape, color, texture, and mass.

PROGRAM RESOURCES

- Learning Masters Book, page 186, or at **myNGconnect.com**

❶ Introduce

Tap Prior Knowledge

- Have students find two objects in the room. Say: **Use your senses to compare the size, shape, color, texture, and mass of the objects.**

Set a Purpose and Read

- Read aloud the heading. Tell students that they will be reading about some properties of pumpkins.

- Have students read pages 22–23.

❷ Teach

Text Feature: Photos and Captions

- Say: **Look at the photo on page 23.** Ask: **How do you know that the pumpkin is large?** (seeing its size compared to the man)

- Explain: **Captions may provide information we wouldn't know by just looking at the photo.** Ask: **What does this caption tell us?** (that the pumpkin was entered in a contest but wasn't the winner)

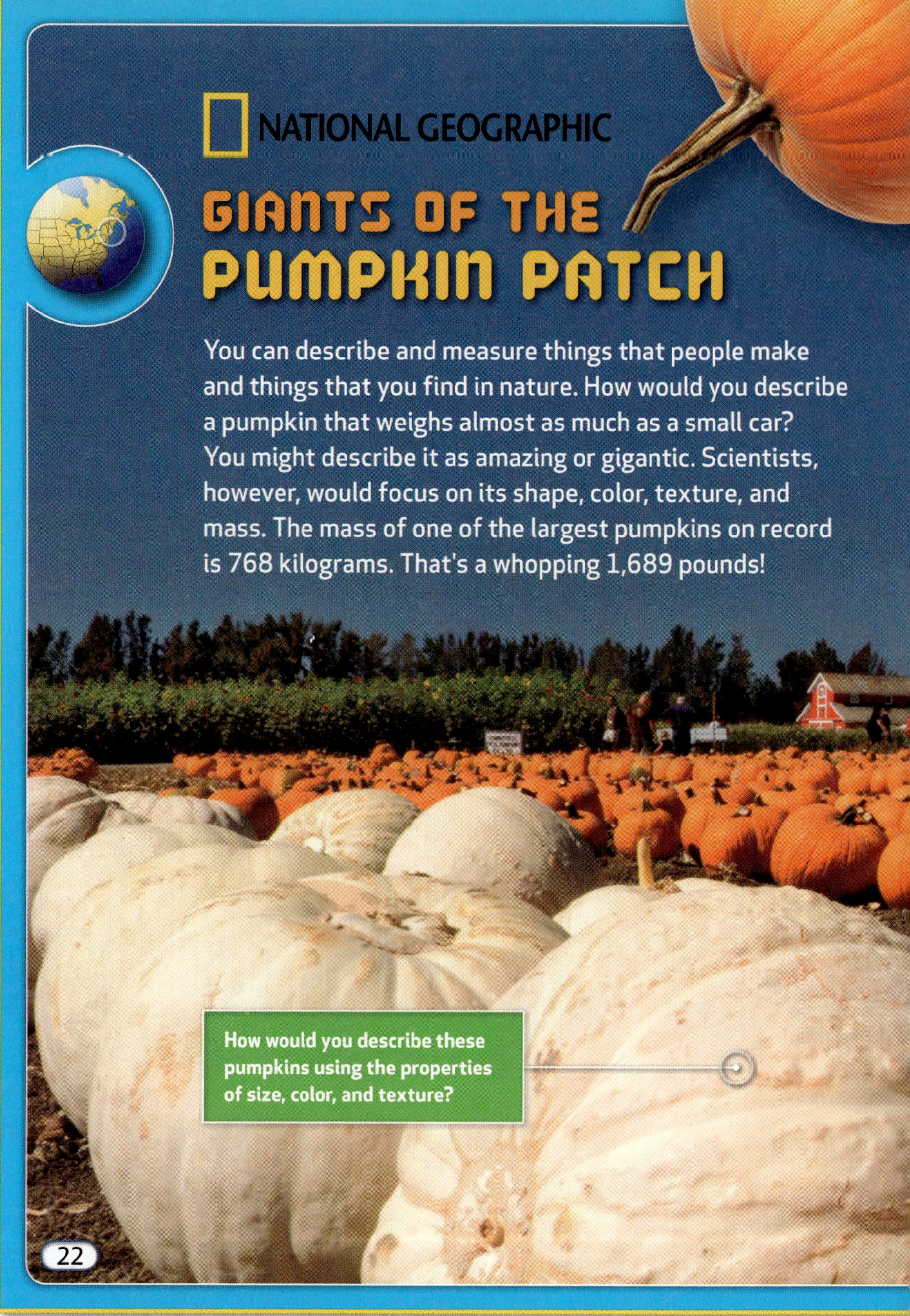

Raise Your SciQ!

Orange Pumpkins Why are pumpkins orange? Pumpkins are orange because of their large amounts of lutein, alpha-carotene, and beta-carotene. These organic compounds are naturally occurring reddish-orange pigments that have known healthful benefits.

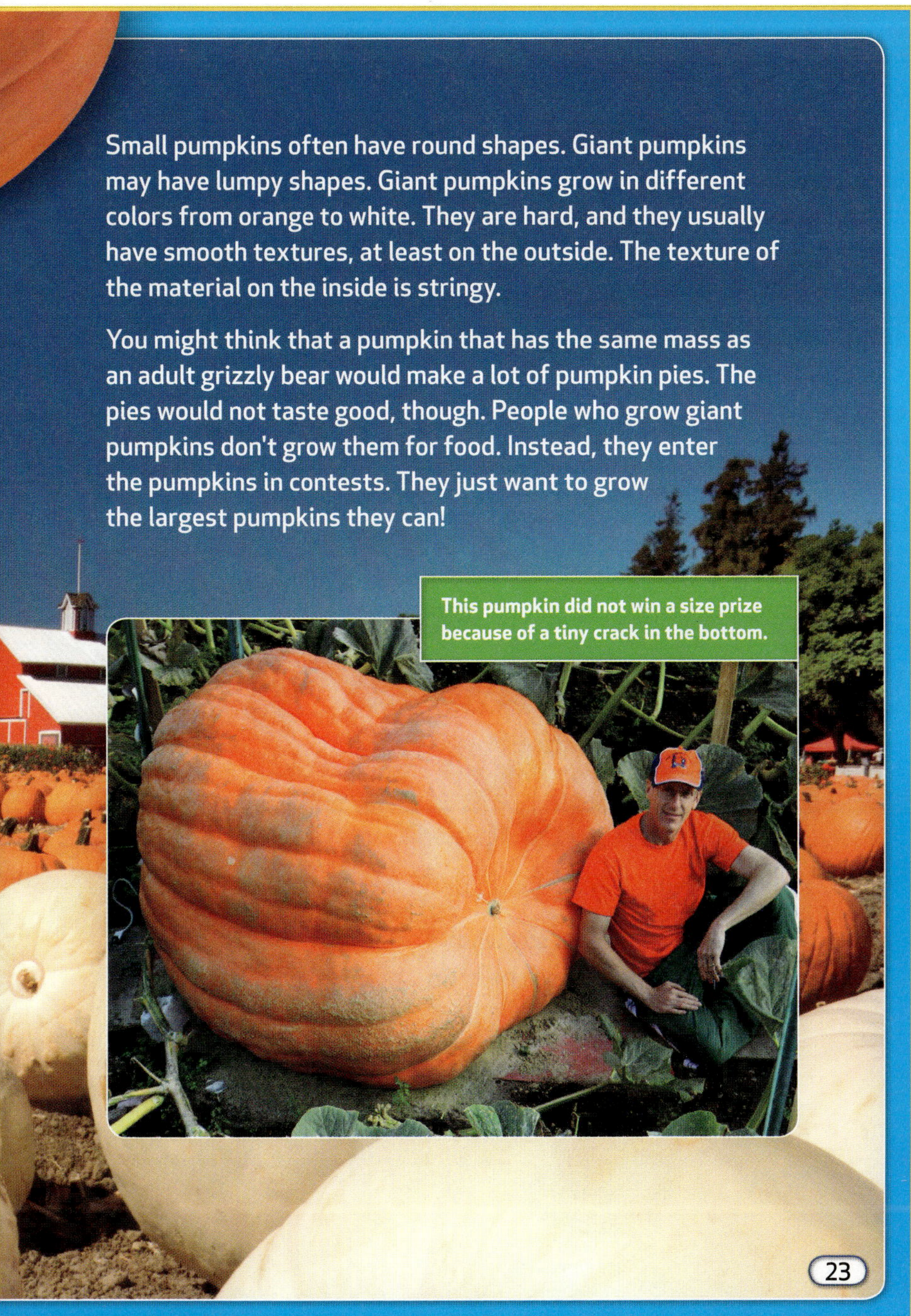

Small pumpkins often have round shapes. Giant pumpkins may have lumpy shapes. Giant pumpkins grow in different colors from orange to white. They are hard, and they usually have smooth textures, at least on the outside. The texture of the material on the inside is stringy.

You might think that a pumpkin that has the same mass as an adult grizzly bear would make a lot of pumpkin pies. The pies would not taste good, though. People who grow giant pumpkins don't grow them for food. Instead, they enter the pumpkins in contests. They just want to grow the largest pumpkins they can!

23

Teach, *continued*

Compare Objects by Their Properties

- Ask: **After reading about the giant pumpkins on page 23, how can you compare their color and texture?** (Most are smooth on the outside and somewhat softer on the inside. Some are orange and some are white.) **What else do you know about the pumpkins' properties?** (They are large in volume and have a large mass.)

- Say: **Recall the definition of matter.** Ask: **What is made of matter in the photos?** (barn, pumpkins, earth, person, trees) **How do you know?** (We can observe properties, such as size, color, texture, shape, and hardness.)

- Say: **Compare the properties of any two pumpkins in the photo.** (Sample response: The pumpkins are both medium-sized, smooth, rounded, and hard on the outside, so they have those properties in common. They are likely different colors, different sizes, and different shapes.)

❸ Assess

1. **Describe** In what ways are pumpkins different from each other? (Pumpkins come in many varieties, shapes, colors, and sizes. Some are for eating. The giant pumpkins are used for competition only.)

2. **Classify** What categories might be used to sort pumpkins? (size, color, shape, weight)

3. **Predict** How does the pumpkin's volume compare with the man's volume? (The pumpkin's volume is greater than the man's volume.)

Share and Compare

How to Describe and Measure Matter

Give students the Learning Master. Have them place the words provided into the correct box. Check their work, and then have them write a short paragraph about an object. Encourage them to use words from all of their completed boxes. Have partners discuss which and how many properties of matter they used to describe their object.

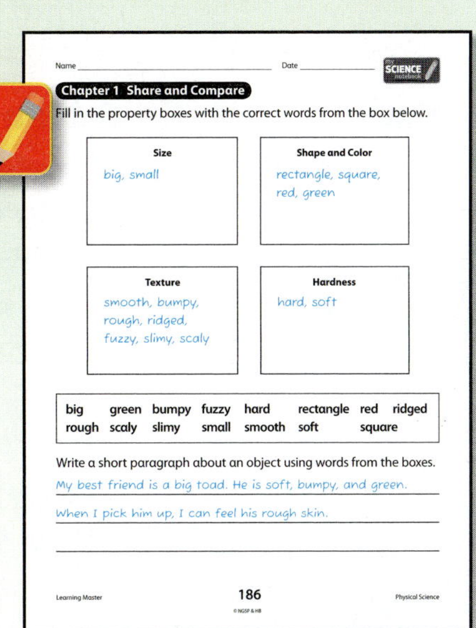

Learning Master 186 or at ☺ myNGconnect.com

Physical Science **T23**

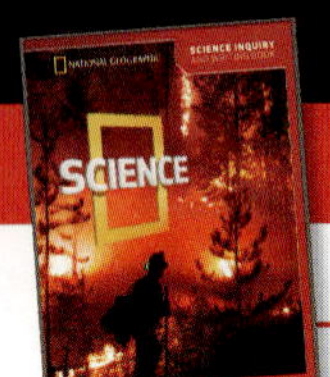

Objectives

Students will be able to:

- Investigate through Directed Inquiry (answer a question; make and compare observations; collect and record data and observations; generate explanations and conclusions based on evidence; share findings; ask questions based on observations to increase understanding).
- Recognize that matter takes up space and has mass, and that two objects cannot occupy the same place at the same time.
- Use a balance and a graduated cylinder to gather information and solve problems.
- Compare the mass and volume of solids and liquids using appropriate mathematical signs.
- Judge whether measurements and computation of quantities are reasonable.

Science Process Vocabulary

measure, predict

PROGRAM RESOURCES

- Science Inquiry and Writing Book: *Physical Science*
- Science Inquiry and Writing Book **eEdition** at ⊘ **myNGconnect.com**
- **Inquiry eHelp** at ⊘ **myNGconnect.com**
- Science Inquiry Kit: *Physical Science*
- Learning Masters Book, pages 187–190, or at ⊘ **myNGconnect.com**
- Inquiry Rubric: Assessment Handbook, page 211, or at ⊘ **myNGconnect.com**
- Inquiry Self-Reflection: Assessment Handbook, page 218, or at ⊘ **myNGconnect.com**

MATERIALS

Kit materials are listed in italics.

graduated cylinder (50 mL); pitcher; water; *marble; rock;* plastic cup (9 oz)*; balance; gram masses*

❶ Introduce

Tap Prior Knowledge

- Remind students that mass is the amount of matter in an object and volume is the amount of space that the object takes up. Also, two objects cannot occupy the same place at the same time. Ask: **Which units can be used to measure mass?** (grams) **Which units can be used to measure volume?** (milliliters, liters, or cubic centimeters)

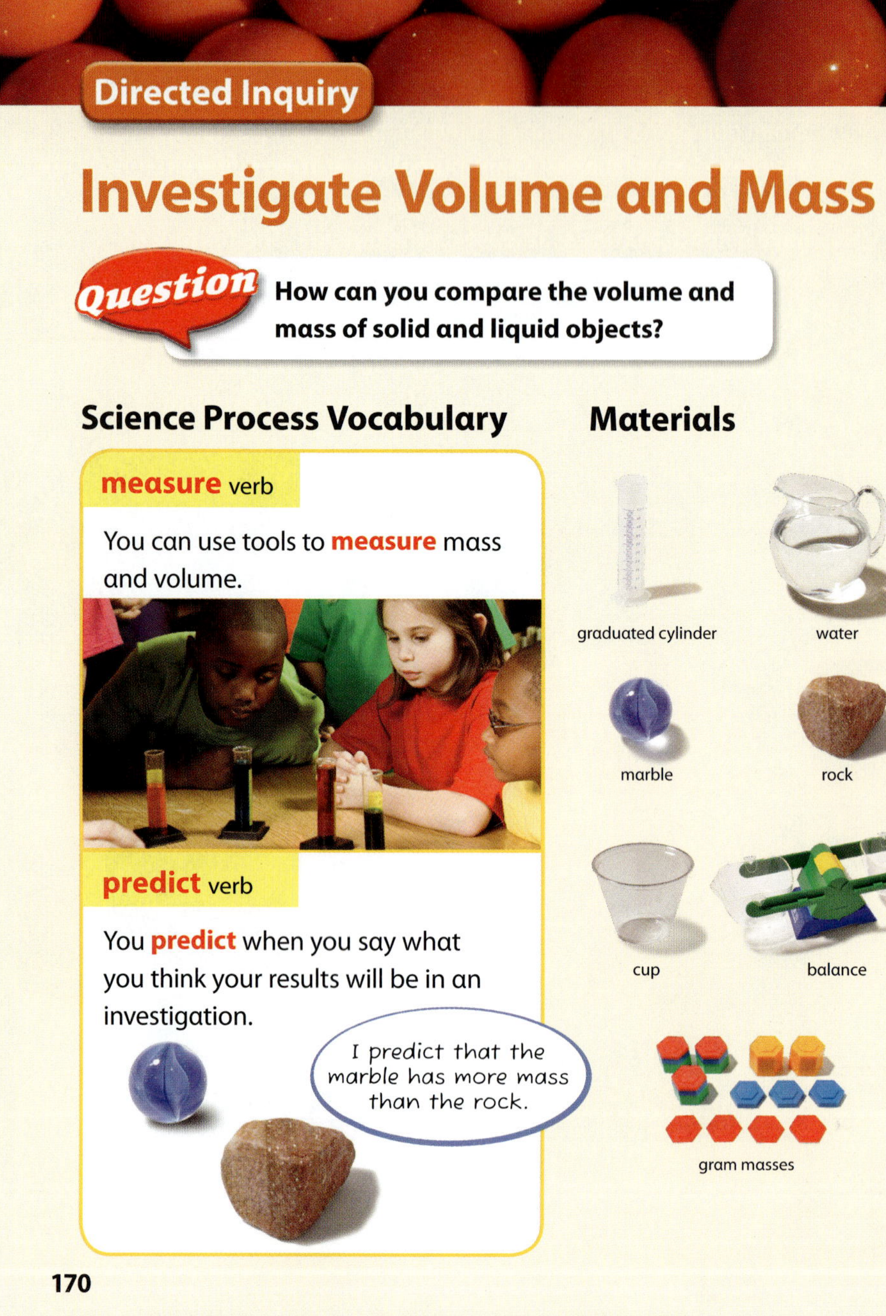

MANAGING THE INVESTIGATION

Time

 20 minutes

Groups

Small groups of 4

Advance Preparation

- Make sure to give students 50-mL graduated cylinders so that they can obtain accurate volume measurements.
- Make sure that the rock provided is small enough to fit into the graduated cylinder.
- Supply each group with 2–5 gram masses in each color.
- Fill pitchers with water. Each group will need about 40 mL of water.

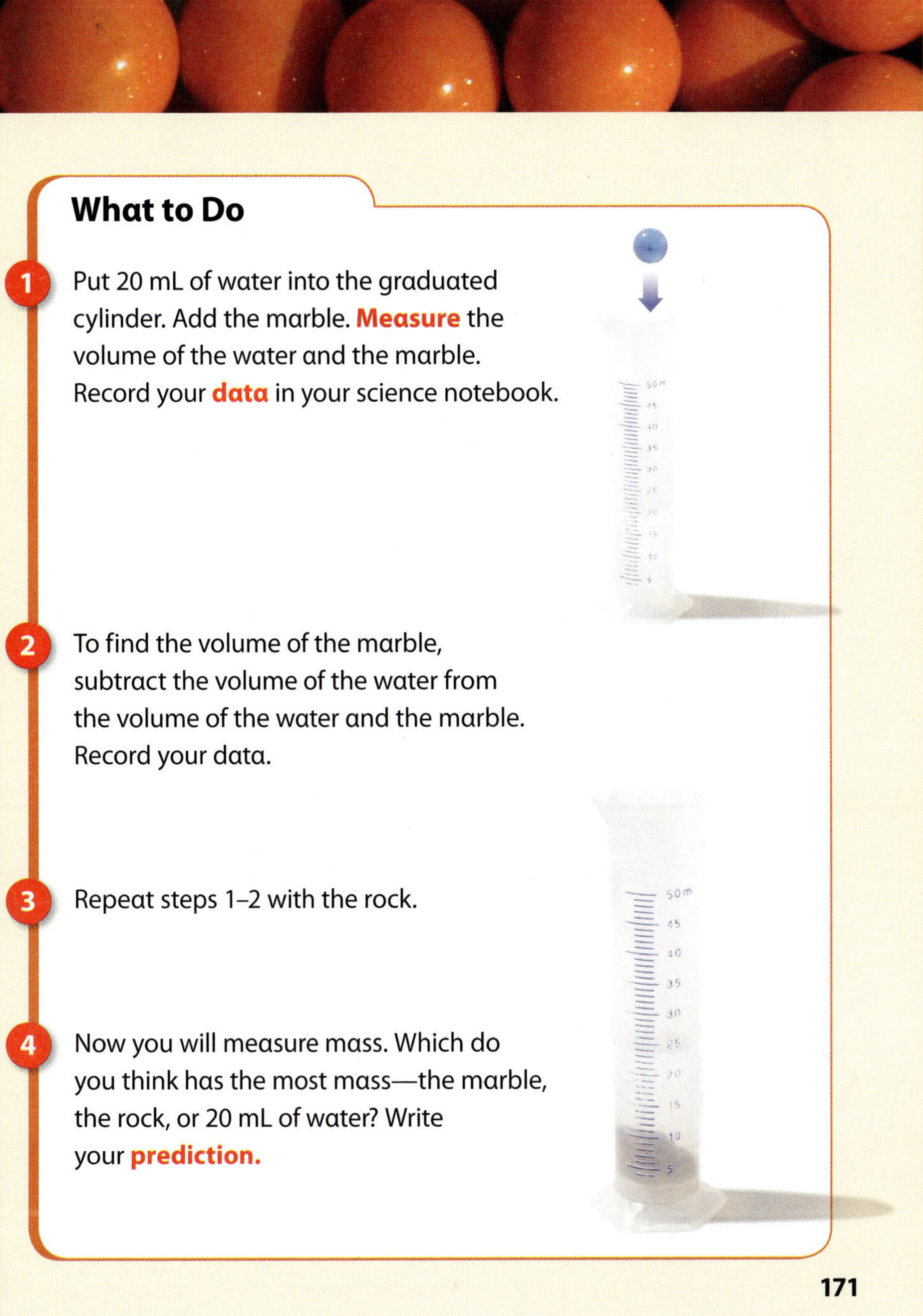

What to Do

1 Put 20 mL of water into the graduated cylinder. Add the marble. **Measure** the volume of the water and the marble. Record your **data** in your science notebook.

2 To find the volume of the marble, subtract the volume of the water from the volume of the water and the marble. Record your data.

3 Repeat steps 1–2 with the rock.

4 Now you will measure mass. Which do you think has the most mass—the marble, the rock, or 20 mL of water? Write your **prediction.**

171

Introduce, continued

Connect to the Big Idea

- Review the Big Idea Question, *How can you describe and measure matter?* Explain to students that this inquiry will help them learn to measure the volume and mass of solids and liquids.

- Have students open their Science Inquiry and Writing Books to page 170. Read the Question and invite students to share ideas about how to measure and compare the volume and mass of solid and liquid objects.

❷ Build Vocabulary

Science Process Words: measure, predict

Use this routine to introduce the words.

1. **Pronounce the Word** Say measure. Have students repeat it in syllables.

2. **Explain Its Meaning** Choral read the sentence. Ask students for another word or phrase that means the same as measure. (find out the length, volume, or mass)

3. **Encourage Elaboration** Ask: **What tool should you use to measure mass?** (balance)

ELL Vocabulary Support

Explain to students that the word measure is a verb, or an action word. The word measurement is a noun, or a thing. Say: **I can measure _____. My measurement is _____.** Have students practice using both forms of the word.

Repeat for the word predict. To encourage elaboration, ask students to make a prediction about the weather tomorrow.

❸ Guide the Investigation

- Distribute materials. Read the inquiry steps on pages 171–172 together with students. Clarify steps if necessary.

- In step 3, tell students to pour the water from the graduated cylinder into a cup so they can retrieve the marble. Then tell students to measure 20 mL of water into the graduated cylinder.

- Before they measure volume or mass, have students hold the marble in one hand and the rock in the other. Ask: **Which feels like it has more mass? Which seems to take up more space?**

Teaching Tips

- Remind students to read the graduated cylinder at eye level. They should measure at the bottom of the meniscus or at the center of the graduated cylinder.

- Have students add the marble and rock carefully to prevent splashing.

- If students have difficulty with the balance, suggest that they use small gram masses, such as 5-g masses or 1-g masses.

What to Expect

- Groups will measure volume and mass of a rock, a marble, and 20 mL of water. They will compare their volume and mass measurements.

Guide the Investigation, continued

- Have students compare the masses and volumes using equations with the appropriate mathematical symbols, such as > for "greater than," < for "less than," and = for "equal to." Have them record their equations in their science notebooks.

- After students record their measurements, ask: **Do your measurements seem reasonable? Why or why not?** Students should give evidence for their answers. If measurements do not seem reasonable, ask students to repeat the steps.

❹ Explain and Conclude

- Guide students to record their data into tables. Groups should compare data and identify which object had the most volume and which had the most mass.

- Ask: **How does using a balance and graduated cylinder help you gather better information?** (The balance and graduated cylinder help me obtain more exact measurements.)

- Ask students to reread their predictions. Have them work in pairs as they evaluate how their understandings have changed.

- If results vary, remind students not to change their data. Instead, they should come up with explanations. For example, the rocks may be different sizes, so they will vary in volume and mass. If students did not measure the water accurately, their mass measurements of the water may vary.

- If students obtain data that is very different from the rest of the class, they may have made errors in their measurements. Ask: **What are some possible measurement errors? What would you do differently next time?**

Answers

1. Answers will vary depending on the rock that is used. The marble should have a volume of about 2 mL.

What to Do, continued

5 Use the balance to find the mass of the cup. Record your data.

6 Place the marble in the cup and find the mass of the cup and marble. Record your data. To find the mass of the marble, subtract the mass of the cup from the mass of the cup and marble. Record your data.

7 Repeat step 6 with the rock and with 20 mL of water.

172

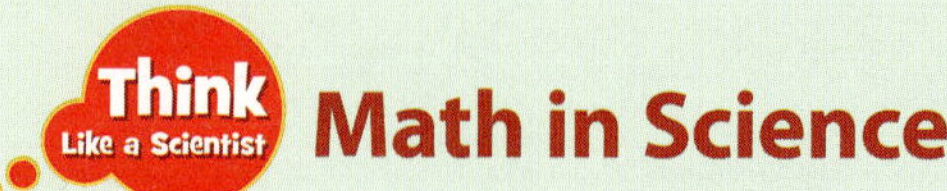

Think Like a Scientist | Math in Science

Volume of Regular Solids Using a graduated cylinder is one way to measure the volume of solid objects. Measurements such as length, width, and height can also be used to calculate volume. For an object such as a cube, shoebox, or a book, multiply length, width, and height to find volume in cubic centimeters.

Record

Write in your science notebook.
Use tables like these.

Volume

Object	Volume of Water (mL)	Volume of Water and Object (mL)	Volume of Object (mL)
Marble	20		
Rock	20		

Mass

Object	Mass of Cup (g)	Mass of Cup and Object (g)	Mass of Object (g)
Marble			
Rock			
Water			

Explain and Conclude

1. **Compare** the volume of the water, marble, and rock. Which had the most volume?
2. Which object had the most mass? Is that what you **predicted?** Explain.
3. **Share** your results with others. Explain any differences.

Think of Another Question

What else would you like to find out about comparing the volume and mass of solids and liquids? How could you find an answer to this new question?

Early balance

173

Explain and Conclude, continued

2. Predictions will depend on the rock that is used.

3. Answers will vary. Students might state that differences are due to incorrect reading of the balance or graduated cylinder, or to differences in the actual mass or volume of the objects.

❺ Find Out More

Think of Another Question

- Students should use their observations to generate new questions. Students may ask: *How can we find the volume of an object that cannot fit into a graduated cylinder?* To do this, students may fill a measuring cup with water and measure how much the water level changes when the object is added.

❻ Reflect and Assess

- To assess student work with the Inquiry Rubric shown below, see Assessment Handbook, page 211, or go online at 🌐 **myNGconnect.com**
- Have students use the Inquiry Self-Reflection on Assessment Handbook, page 218, or at 🌐 **myNGconnect.com**

Inquiry Rubric at 🌐 myNGconnect.com	Scale			
The student used a graduated cylinder and water to **measure** the volume of a marble, a rock, and an amount of water.	4	3	2	1
The student made a **prediction** about whether the marble, the rock, or the water had the most mass.	4	3	2	1
The student used a balance to measure the mass of the marble, the rock, and the water.	4	3	2	1
The student collected **data** in a table and then **shared** and **compared** results with others.	4	3	2	1
The student explained differences between groups and identified possible errors in measurements.	4	3	2	1
Overall Score	4	3	2	1

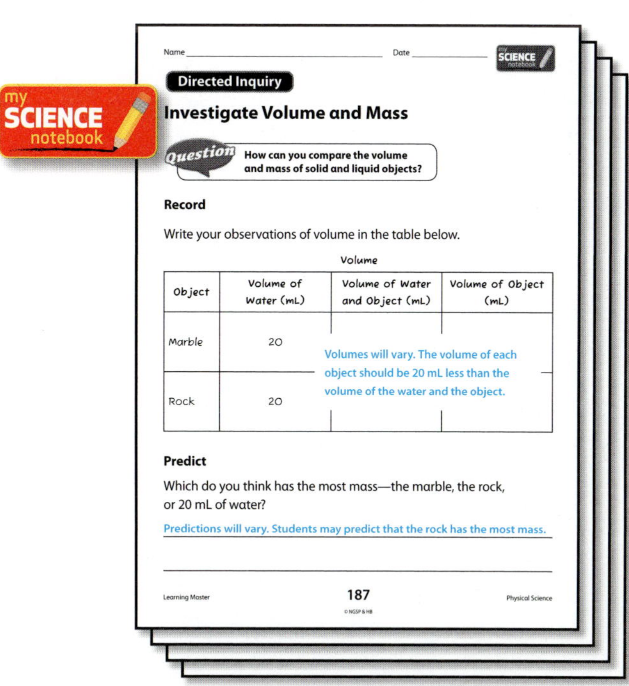

Learning Masters 187–190
or at 🌐 myNGconnect.com

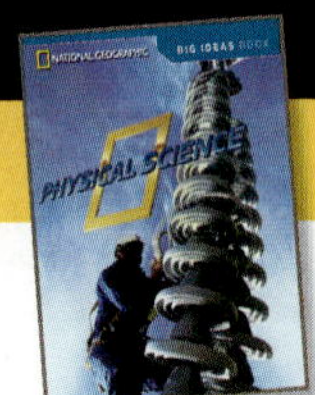

PROGRAM RESOURCES
- Chapter 1 Test, Assessment Handbook, pages 88–90, or at 🌐 **myNGconnect.com**
- NGSP ExamView CD-ROM

❶ Sum Up the Big Idea

- Display the chart from page T8–T9. Read the Big Idea Question. Then add new information to the chart.

- Ask: **What did you find out in each section? Is this what you expected to find?**

How Can You Describe and Measure Matter?

Properties of Matter

Jenny: You can observe matter's size, shape, color, texture, and hardness.

Measuring Mass

Kanika: Measuring mass is a good way to compare solids. You can also compare the mass of liquids.

Measuring Volume

Alice: Measuring and comparing volumes tells how much space things take up.

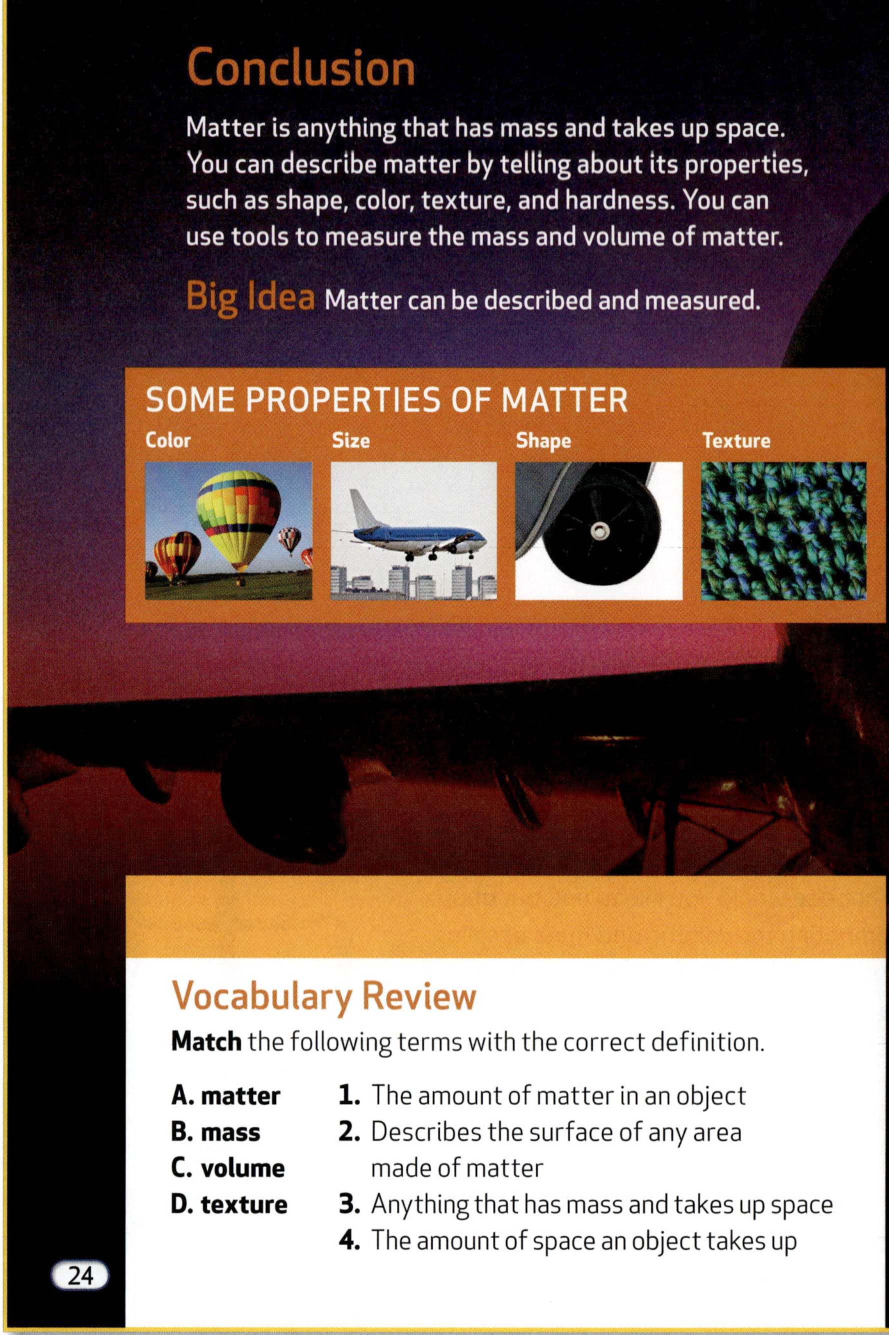

Review Academic Vocabulary

Academic Vocabulary Tell students that scientists use special words when describing their work to others. For example:

matter texture mass volume

Have students write the list in their science notebook and use each word in a sentence that tells how to describe and measure matter. They should then share their sentences with a partner.

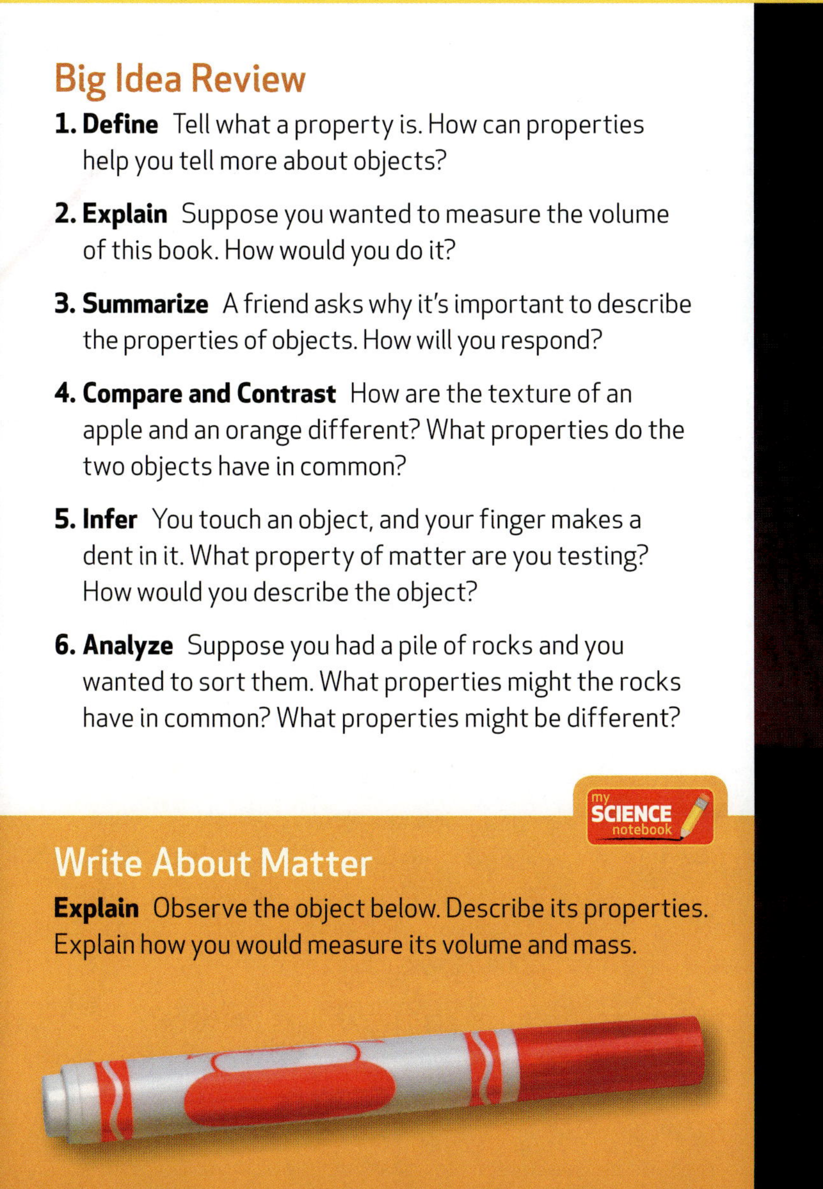

❷ Discuss the Big Idea

Notetaking 

Have students write in their science notebook to show what they know about the Big Idea. Have them:

1. Write about how matter takes up space and has measurable mass.

2. Describe how properties of matter, such as size and shape, can be observed and compared.

3. Explain how to measure the mass and volume of liquids and solids.

❸ Assess the Big Idea

Vocabulary Review

1. B   2. D   3. A   4. C

Big Idea Review

1. A property is something you can observe about an object. Properties can tell you about an object's size, color, shape, texture, and hardness.

2. Take a ruler. Use it to measure the length, width, and height of the book. Multiply the three measurements together. The result is the volume of this book.

3. Describing the properties of objects helps us talk about things, find things, and decide if things will work.

4. An apple is smooth while an orange is bumpy. The shape of both is almost round and their mass is similar.

5. The property being tested for is hardness. This particular object is soft because your finger makes a dent in it.

6. Similar properties: size, shape, texture, color, hardness in any order; Different properties: mass, volume in any order

Write About Matter 

The marker is made of matter. The marker is hard, smooth, light weight, and red and white in color. To measure its volume, I would put water in a beaker and note its volume. Then I would insert the marker and note the new volume. The volume of the marker would be the volume of the water alone subtracted from the combined volume.

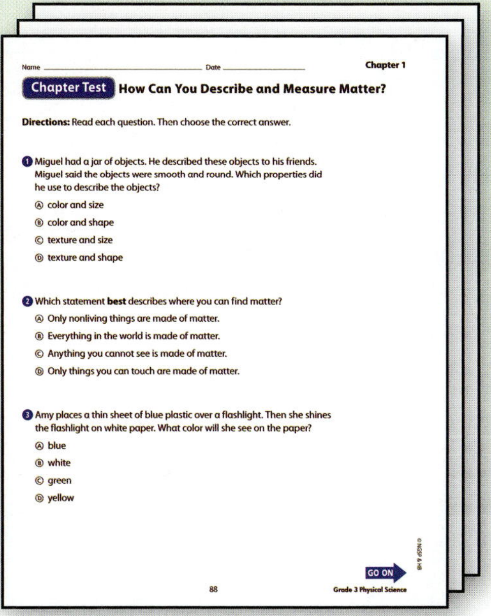

Assess Student Progress
Chapter 1 Test

Have students complete the Chapter 1 Test to assess their progress in this chapter.

Chapter 1 Test, Assessment Handbook, pages 88–90, or at 🌐 **myNGconnect.com**

or NGSP ExamView CD-ROM

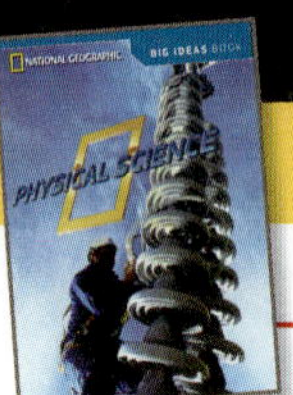

Objectives

Students will be able to:

• Describe the biographies of and contributions made by various scientists and inventors from diverse gender and ethnic backgrounds.

PROGRAM RESOURCES

• Big Ideas Book: *Physical Science*
• Big Ideas Book: *Physical Science* **eEdition** at **myNGconnect.com**
• **Digital Library** at **myNGconnect.com**

❶ Introduce

Tap Prior Knowledge

• Ask: **What is the hardest matter that you know of?** (Possible answers: rocks, diamonds) List student responses on the board. Then have them choose which item on the list they think is hardest.

Set a Purpose and Read

• Read aloud the heading. Say: **Physical chemists study matter and its properties.**

• Explain: **These pages show an interview with Rod Ruoff. The interviewer asked him questions about his career in physical chemistry.**

• Have students read pages 26–27.

❷ Teach

Describe How Physical Chemists Work

• Say: **Physical chemists study properties.** Ask: **What properties do you think they study when they compare diamonds and the "lead" in pencils?** (hardness and texture, and possibly shape, color, and size)

• Ask: **Why do you think it is important for scientists to write reports about their work?** (because scientists share information with other scientists)

• Ask: **Why is their work important?** (Physical chemists create new, better solutions to people's problems. Sometimes a problem exists because the right material hasn't been invented for building a solution.)

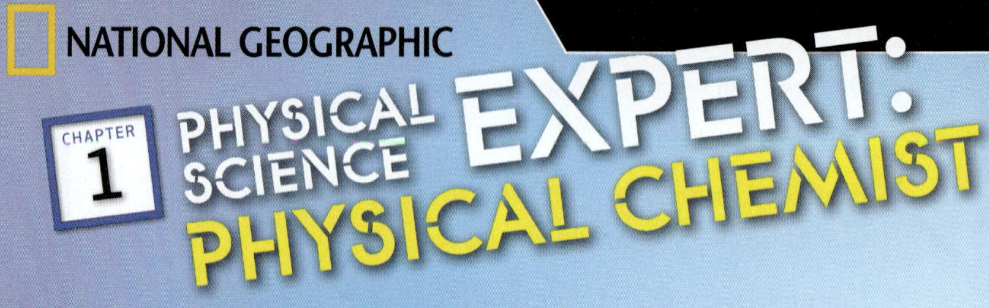

Rod Ruoff is a physical chemist. He works at the University of Texas at Austin. He also started a company where he works with his team to research new ways of making energy. The team uses materials with interesting properties.

NG Science: What do you currently study?

Rod Ruoff: My team and I study properties of materials. We think about how these materials can make society better.

NG Science: When did you first know you wanted to be a physical chemist?

Rod Ruoff: Actually, I wanted to be either a professional hockey player or soccer player! In college, I took a physical chemistry course. I became fascinated with looking at different materials and thinking about new ways to use them.

NG Science: What type of research have you done?

Rod Ruoff: At first, I studied tiny pieces of matter to learn about how they stick together. Now I mostly research materials made of carbon. Diamonds are made of carbon, and so is graphite. Graphite is the same material as the "lead" in your pencil.

26

Nanotech Short for *nanotechnology*, nanotech involves controlling matter on an atomic and molecular level. Scientists try to create new materials and devices. This is a primary focus of Rud Ruoff's research at the University of Texas. Nanotechnology has created new materials that are harder and stronger than conventional materials. One goal is to create tinier computers.

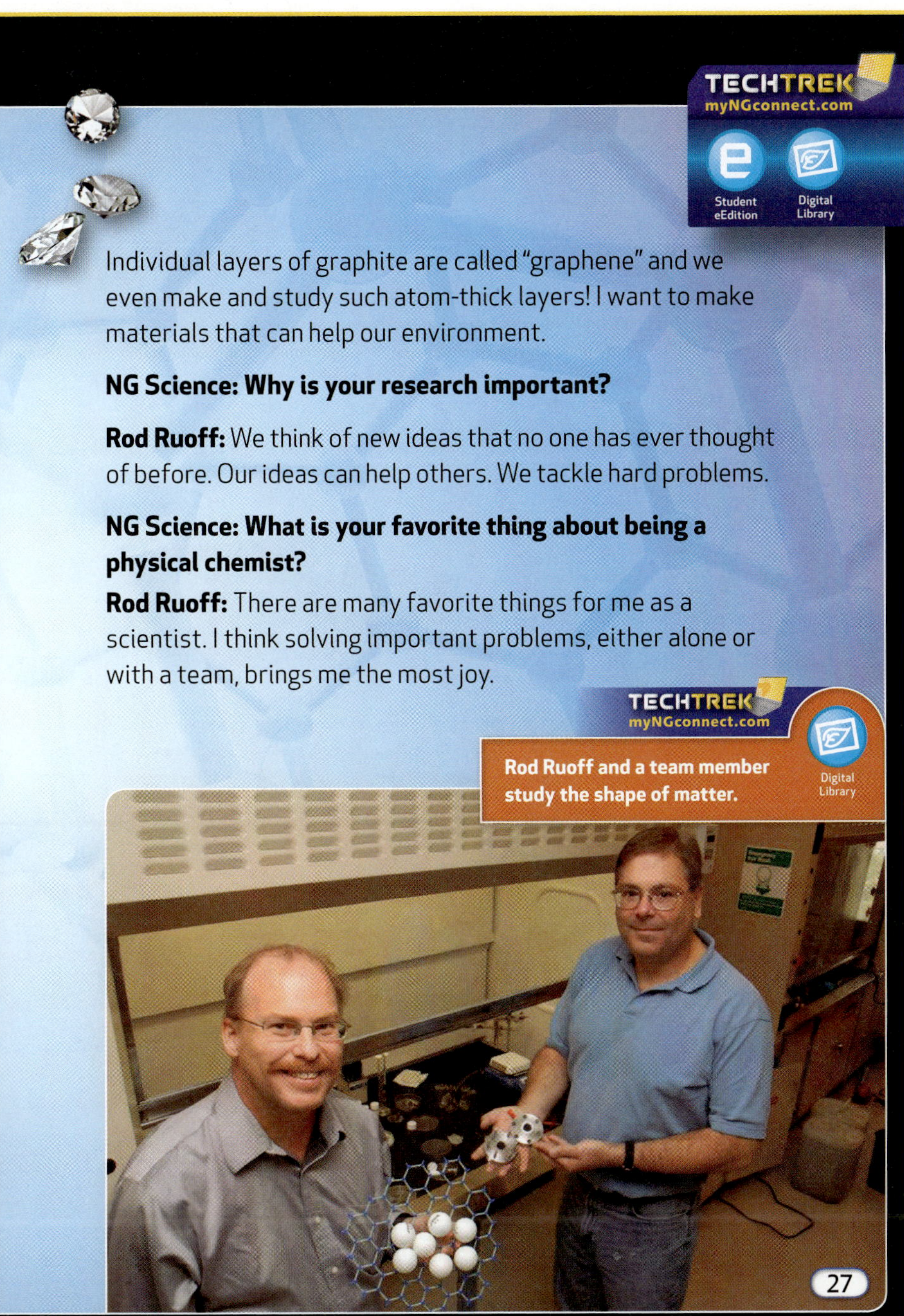

Individual layers of graphite are called "graphene" and we even make and study such atom-thick layers! I want to make materials that can help our environment.

NG Science: Why is your research important?

Rod Ruoff: We think of new ideas that no one has ever thought of before. Our ideas can help others. We tackle hard problems.

NG Science: What is your favorite thing about being a physical chemist?

Rod Ruoff: There are many favorite things for me as a scientist. I think solving important problems, either alone or with a team, brings me the most joy.

27

Differentiated Instruction

ELL Language Support for Comparing Materials

BEGINNING	INTERMEDIATE	ADVANCED
Help students learn words to describe physical chemists by asking either/or questions such as: **Do physical chemists study only diamonds or other materials too?**	Help students describe physical chemists using Academic Language Frames. For example: • *Physical chemists study _____.* • *They find new ways to use _____.*	Help students describe physical chemists using Academic Language Stems. For example: • *Physical chemists study …* • *Rod Ruoff studies materials such as …*

Teach, continued

Find Out More

• Say: **A laboratory contains scientific measurement equipment and materials that scientists use to perform experiments. This is the kind of place where a physical chemist works.**

• Ask: **Is this a career you would find interesting? Why or why not?** (Accept all answers, positive or negative.)

• Encourage interested students to research physical chemistry.

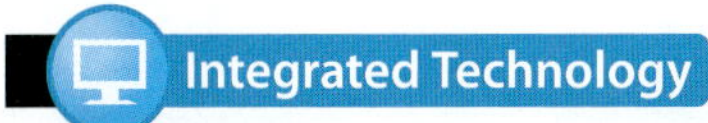

Digital Library

myNGconnect.com

Have students use the **Digital Library** to find photos related to physical chemistry.

Integrated Technology

Whiteboard Presentation Students can use the photos and information to make a presentation about what physical chemists do and the materials and equipment with which they work. Help them show the presentation on a whiteboard.

❸ Assess

1. **Recall** What does physical chemist Rod Ruoff study? (He studies the properties of different materials. He has studied carbon materials, such as diamonds, and now studies new ways to make energy.)

2. **Explain** Why is Rod Ruoff's research important? (He researches new ideas and materials. His research can help others.)

3. **Infer** How might Rod Ruoff's work benefit society and the world? (Rod Ruoff is interested in new ways of making energy. This new way of making energy might cost less or be more environmentally friendly than current methods of making energy.)

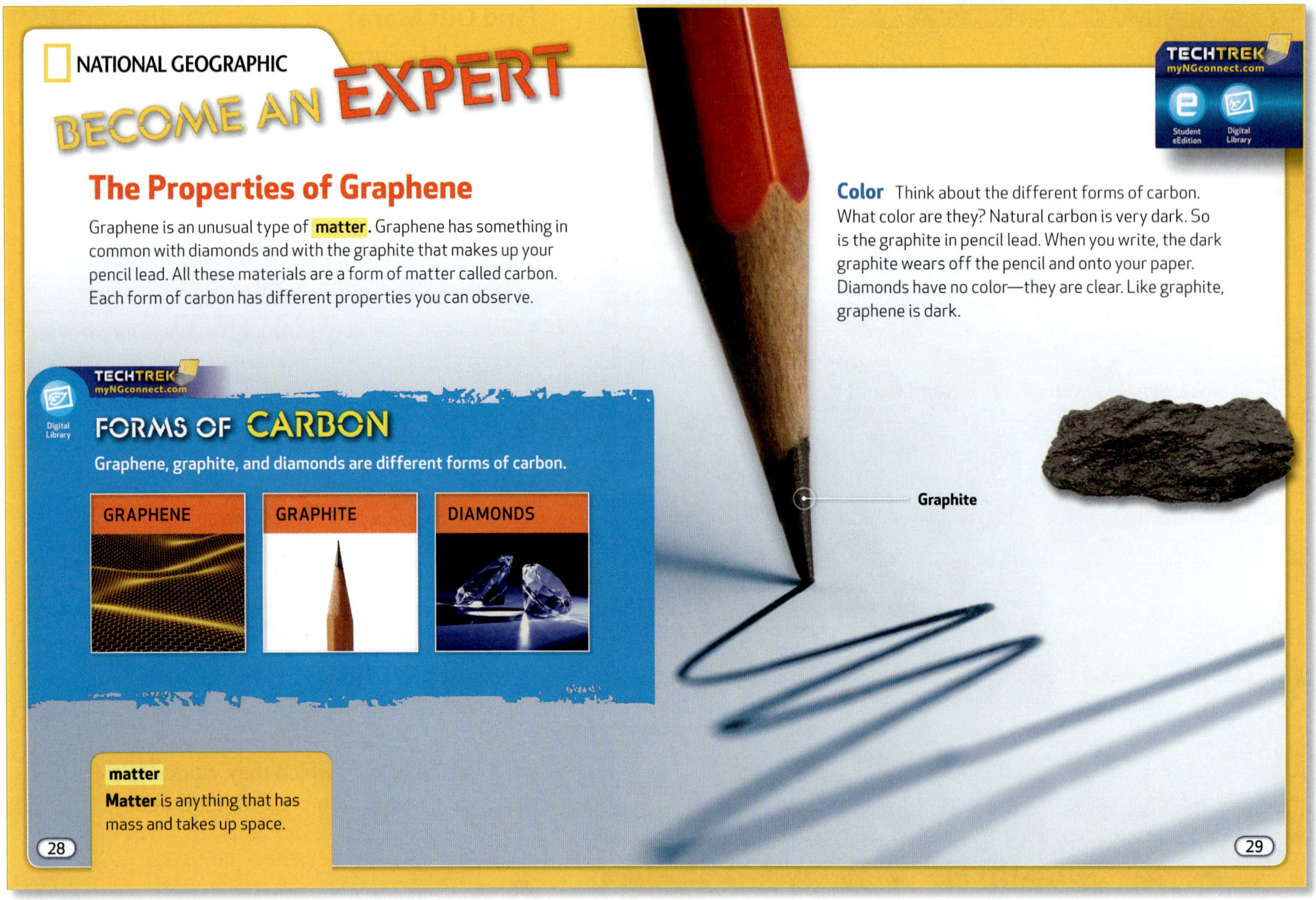

NATIONAL GEOGRAPHIC
BECOME AN EXPERT

The Properties of Graphene

Graphene is an unusual type of **matter**. Graphene has something in common with diamonds and with the graphite that makes up your pencil lead. All these materials are a form of matter called carbon. Each form of carbon has different properties you can observe.

TECHTREK
myNGconnect.com
Digital Library

FORMS OF CARBON

Graphene, graphite, and diamonds are different forms of carbon.

GRAPHENE

GRAPHITE

DIAMONDS

matter
Matter is anything that has mass and takes up space.

28

Color Think about the different forms of carbon. What color are they? Natural carbon is very dark. So is the graphite in pencil lead. When you write, the dark graphite wears off the pencil and onto your paper. Diamonds have no color—they are clear. Like graphite, graphene is dark.

TECHTREK
myNGconnect.com
Student eEdition
Digital Library

Graphite

29

PROGRAM RESOURCES

- Big Ideas Book: *Physical Science*
- Big Ideas Book: *Physical Science* **eEdition** at ⊘ **myNGconnect.com**
- **Digital Library** at ⊘ **myNGconnect.com**

Access Science Content

Compare Three Types of Carbon

- Have students observe the images on pages 28–29. Point out the image of the pencil mark on the paper. Say: **You might call the dark material in a pencil *lead*, but it is actually a type of** **matter** **made entirely of carbon atoms. It does not contain any lead.**

- Point out the images of the three forms of carbon. Ask: **How are these three types of** **matter** **similar?** (They are all made of carbon.) Explain: **Carbon is found in nature.**

- Ask: **How is the color of each type of** **matter** **different?** (Graphene and graphite are dark, and diamonds are colorless.)

Digital Library

⊘ **myNGconnect.com**

Have students use the **Digital Library** to find photos of graphene, graphite, and diamonds.

Making and Recording Observations

Have students study the images and text on pages 28–29. Ask them to create a table called Carbon **Matter** in their science notebook. They should list the different types of carbon and each one's properties, such as texture, color, hardness, shape, and size. Have students create a column to write additional notes about each carbon form.

Access Science Content

Compare Size and Strength

- Have students read pages 30–31.

- Direct students to the small photo on page 30 and read the caption. Point out the grains of rice. Explain: **You can measure the length of a grain of rice, which is about 5 millimeters. That means that a grain of rice is about 5 million nanometers long.** Ask: **Do you think you could see graphene without magnifying it?** (No, it's too small.)

- Direct students to the photos on page 31. Ask: **What is the beam made of?** (steel) **What is happening to the paper?** (It is being ripped.) **Could you tear graphene like you tear paper?** (No, it is stronger than steel.) **What do these two images tell us about graphene?** (It is very strong.) Say: **It would be easier to tear the steel beam than to tear graphene.** Ask: **What value do you think something so strong and so thin has?** (Sample response: It might be used as protective clothing or to strengthen plastics or metal used in airplanes or cars.)

Assess

1. **Name** What is a single layer of carbon atoms that is almost 200 times as strong as steel? (graphene)

2. **Interpret Charts** What are three forms of carbon shown on the chart on page 28? (graphite, graphene, diamond)

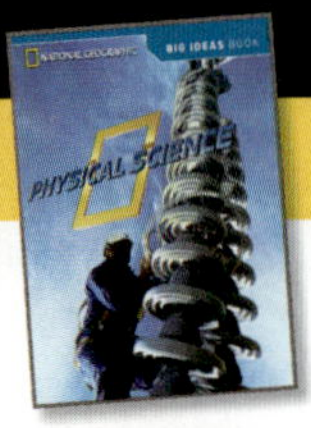

BECOME AN EXPERT

NATIONAL GEOGRAPHIC

Hardness The graphite form of carbon is soft. The diamond form of carbon, though, is one of the hardest materials on Earth. Graphene is a hard material, too. Adding graphene to materials can make them stiff. If you hold a sheet of paper on the side by its edges, it will bend toward the floor. A sheet of graphene is much thinner than a sheet of paper. If you could hold a sheet of graphene, it would be stiff. Researchers are looking for ways to make materials stronger and stiffer by adding graphene to them.

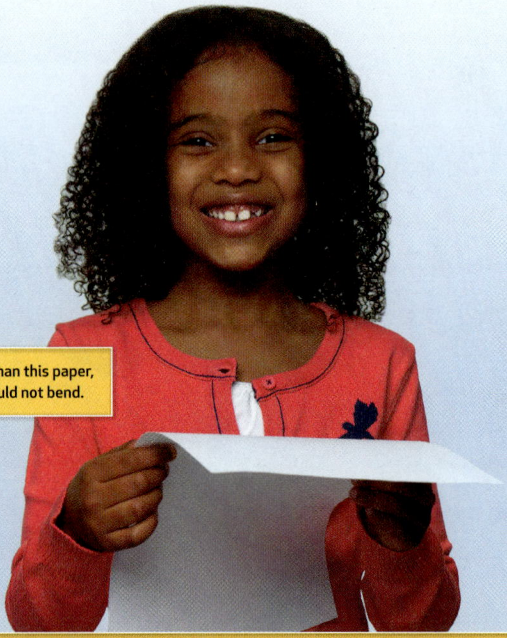

Graphene is much thinner than this paper, but a sheet of graphene would not bend.

Shape and Texture What makes graphene so strong? The property of shape is part of graphene's super strength. If you look at a very magnified view of graphene, you see it looks like chicken wire. Each of these shapes works together to form a strong network. The **texture** of a sheet of graphene is smooth.

Someone can see through chicken wire by looking through the holes.

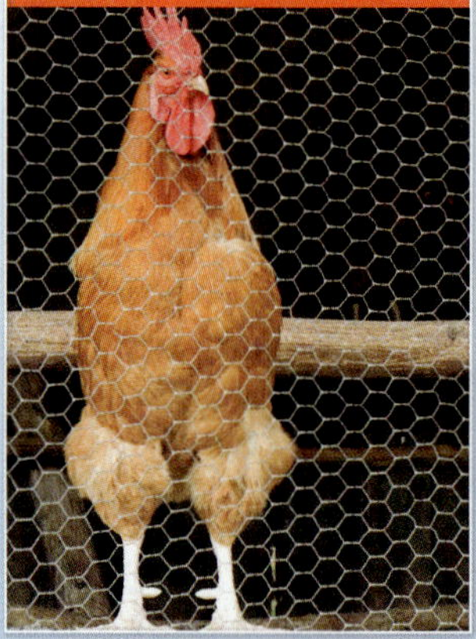

Graphene looks a lot like chicken wire.

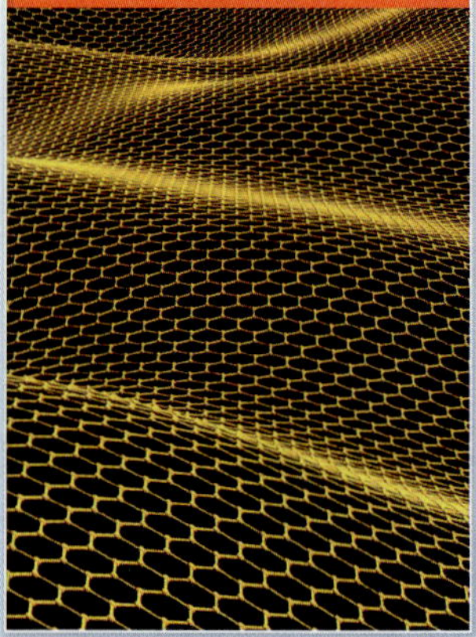

texture
Texture describes the surface of any area made up of matter.

32

33

Access Science Content

Describe the Hardness and Texture of Graphene

- Have students observe the image on page 32. Say: **Compare the hardness of graphite and graphene.** (Graphene is a lot harder than graphite.) **What is the texture of graphene?** (It is smooth.)
- Have students read pages 32–33. Ask: **What value do scientists think graphene has?** (Graphene can be added to make something stronger.)

Differentiated Instruction

ELL Language Support for Describing Graphene

BEGINNING

Help students describe graphene with either/or questions, such as: **Is graphene smooth or rough? Is graphene hard or soft?**

INTERMEDIATE

Use Academic Language Frames to help students describe graphene.

- *The texture of graphene is _____.*
- *Graphene is useful because it is _____.*

ADVANCED

Use Academic Language Stems to help students describe graphene.

- *Graphene is . . .*
- *Scientists think graphene is valuable because . . .*

Notetaking

my SCIENCE notebook

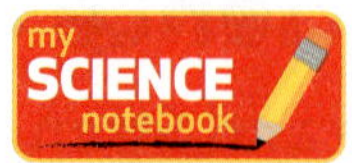

As students read pages 32–33, ask them to write down new words in their science notebook. Have them record the definition given on the page. In addition, have them write a descriptive sentence in their own words. Encourage them to draw a picture, when possible, to help them better understand the word.

Access Science Content

Measure Graphene's Mass and Volume

- Have students read page 34. Say: **Mass is measured in grams. If you place two solids in a balance scale, the one with more matter hangs lower. A book has more matter than a teaspoon of sugar.** Ask students to look back at the sheet of graphene on page 33. Ask: **What would happen if the 1-meter square sheet of graphene and a teaspoon of salt were placed in a balance scale?** (The salt would hang lower because it is heavier, or has more mass.)

- Remind students: **To find the volume of a solid, you multiply its length times width times height.** Have students look at the image on page 34. Say: **A layer of graphene large enough to cover this field would have a length of 110 meters and a width of 49 meters. 110 times 49 equals 5,390. That is a large number, but we would have to multiply it by the height of the graphene to get its volume. The height is less than a nanometer, which is tiny.**

- Have students read page 35. Ask: **Why is graphene able to let light through its surface?** (It is very thin.) Say: **Graphene could be used to create paper-thin computer screens.** Ask students to think about books or news being read on graphene.

Assess

1. **Recall What is the texture of a sheet of graphene?** (smooth)

2. **Interpret Why are graphene's mass and volume so very small?** (Graphene is so thin, less than a nanometer, that even a large layer of it will still not have much mass or volume.)

Share and Compare

Turn and Talk

Ask students to turn to partners and talk about what they learned about the properties of graphene. Prompt students by asking:

1. **Recall** **Name three different forms of carbon, and list a few properties of each form.** (graphene, graphite, diamond; Diamonds are hard and clear. Graphite is dark. Graphene is dark, thin, and strong.)

2. **Explain** **How are the properties of graphene special?** (Graphene is incredibly thin, stiff, and strong, and it allows light to go through it. It allows energy to flow through it. It is almost 200 times as strong as steel.)

3. **Predict** **How do you expect graphene to be used in technology in the future?** (Answers will vary, but students should be able to explain how the strong properties of graphene could improve the technology.)

Read

Emphasize that the pages students select should be the ones they found most interesting. Allow time for students to practice reading the pages. Tell students to think about why they found those pages most interesting before talking about them with others.

Write

Have students write a conclusion that tells what important ideas they learned about the properties of graphene. Tell them to write the Big Idea in their own words. Have students share their conclusion with a classmate and look for what is unique about graphene.

Draw

Have students think about what graphene might look like if it were magnified even more than it was in the photos of this book. Ask students to draw what they think graphene would look like. Ask them to combine their drawings with those of their classmates.

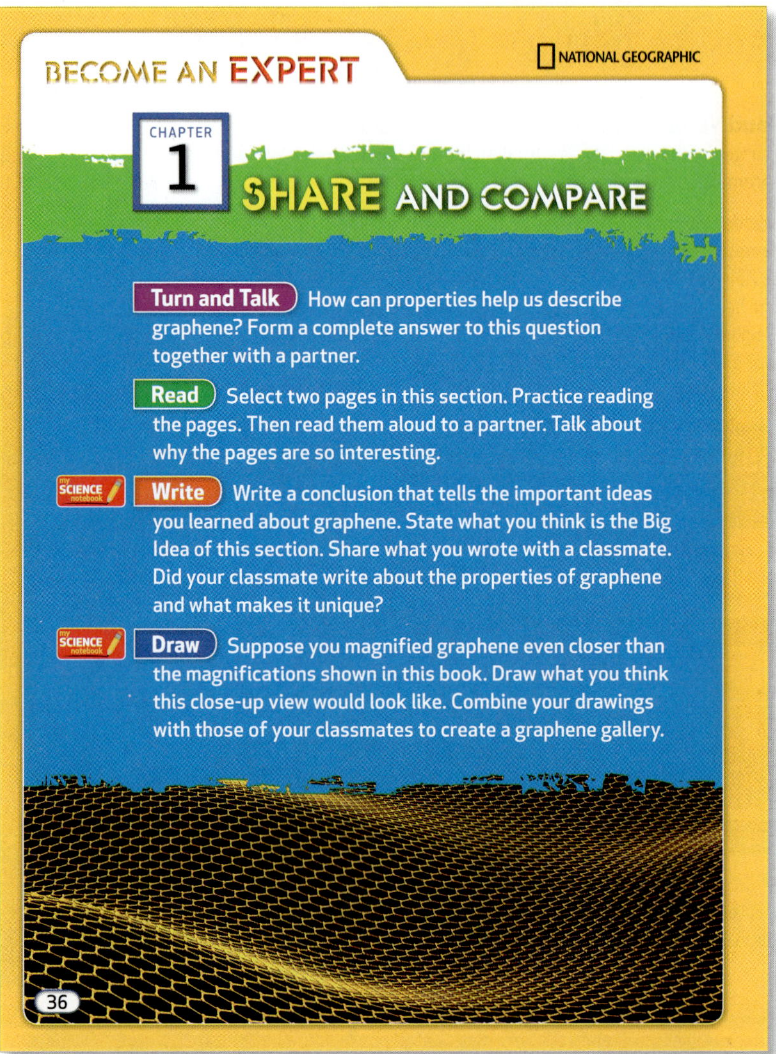

❯ Sum Up

Tell students that to sum up a text helps them pull together the ideas of the text. Remind students that they summed up the Become an Expert lesson in the Write section of Share and Compare. Have students take turns reading the conclusions they wrote to a partner. Invite them to compare and contrast their conclusions.

PHYSICAL SCIENCE

Chapter 2 Contents

What Are States Of Matter?

States of Matter

Matter Changes State

Measure Temperature

After reading Chapter 2, you will be able to:

- Recognize that all objects are made of matter and that matter takes up space and has mass. **STATES OF MATTER**

- Describe, classify, and explain the properties of solids, liquids and gases and recognize that water can exist in all three states. **STATES OF MATTER**

- Recognize that air is a substance that surrounds us, has mass, and takes up space. **STATES OF MATTER**

- Understand that the properties of an object are dependent on the conditions of the present surroundings in which the object exists, such as temperature. **MATTER CHANGES STATE**

- Recognize that heating and cooling (temperature change) may cause changes in the properties of materials such as water, including phase changes in the state of matter. **MATTER CHANGES STATE**

- Measure and compare the temperature of water when it exists as a solid to its temperature when it exists as a liquid. **MATTER CHANGES STATE**

- **Science in a Snap!** Understand that the properties of an object are dependent on the conditions of the present surroundings in which the object exists, such as temperature. **MATTER CHANGES STATE**

37

❯ Preview and Predict

Tell students that they can preview a text by looking over it to get an idea of what the text is about and how it is organized. Tell students that predicting is forming ideas about what they will read. Students should confirm their ideas with others.

Have students preview and make predictions about the chapter. Remind them to use section heads, vocabulary words, pictures, and their background knowledge to make predictions.

CHAPTER 2 □ What Are States of Matter?

LESSON	PACING	OBJECTIVES
Directed Inquiry **1** *Investigate Water and Temperature* pages T37e–T37h **Science Inquiry and Writing Book** pages 174–177	**15** minutes	Investigate through Directed Inquiry (answer a question; make and compare observations; collect and record data and observations; generate explanations and conclusions based on evidence; share findings; ask questions based on observations to increase understanding). Recognize that matter can exist as a solid, liquid, or gas. Describe changes in the physical properties of water when it changes state through heating and cooling. Identify the evidence used in an explanation. Question the explanation from others, seeking clarification and comparing them with their own explanations and understandings. Keep records of investigations and observations and do not change records that are different from someone else's work.
2 **Big Idea Question and Vocabulary** pages T38-T39–T40-T41 **States of Matter** pages T42–T45	**35** minutes	Recognize that water, as an example of matter, can exist as a solid, liquid, or gas. Recognize that all objects are made of matter and that matter takes up space and has mass. Understand that two objects cannot occupy the same place at the same time. Describe, classify, and explain the properties of solids, liquids, and gases. Recognize that air is a substance that surrounds us, has mass, and takes up space.
3 **Matter Changes State** pages T46–T51	**30** minutes	Understand that the properties of an object are dependent on the conditions of the present surroundings in which the object exists, such as temperature. Recognize that heating and cooling (temperature change) may cause changes in the properties of materials such as water, including phase changes in the state of matter. Describe the changes in the physical properties of water when it is frozen or melted. Measure and compare the temperature of water when it exists as a solid to its temperature when it exists as a liquid. Recognize that liquid water evaporates, changing into a gas as it moves into the air.
4 **Measure Temperature** pages T52–T53	**20** minutes	Understand that the properties of an object are dependent on the conditions of the present surroundings in which the object exists, such as temperature.
5 NATIONAL GEOGRAPHIC **Fog Fences in Central America** pages T54–T55	**20** minutes	Recognize that heating and cooling (temperature change) may cause changes in the properties of materials such as water, including phase changes in the state of matter.

VOCABULARY	RESOURCES	ASSESSMENT
observe **compare**	Science Inquiry and Writing Book: *Physical Science* Science Inquiry Kit: *Physical Science* Directed Inquiry: Learning Masters 191–194	Inquiry Rubric: Assessment Handbook, page 212 Inquiry Self-Reflection: Assessment Handbook, page 219 Reflect and Assess, page T37h
state of matter **solid** **liquid** **gas**	Vocabulary: Learning Master 195 Big Ideas Book: *Physical Science*	Assess, page T45
evaporation **condensation**	Science Inquiry and Writing Book: *Physical Science*	Assess, page T51
		Assess, page T53
	Share and Compare: Learning Master 196	Assess, page T55

TECHNOLOGY RESOURCES

STUDENT RESOURCES

🔵 **myNGconnect.com**

- **Student eEdition**
- Big Ideas Book
- Science Inquiry and Writing Book
- Explore on Your Own Books
- **Read with Me**
- **Vocabulary Games**
- **Digital Library**
- **Enrichment Activities**

National Geographic Kids

National Geographic Explorer!

TEACHER RESOURCES

🔵 **myNGconnect.com**

- **Teacher eEdition**
- Teacher's Edition
- Science Inquiry and Writing Book
- Explore on Your Own Books
- Online Lesson Planner
- National Geographic Unit Launch Videos
- Assessment Handbook
- **Presentation Tool**
- **Digital Library**

NGSP ExamView CD-ROM

▶▶▶

IF TIME IS SHORT...
FAST FORWARD.

CHAPTER 2 □ What Are States of Matter?

LESSON	PACING	OBJECTIVES
6 **Think Like a Scientist** *Math in Science: Measuring Temperature* pages T55a–T55d **Science Inquiry and Writing Book** pages 178–181	**10** minutes	Develop strategies and skills for information-gathering and problem-solving using appropriate scientific tools and technologies. Conduct an investigation in which temperature is measured to the nearest degree Celsius. Make observations and collect data using scientific process skills. Compare amounts and measurements made by others and seek reasons to explain the differences. Express quantities using appropriate number formats.
Guided Inquiry **7** *Investigate Melting* pages T55e–T55h **Science Inquiry and Writing Book** pages 182–185	**40** minutes	Investigate through Guided Inquiry (answer a question; make and compare observations; collect and record data and observations; generate explanations and conclusions based on evidence; share findings; ask questions based on observations to increase understanding; adjust explanations based on findings and new ideas). Measure and compare the temperature of liquid and solid water. Plan and conduct a fair test. Construct simple charts and tables from data and observations and summarize that information to answer scientific questions.
8 **Conclusion and Review** pages T56–T57	**15** minutes	
9 NATIONAL GEOGRAPHIC **PHYSICAL SCIENCE EXPERT** *Ice Sculptor* pages T58–T59 NATIONAL GEOGRAPHIC **BECOME AN EXPERT** *Sweden: Ice Hotel Construction* pages T60-T61–T68	**35** minutes	Understand that scientists and inventors come from different gender and ethnic backgrounds and describe the work they have contributed to science and technology. Describe how water undergoes changes through heating and cooling using familiar terms such as melting and freezing. Describe, classify, and explain the properties of solids, liquids, and gases. Understand that the properties of an object are dependent on the conditions of the present surroundings in which the object exists, such as temperature.

FAST FORWARD ▶▶▶
ACCELERATED PACING GUIDE

DAY 1 🕒 **15** minutes

Directed Inquiry

Investigate Water and Temperature, page T37e

DAY 2 🕒 **35** minutes

NATIONAL GEOGRAPHIC **PHYSICAL SCIENCE EXPERT** *Ice Sculptor,* page T58

NATIONAL GEOGRAPHIC **BECOME AN EXPERT** *Sweden: Ice Hotel Construction,* page T60–T61

VOCABULARY	RESOURCES	ASSESSMENT
	Science Inquiry and Writing Book: *Physical Science* Think Like a Scientist: Learning Masters 197–198	Assess, page T55d
experiment **variable**	Science Inquiry and Writing Book: *Physical Science* Science Inquiry Kit: *Physical Science* Guided Inquiry: Learning Masters 199–202	Inquiry Rubric: Assessment Handbook, page 212 Inquiry Self-Reflection: Assessment Handbook, page 220 Reflect and Assess, page T55h
		Assess the Big Idea, page T57 Chapter 2 Test: Assessment Handbook, pages 92–95 NGSP ExamView CD-ROM
		Assess, pages T59, T62–T63, T66–T67

TECHNOLOGY RESOURCES

STUDENT RESOURCES

myNGconnect.com

Student eEdition

Big Ideas Book

Science Inquiry and Writing Book

Explore on Your Own Books

Read with Me

Vocabulary Games

Digital Library

Enrichment Activities

National Geographic Kids

National Geographic Explorer!

TEACHER RESOURCES

myNGconnect.com

Teacher eEdition

Teacher's Edition

Science Inquiry and Writing Book

Explore on Your Own Books

Online Lesson Planner

National Geographic Unit Launch Videos

Assessment Handbook

Presentation Tool

Digital Library

NGSP ExamView CD-ROM

DAY 3 🕐 40 minutes

Guided Inquiry

Investigate Melting, page T55e

Objectives

Students will be able to:

- Investigate through Directed Inquiry (answer a question; make and compare observations; collect and record data and observations; generate explanations and conclusions based on evidence; share findings; ask questions based on observations).
- Recognize that matter can exist as a solid, liquid, or gas.
- Describe changes in the physical properties of water when it changes state through heating and cooling.
- Identify the evidence used in an explanation.
- Question the explanation from others, seeking clarification and comparing them with their own understandings.
- Keep records of investigations and observations and do not change records that are different from someone else's work.

Science Process Vocabulary

observe, compare

PROGRAM RESOURCES

- Science Inquiry and Writing Book: *Physical Science*
- Science Inquiry and Writing Book **eEdition** at **myNGconnect.com**
- **Inquiry eHelp** at **myNGconnect.com**
- Science Inquiry Kit: *Physical Science*
- Learning Masters Book, pages 191–194, or at **myNGconnect.com**
- Inquiry Rubric: Assessment Handbook, page 212, or at **myNGconnect.com**
- Inquiry Self-Reflection: Assessment Handbook, page 219, or at **myNGconnect.com**

MATERIALS

Kit materials are listed in italics.

2 resealable plastic bags; masking tape; *graduated cylinder (100 mL);* pitcher; water

❶ Introduce

Tap Prior Knowledge

- Ask students to tell about water in its different states. Guide them to use *melting, freezing,* and *boiling* as they describe the different states.

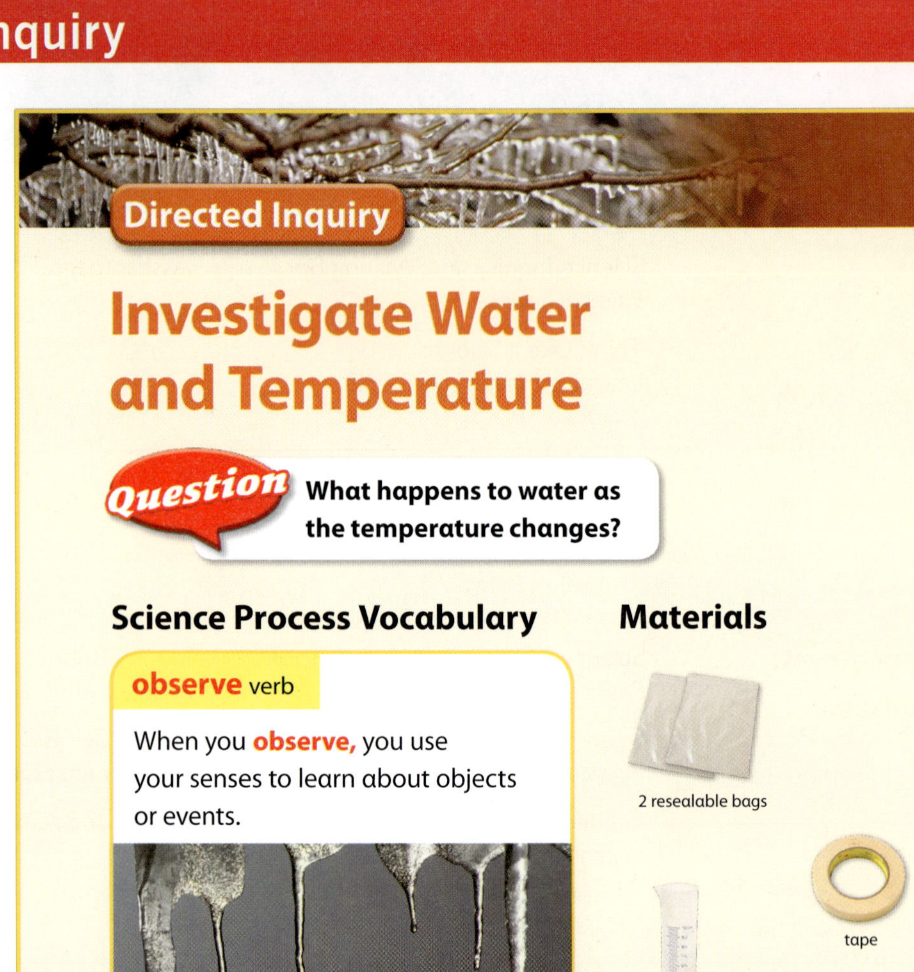

Investigate Water and Temperature

Question What happens to water as the temperature changes?

Science Process Vocabulary

observe verb

When you **observe,** you use your senses to learn about objects or events.

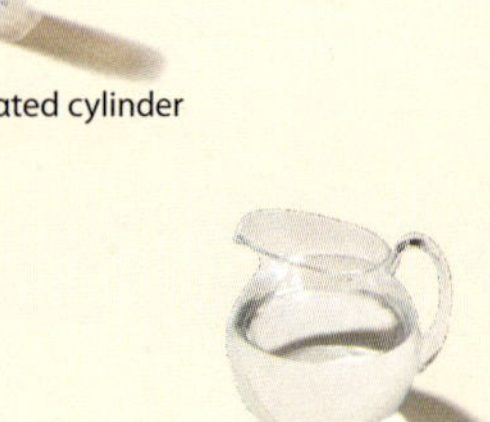

compare verb

When you **compare,** you tell how objects or events are alike and different.

Materials

2 resealable bags

tape

graduated cylinder

water

174

MANAGING THE INVESTIGATION

Time

15 minutes per day for 4 days

Groups

Small groups of 4

Advance Preparation

- Fill pitchers with water. Each group will need at least 200 mL of water.

Teaching Tips

- In steps 4 and 5, have students place paper towels or newspapers under the bags to absorb moisture from condensation or leaks.
- In step 5, place the bags under a lamp if it is not sunny outside.

What to Do

1 Label the plastic bags **Bag 1** and **Bag 2.** Use the graduated cylinder to **measure** 100 mL of water. Pour the water into one bag. Seal the bag. Repeat with the other bag.

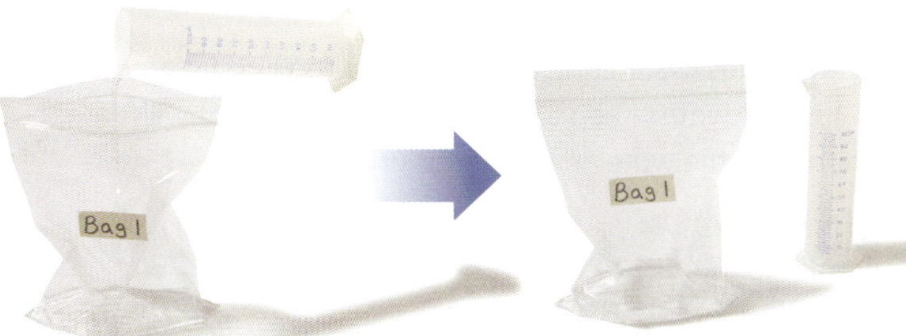

2 Carefully place both bags in the freezer. **Predict** what will happen to the water in the bags. Record your prediction in your science notebook.

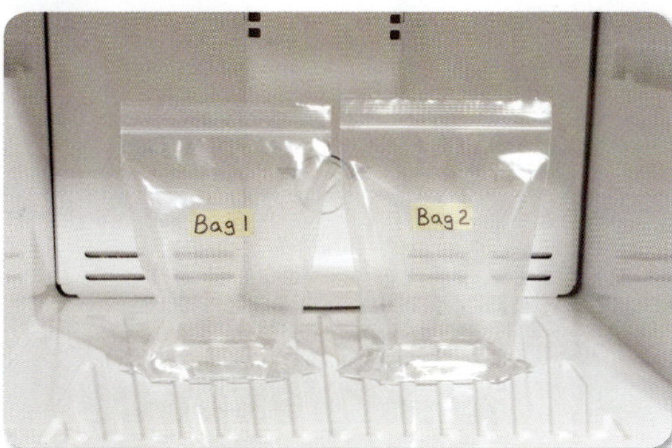

175

Teaching Tips

- When water is poured into a graduated cylinder, the surface of the water often forms a curve, or a meniscus. This is due to attraction of water molecules to the glass. The proper way to determine the volume is to read the value at the bottom of the meniscus. In some types of plastic graduated cylinders, a meniscus will not be evident.

What to Expect

- Students will observe water melting, freezing, condensing, and evaporating. Water will evaporate and condense in the sealed bag and evaporate from the open bag.

Introduce, continued

Connect to the Big Idea

- Review the Big Idea Question, *What are states of matter?* Explain to students that this inquiry will help them learn how water changes from one state of matter to another.

- Have students open their Science Inquiry and Writing Books to page 174. Read the Question and invite students to tell what happens to water when the temperature changes.

❷ Build Vocabulary

Science Process Words: observe, compare

Use this routine to introduce the words.

1. **Pronounce the Word** Say **observe**. Have students repeat it in syllables.

2. **Explain Its Meaning** Choral read the sentence. Ask students for another word or phrase that means the same as **observe**. (watch carefully)

3. **Encourage Elaboration** Ask: **What senses will you use to make observations in this investigation?** (sight, touch)

   ELL **Use Language Frames**
   Write *I observe* _____. Show students a cup of water. Ask: **What do you observe about the water?** (I observe that it is a liquid.) Remind students to use the language frame to respond.

Repeat for the word **compare**. To encourage elaboration, hold up a pencil and a pen and ask students to **compare** the two objects.

❸ Guide the Investigation

- Distribute materials. Read the steps on pages 175–176 together with students. Move from group to group and clarify steps if necessary.

- In step 1, one student may hold the bag open while another student pours the water. Students should check that the bags are tightly sealed.

- Remind students to read the volume measurement at the bottom of the meniscus or the center of the graduated cylinder.

- Students should write predictions in their science notebooks. Predictions, data, and conclusions should be clearly labeled.

Guide the Investigation, continued

- In steps 3 and 4, students should compare their observations of solid ice and liquid water. Ask: **When you turn the bags, what happens to the solid ice?** (It keeps its shape.) **What happens to the liquid water?** (It changes shape and takes the shape of the bag.)

- Review with students the meanings of *evaporation* and *condensation*. Then ask: **In which bag did the water evaporate into the air?** (Bag 1) **In which bag did the water evaporate and condense?** (Bag 2) **How can you tell?** (There was less water in Bag 1, and there were water droplets in Bag 2.)

❹ Explain and Conclude

- Guide students as they record their observations in a table. Be sure students draw the table large enough to include detailed observations.

- Ask students to reread their predictions. They should work in pairs as they evaluate how their understanding has changed.

- Encourage students to describe the physical properties of the water as it changes state, such as its shape and whether it is visible. Then encourage them to make inferences and generalizations about how water changes when temperature changes.

- Ask students to identify evidence used in their explanations. Ask: **What evidence did you use to support what happens to water when the temperature changes?**

- Discuss how questioning explanations helps students to clarify their own understanding and observations. Encourage students to question and compare observations and results.

- Discuss why students should not change their observations or results if their data varies from others. Instead, explain that they should come up with an explanation for their results.

Answers

1. Answers will vary, but students may have predicted that the water in both bags on the desk would melt and that the water in the open bag would evaporate.

What to Do, continued

3 The next day, take the bags out of the freezer. Turn the bags in different directions. **Observe** the shape of the water. Record your observations.

4 Put the bags on your desk. Wait 30 minutes. Turn the bags in different directions. Record your observations.

5 Put the bags in sunlight. Open Bag 1, being careful not to spill the water. Keep Bag 2 sealed. Predict what will happen to the water in both bags after 3 days.

6 Observe the bags every day for the next 3 days. Record your observations.

176

NATIONAL GEOGRAPHIC ## Raise Your SciQ!

Condensation Condensation occurs when a gas cools and turns into a liquid. Students may observe water droplets forming on the outside of the plastic bags filled with ice. This happens because the ice cools the water vapor in the air around the bag. Water droplets may also form on the inside of Bag 2 after the water has been heated by the sun for a few days. Water in the bag evaporates, but it cannot escape because the bag is sealed. When the air cannot hold any more water vapor, some must condense into droplets.

Record

Write or draw in your science notebook.
Use a table like this one.

Observations of Bags with Water

	Bag 1	Bag 2
After 1 day in the freezer		
After 30 minutes on the desk		

Explain and Conclude

1. Did your results support your **predictions?** Explain.

2. **Compare** your results with the results of other groups. What patterns do you see?

3. Use your results from steps 2–6 to **conclude** what happens to water when the temperature goes down. What happens to frozen water when the temperature goes up?

Think of Another Question

What else would you like to find out about what happens to water as the temperature changes? How could you find an answer to this new question?

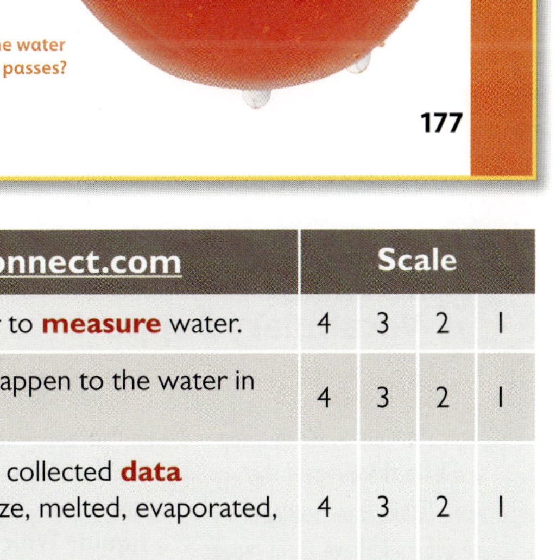
What will happen to the water on this tomato as time passes?

177

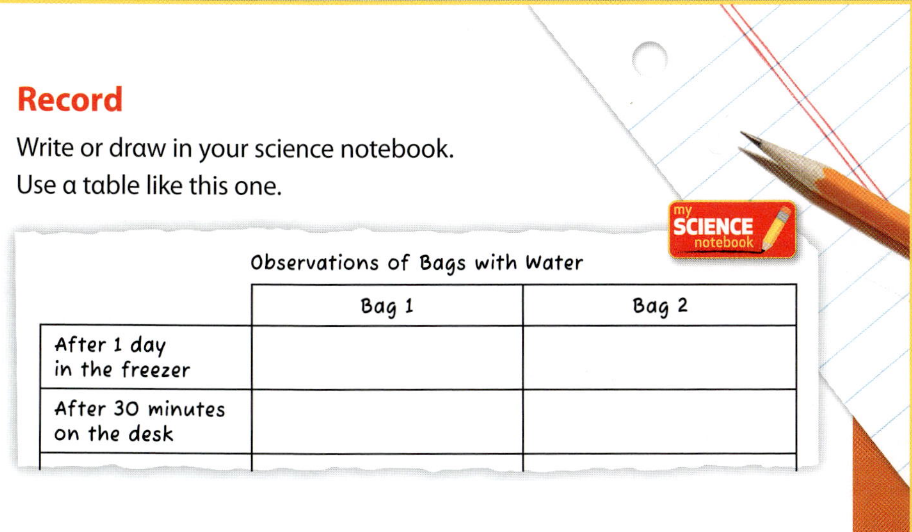

Explain and Conclude, continued

2. Answers will vary. Students should notice patterns in how the water changes in each step.

3. Possible answer: When the temperature goes down, water freezes and becomes solid. When the temperature goes up, frozen water melts and becomes liquid. It may evaporate and become a gas. In a closed plastic bag, some water vapor may condense and become a liquid.

❺ Find Out More

Think of Another Question

- Students should use their observations to generate new questions that arise from their investigations. Discuss what students could do to find answers.

❻ Reflect and Assess

- To assess student work with the Inquiry Rubric shown below, see Assessment Handbook, page 212, or go online at ⊙ **myNGconnect.com**

- Score each item. Decide on one overall score.

- Have students use the Inquiry Self-Reflection on Assessment Handbook, page 219, or at ⊙ **myNGconnect.com**

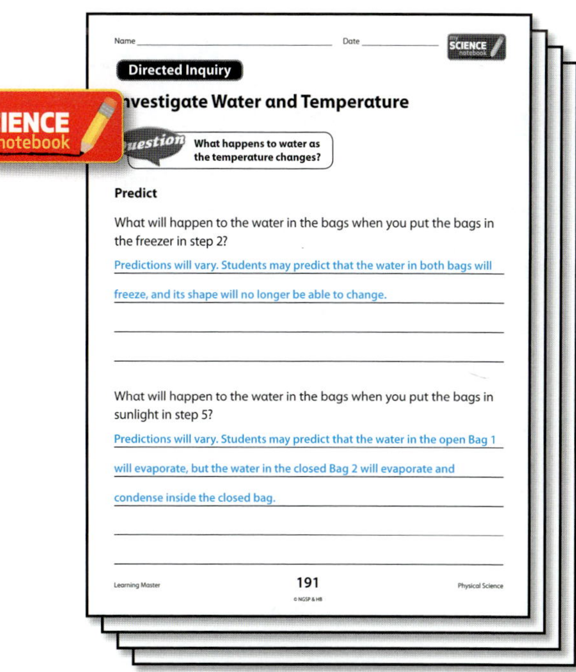

Inquiry Rubric at ⊙ myNGconnect.com	Scale			
The student used a graduated cylinder to **measure** water.	4	3	2	1
The student **predicted** what would happen to the water in the bags.	4	3	2	1
The student made **observations** and collected **data** about how the water changed as it froze, melted, evaporated, and condensed.	4	3	2	1
The student **compared** ideas and **shared** results.	4	3	2	1
The student used the results to draw **conclusions** about what happens to frozen water when temperatures change.	4	3	2	1
Overall Score	4	3	2	1

Learning Masters 191–194
or at ⊙ myNGconnect.com

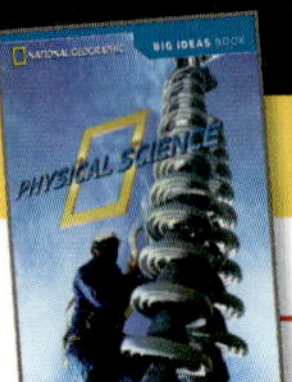

Objectives

Students will be able to:

- Recognize that water, as an example of matter, can exist as a solid, liquid, or gas.

Science Academic Vocabulary

states of matter, solid, liquid, gas, evaporation, condensation

PROGRAM RESOURCES

- Big Ideas Book: *Physical Science*
- Big Ideas Book: *Physical Science* **eEdition** at ⊘ **myNGconnect.com**
- **Vocabulary Games** at ⊘ **myNGconnect.com**
- **Digital Library** at ⊘ **myNGconnect.com**
- **Enrichment Activities** at ⊘ **myNGconnect.com**
- **Read with Me** at ⊘ **myNGconnect.com**
- Learning Masters Book, page 195, or at ⊘ **myNGconnect.com**

❶ Introduce

Tap Prior Knowledge

- Briefly review the *Physical Science* video. Ask students to explain how the scientist studies the way substances change. Encourage them to offer what they already know about how water changes.

❷ Focus on the Big Idea

Big Idea Question

- Read the Big Idea Question aloud and have students echo it.

- Preview pages 42–45, 46–51, and 52–53, linking the headings with the Big Idea Question.

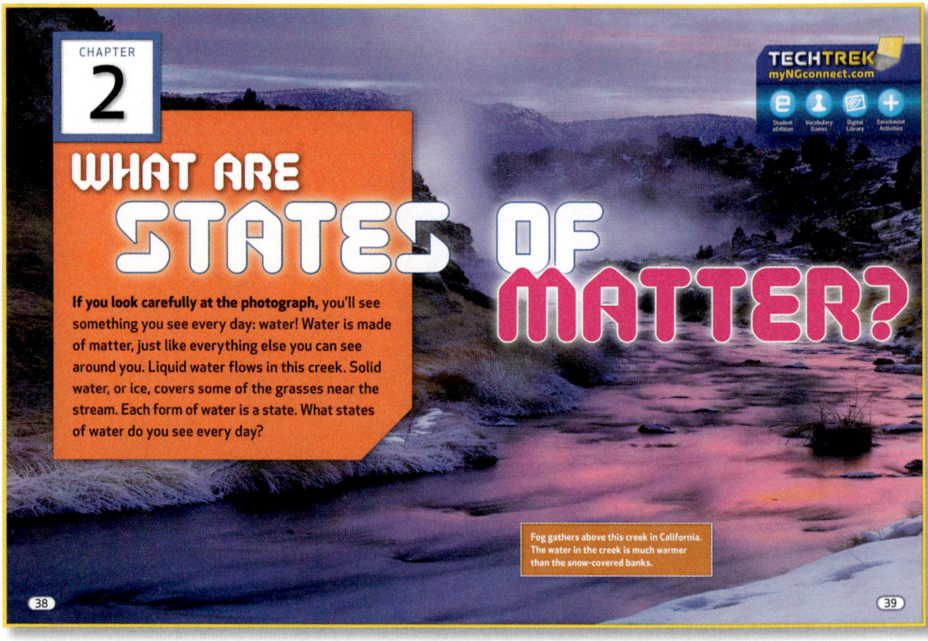

Differentiated Instruction

ELL Vocabulary Support

BEGINNING	INTERMEDIATE	ADVANCED
Show students water in a cup and have them repeat the word *liquid*. Show students an ice cube and have them repeat the word *solid*.	Show students water in a cup and a solid such as an eraser, and ask, **Which is a liquid? Which is a solid?** Students should point and respond verbally: *Water is a liquid. The eraser is a solid.*	Have students name several liquids (e.g., *water, juice, milk*) and several solids (e.g., *wood, metal, paper*).

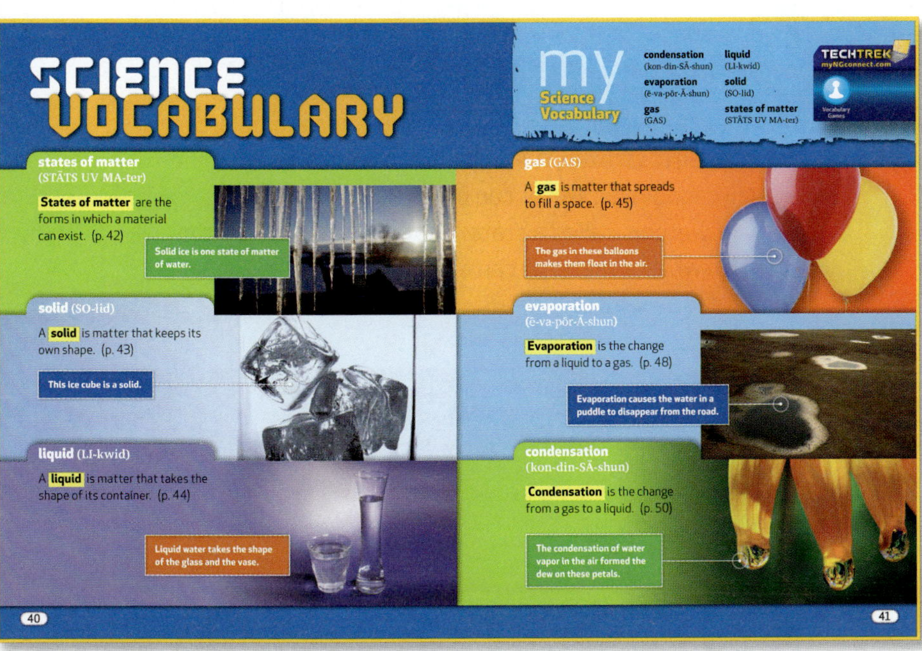

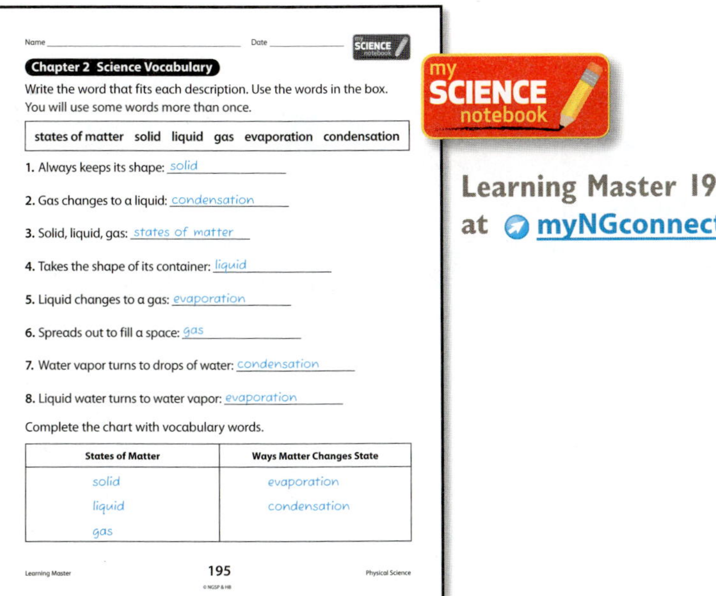

**Learning Master 195 or
at myNGconnect.com**

Focus on the Big Idea, continued

- With student input, post a chart that displays the headings. Have students orally share what they expect to find in each section. Then read page 38 aloud.

❸ Teach Vocabulary

Have students look at pages 40–41, and use this routine to teach each word. For example:

1. **Pronounce the Term** Say **states of matter** and have students repeat it.

2. **Explain Its Meaning** Read the term, definition, and sample sentence, and use the photo to explain the term's meaning. Point out the icicles: **Water can be in different states of matter.**

3. **Encourage Elaboration** Have students look at the icicles. Ask: **In what state of matter is the water?** (solid state)

- Repeat for the words **solid**, **liquid**, **gas**, **evaporation**, and **condensation** using the following Elaboration Prompts:

- **If you put the solid in a cup, what shape would the solid have?** (the same shape as before)

- **If you poured the liquid into a cup, what shape would the liquid have?** (the cup's shape)

- **If you pop a balloon, what happens to the gas inside?** (It spreads out into the air.)

- **To what state does evaporation lead?** (gas)

- **To what state does condensation lead?** (liquid)

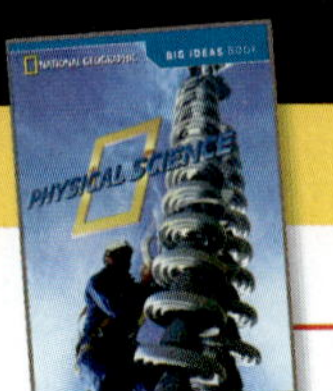

Objectives
Students will be able to:

- Recognize that all objects are made of matter and that matter takes up space and has mass.
- Understand that two objects cannot occupy the same place at the same time.
- Describe, classify, and explain the properties of solids, liquids, and gases.
- Recognize that water, as an example of matter, can exist as a solid, liquid, or gas.

Science Academic Vocabulary
states of matter, solid

PROGRAM RESOURCES
- **Digital Library** at **myNGconnect.com**

❶ Introduce

Tap Prior Knowledge
- Ask students to share what they know about liquid water, snow, and ice. Have students describe any bodies of water located where they live. Ask about the differences in the water during different temperatures.

Set a Purpose and Read
- Read the heading. Tell students that **"states of matter"** means the same as "forms of matter" and that they will read about the different forms that matter can be in.
- Have students read pages 42–45.

❷ Teach

Academic Vocabulary: *states of matter, solid*
- Call attention to the term **states of matter**. Say: **Matter makes up every substance that exists, including air, water, rocks, and you. Matter can be in different forms, or states.** Review the concept that matter takes up its own space.
- Pronounce the word **solid** and write it. Have a volunteer read the sentence that defines it. Say: **Solid is a state of matter that keeps its shape.**

States of Matter

Take a close look at the photo. What do you see? You see matter! Everything around you is made of matter. Matter has mass and takes up space. Each piece of matter takes up its own space. That's why only one fishing pole, for example, can be in the same place at the same time.

Matter can have different forms. Water, for example, can flow. When it's frozen, it's hard. There is even water in the air you cannot see. Water can be a solid, liquid, or a gas. Each form of water is called a state. **States of matter** are the forms in which a material can exist.

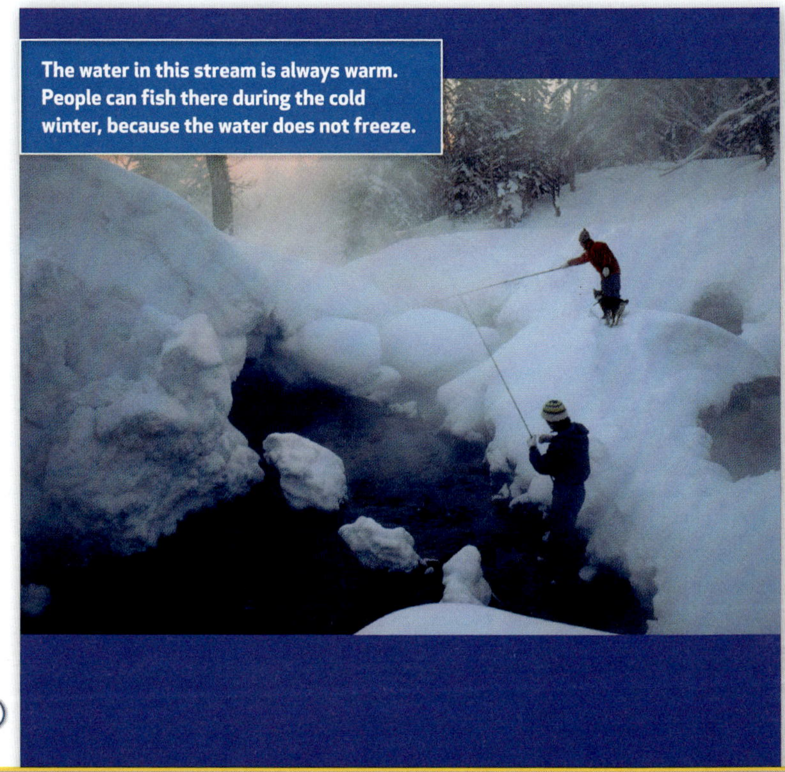

42

NATIONAL GEOGRAPHIC Raise Your SciQ!

Particles in Solids Particles in a solid—its atoms or molecules—vibrate, but they do not move about freely. The particles are packed together tightly, held by attractive forces. As a result, solids are rigid and hold their shape.

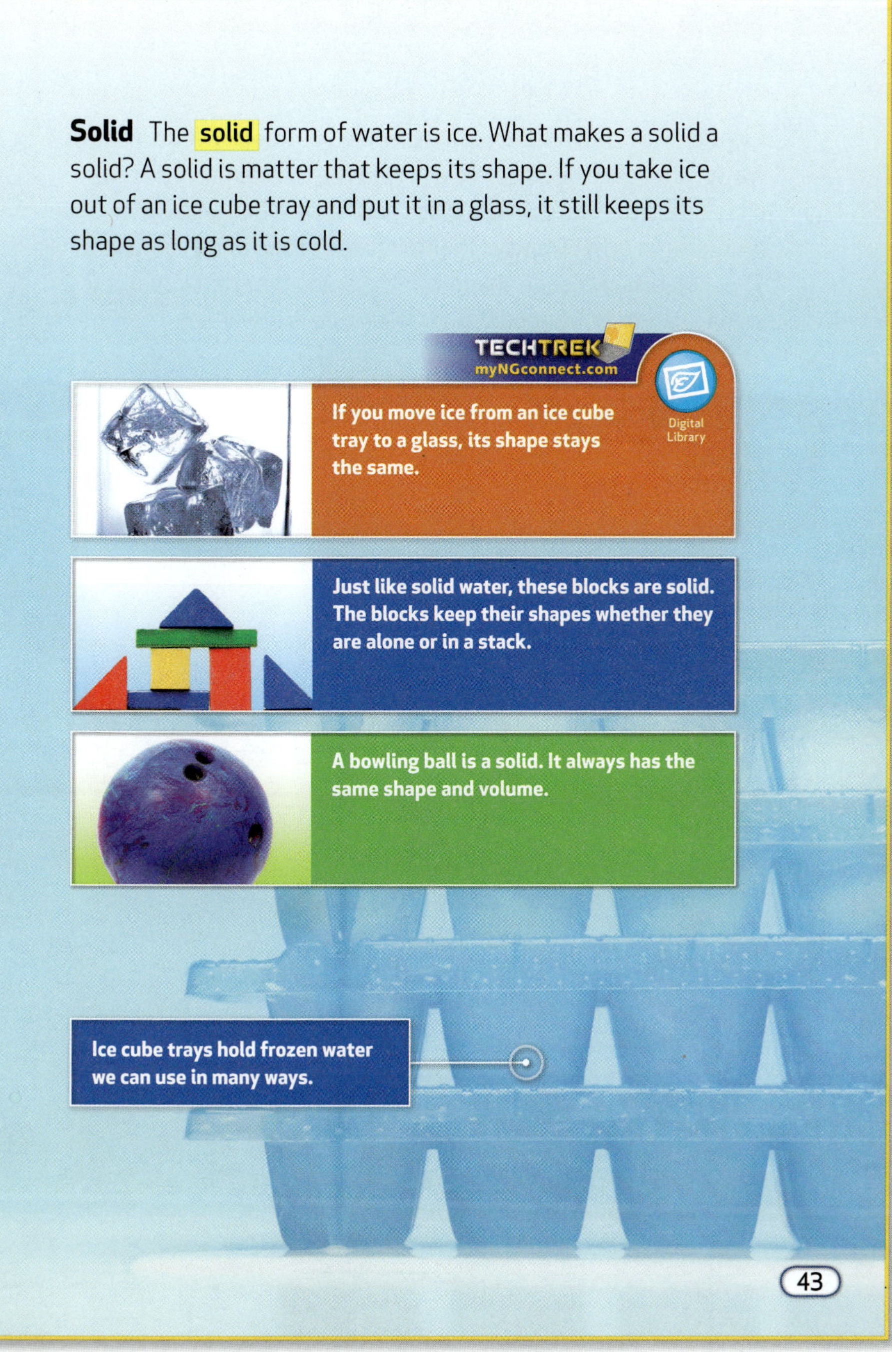

Solid The **solid** form of water is ice. What makes a solid a solid? A solid is matter that keeps its shape. If you take ice out of an ice cube tray and put it in a glass, it still keeps its shape as long as it is cold.

Differentiated Instruction

 Language Support for Describing Solids

BEGINNING	INTERMEDIATE	ADVANCED
Ask questions to help students describe the solid state. For example: **Does a solid keep its shape in a container? Which is solid—air or ice?**	Provide these Academic Language Stems to help students describe solids: • *A solid keeps its ______.* • *______ is a solid.* • *______ is not a solid.*	Help students describe solids with Academic Language Stems. For example: • *When a solid is put in a container, it will . . .* • *______ is a solid because . . .*

Teach, continued

Recognize the States of Matter

- Explain that water is a substance that is very common on Earth and that water has different forms, or states. Ask: **Where have you seen water?** (lakes, puddles, rain, tap water, bottled water) **Where do you see water that is solid, or frozen?** (icicles, ice cubes)

- Explain that water is also present in the air, but it is an invisible gas. Say: **You can't see water that's in the air, but you can sometimes feel it. For example, on hot summer days, the air can feel sticky and uncomfortable. That's because there's a lot of water in the air.**

Describe Solids

- Say: **A solid is matter that keeps its shape. If you put an ice cube tray filled with water in the freezer, the water freezes and forms ice.** Ask: **What do ice cubes feel like?** (hard, cold) **What happens to the shape of the ice cubes if you take them out of the tray and put them in an empty glass?** (Their shape stays the same.) **Ice is water in its solid form, or state. How can you tell that ice is a solid?** (It keeps its shape.)

- Demonstrate the properties of a **solid** by placing a ball or other small object in a plastic cup. Then move the object to a container of a different size. Ask students to describe what they observe about the shape of the **solid**.

- Explain: **Remember that volume is the amount of space something takes up. A solid always takes up the same amount of space. Its volume stays the same.**

- Ask students to name some **solids** other than ice.

 **Digital Library**

 **myNGconnect.com**

Have students use the **Digital Library** to find other photos of water in its **solid** state.

Integrated Technology

Use Digital Cameras Have students take photos of water in its **solid** and liquid states. If possible, have students prepare a digital slide show to share with small groups.

LESSON 2 □ States of Matter

Objectives

Students will be able to:

- Describe, classify, and explain the properties of solids, liquids, and gases.
- Recognize that water, as an example of matter, can exist as a solid, liquid, or gas.
- Recognize that air is a substance that surrounds us, has mass, and takes up space.

Science Academic Vocabulary

liquid, gas

PROGRAM RESOURCES

- **Enrichment Activities** at ⊘ **myNGconnect.com**
- **Digital Library** at ⊘ **myNGconnect.com**

Teach, continued

Academic Vocabulary: *liquid, gas*

- Call attention to the word **liquid**. Pronounce it together and have a volunteer read the definition. Say: **Liquid is a state of matter.** Have students look through pages 42–45 to identify **liquids** in the photos.

- Point out the word **gas**. Pronounce it together and have a volunteer read the definition. Say: **Gas is a state of matter. You can't see a gas—gases are invisible.**

- Ask: **What are the three states of matter?** (solid, liquid, gas)

Describe Liquids

- Have students reread page 44.

- Ask: **How can you tell if matter is a liquid?** (Its shape changes to take the shape of the container.) Tell students to compare the containers in the photo. Ask: **How does the shape of the water change?** (The shape of the water changes into the shape of the container.)

- Say: **Remember that volume is the amount of space something takes.** Ask: **How are solids and liquids alike?** (They keep the same volume.)

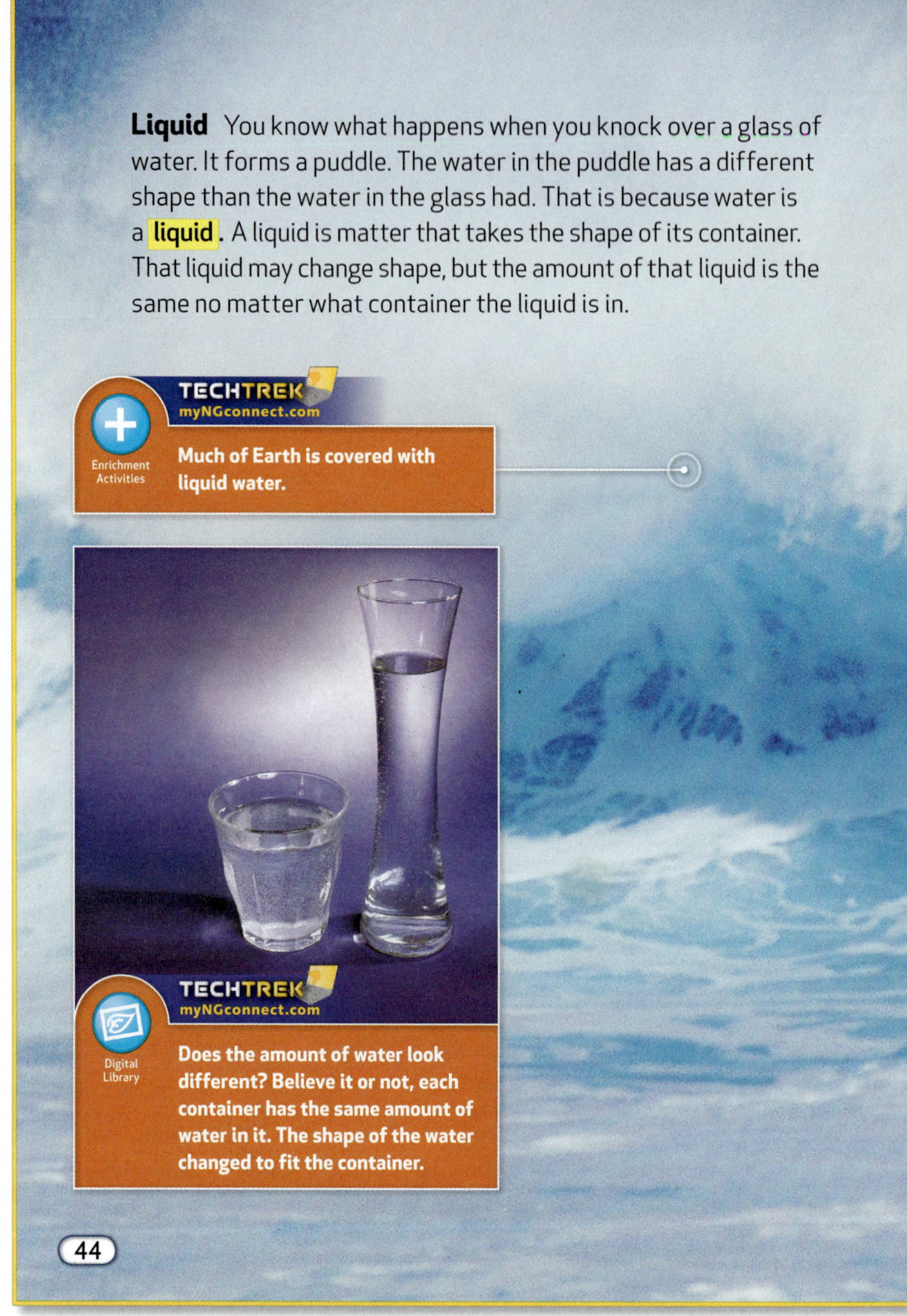

Liquid You know what happens when you knock over a glass of water. It forms a puddle. The water in the puddle has a different shape than the water in the glass had. That is because water is a **liquid**. A liquid is matter that takes the shape of its container. That liquid may change shape, but the amount of that liquid is the same no matter what container the liquid is in.

Raise Your SciQ!

Particles in Liquids and Gases Unlike the particles in solids, liquid particles are not held close together without freedom to move about. They are farther apart and move around each other randomly. Yet they are still bound together. As a result, a liquid will assume the shape of its container without changing its volume.

Particles in a gas are much farther apart than particles in a liquid. Gas particles move about more freely and rapidly. As a result, a gas can expand to fill a container or be compressed to fit inside a smaller one. A gas can change both its shape and its volume.

Gas If you have ever watched a pot of water heat up on the stove, you may have noticed bubbles in the water. Those bubbles show that a **gas** is forming. A gas is matter that spreads to fill a space. The gas form of water is water vapor. Like most gases, water vapor is invisible. Air is a gas, too. Even though you cannot see air, it is still matter. It has a small amount of mass, and it takes up space.

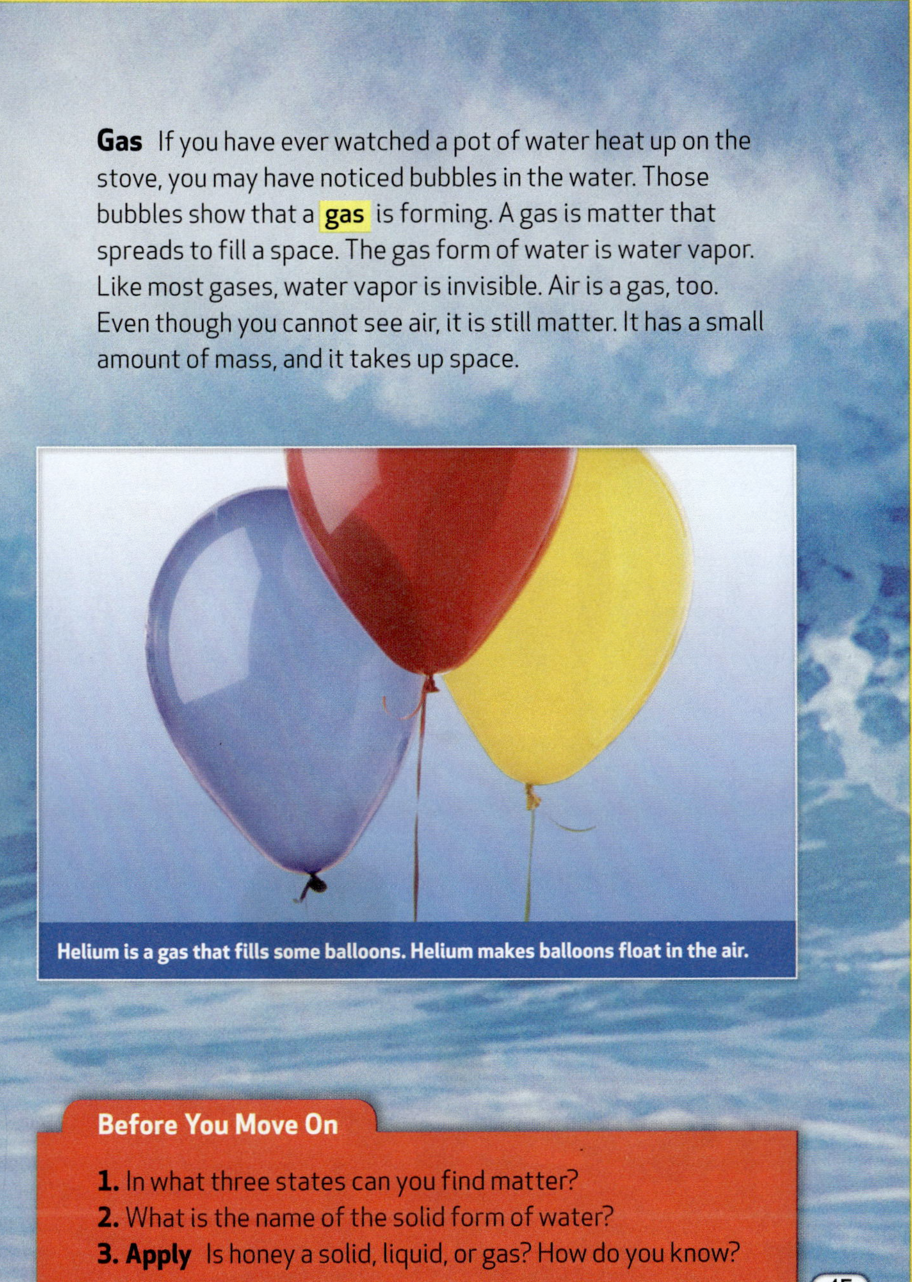
Helium is a gas that fills some balloons. Helium makes balloons float in the air.

Before You Move On

1. In what three states can you find matter?
2. What is the name of the solid form of water?
3. **Apply** Is honey a solid, liquid, or gas? How do you know?

45

Differentiated Instruction

Extra Support

Have students manipulate substances in different states of matter and describe their observations. They may blow a bubble and pop it or inflate and then deflate a balloon; pour the same amount of water into several differently shaped containers; and place a solid in different containers.

Challenge

Display or show a picture of a glass of soda with ice cubes. Ask students to identify the states of matter.

Teach, continued

Describe Gases

- Have students reread page 45. Say: **You cannot see a gas, but gases are all around you.** Tell students to breathe in and blow out. Ask: **What went into your lungs as you breathed in?** (air) **What is air?** (gases) **Can you see air?** (No, it's invisible.)

- Explain: **Water in the form of a gas is called water vapor. You cannot see this gas, but you can see evidence of it.** Ask: **How can you tell that a gas is forming when you see a pan of water start to boil on the stove?** (The gas causes bubbles to form in the water.)

Make a Compare-Contrast Chart

Have students make a three-column chart with the headings *Solid*, *Liquid*, and *Gas*. Tell them to list properties and an example for each state of matter.

Enrichment Activities

🌐 **myNGconnect.com**

Have students use the **Enrichment Activities** to view and interact with various scenes in which water occurs as a solid, a **liquid**, and a **gas**.

Digital Library

🌐 **myNGconnect.com**

Have students use the **Digital Library** to find other photos of water in its **liquid** state.

Integrated Technology

Computer Presentation Students can use the photos to make a whiteboard presentation explaining water in its three states—solid, **liquid**, and **gas**.

❸ Assess

» Before You Move On

1. **List** In what three states can you find matter? (as a liquid, solid, and gas)

2. **Define** What is the name of the solid form of water? (ice)

3. **Apply** Is honey a solid, **liquid**, or **gas**? How do you know? (It is a liquid. It flows and takes the shape of its container.)

Objectives

Students will be able to:

- Understand that the properties of an object are dependent on the conditions of the present surroundings in which the object exists, such as temperature.
- Recognize that heating and cooling (temperature change) may cause changes in the properties of materials such as water, including phase changes in the state of matter.
- Describe the changes in the physical properties of water when it is frozen or melted.
- Measure and compare the temperature of water when it exists as a solid to its temperature when it exists as a liquid.

PROGRAM RESOURCES

- **Digital Library** at 🌐 **myNGconnect.com**

❶ Introduce

Tap Prior Knowledge

- Have students study the photo. Ask: **Have you ever seen ice skaters? How did they move? What could you tell about the ice?** (It had a hard, smooth surface; it was slippery.)

Preview

- Read the heading. Then preview the photos and read the subheads on pages 46–51. Tell students that they will read about how matter changes state.

❷ Teach

Recognize That Temperature Can Cause Water to Change State

- Have students read pages 46–47.
- Point out the background photo. Ask: **How can you tell that it's cold where the ice skaters are?** (because they are dressed warmly; because there is ice) Explain: **The temperature must be below 0°C for water to remain in its solid state, as ice.**
- Ask: **What causes water to change from one state to another, for example, from a liquid to a solid?** (change in temperature)

Matter Changes State

Walking on water is impossible, right? If you have ever skated on ice, though, you have almost done just that! What state is the water in the photo of the ice skaters?

You can find water in all three states in your kitchen. You can even change it from one state to another. Changes of state can happen when the temperature changes.

46

NATIONAL GEOGRAPHIC — Raise Your SciQ!

Temperature and State of Matter Temperature can change the state of matter by changing the motion of its particles. The particles of a liquid such as water move about more freely than those of a solid. When a liquid is cooled, energy is subtracted. Its particles begin to slow down. At the freezing point, they come together and vibrate in place, forming a solid.

When a solid is heated, the added energy changes it to a liquid by causing its particles to move faster and more freely apart. If heated to boiling, the high energy causes the particles to move even faster and much farther apart, forming a gas.

Freezing and Melting Ice forms when water freezes. Freezing is the change from a liquid to a solid. Freezing happens at low temperatures. It happens when water cools to its freezing point, which is 0°C (32°F).

Melting is the opposite of freezing. Melting is the change from a solid to a liquid. When the temperature rises above 0°C, ice changes to water. Other substances melt, too. If you have ever heated butter in a pan on the stove, you've probably seen a solid melt.

When the temperature rises to 0°C and climbs higher, icicles drip, get smaller, and soon disappear.

47

Math in Science

Temperature Scales Explain that *C* in a temperature reading refers to Celsius, a scale for measuring temperature. Point out the thermometer on the page and note that Celsius temperatures are shown on the right. Explain that the left side of the thermometer shows a different scale for measuring temperature, labeled *F* for Fahrenheit. Give students practice in reading both scales on the thermometer. Ask: **What is the temperature in Fahrenheit when it is 0°C?** (32°F)

Teach, continued

Describe Freezing and Melting

- Say: **Liquids freeze at a certain temperature.** Ask: **At what temperature will water freeze?** (0°C) **How does the state of water change during freezing?** (It changes from a liquid to a solid, ice.)

- Explain: **Solids melt at a certain temperature.** Ask: **At what temperature will ice begin to melt?** (when it reaches above 0°C) **What does ice become when it melts?** (liquid water)

- Say: **Now sum up these changes in the states of water.** Ask: **What is freezing?** (changing from liquid water to ice) **What is melting?** (changing from a solid, ice, to a liquid)

Text Feature: Photo and Caption

- Have students read the caption to the photo of icicles and the thermometer. Ask: **What will happen to these icicles if the temperature rises a few degrees Celsius?** (The icicles will start to melt, or change from solid water, or ice, to liquid water.) **What does the thermometer show you?** (The temperature is just about 0°C, which is the freezing and melting point of water.)

 **Digital Library**

myNGconnect.com

Have students use the **Digital Library** to find other images of icicles melting.

Integrated Technology

Make Sound Files Have students work in small groups to create a narration that explains the freezing and melting process. Students should record their narration as a sound file.

Make Inferences

Ask: **What must the temperature have been at one point in the location where the ice skaters are skating?** (0°C) **How do you know?** (because the water froze into ice)

LESSON 3 □ Matter Changes State

Objectives

Students will be able to:

- Understand that the properties of an object are dependent on the conditions of the present surroundings in which the object exists, such as temperature.
- Recognize that heating and cooling (temperature change) may cause changes in the properties of materials such as water, including phase changes in the state of matter.
- Recognize that liquid water evaporates, changing into a gas as it moves into the air.

Science Academic Vocabulary

evaporation

PROGRAM RESOURCES

- **Digital Library** at 🌐 **myNGconnect.com**

Teach, continued

Academic Vocabulary: *evaporation*

- Write the words **evaporate** and **evaporation** and sound out the syllables as you say them. Underline the word *vapor* in **evaporation**. Explain: **The word *vapor* is another word for gas. The gaseous state of water is called *water vapor*.**

- Have a volunteer read the definition. Explain that **evaporation** is another way in which water and other liquids change state.

Describe Evaporation

- Have students read page 48.

- Ask students to describe how the air feels on a warm, humid day. Explain: **Water vapor forms when water changes from a liquid to a gas. This change in state is called evaporation. Warm air can hold more water vapor than cool air, so warm days often feel more humid than cool days.**

- Ask: **Why do the puddles dry up?** (The water evaporates.) Explain: **Heat energy is needed for water to evaporate. That is why puddles dry up faster when the sun is shining on them.**

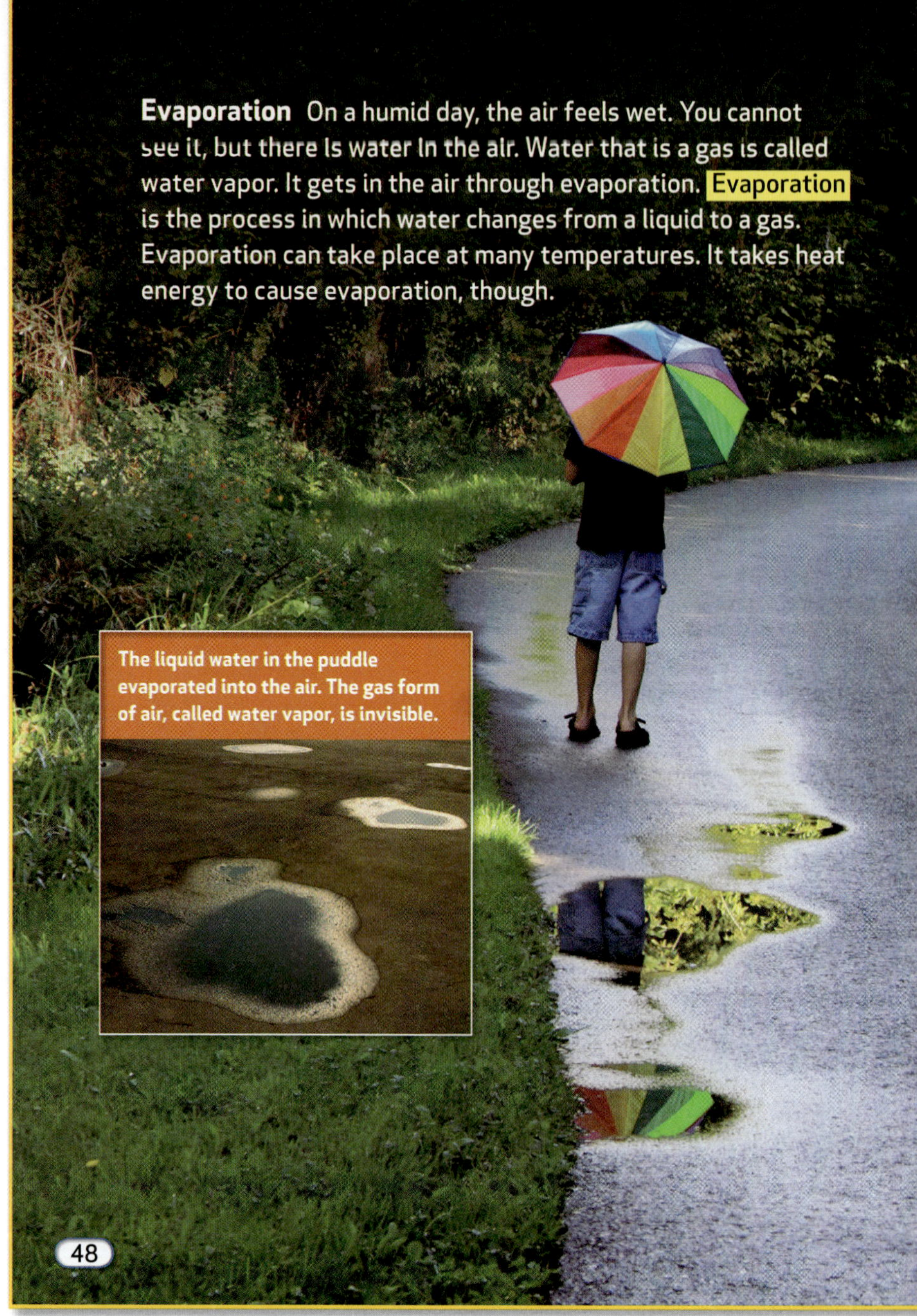

Evaporation On a humid day, the air feels wet. You cannot see it, but there Is water In the alr. Water that is a gas is called water vapor. It gets in the air through evaporation. **Evaporation** is the process in which water changes from a liquid to a gas. Evaporation can take place at many temperatures. It takes heat energy to cause evaporation, though.

The liquid water in the puddle evaporated into the air. The gas form of air, called water vapor, is invisible.

48

Differentiated Instruction

Extra Support

Help students understand evaporation by showing how water droplets evaporate. Place a small drop of water on a piece of paper. Observe how the paper absorbs the water droplet and then how the wet paper dries.

Challenge

Vary the size of the droplets on the paper. Make some larger than others. Move the paper to sunnier or warmer places in the classroom and observe how long it takes the paper to completely dry.

Boiling Look at the pot of water boiling on the stove. What do you observe? You probably notice the bubbles that show that water is changing from a liquid to a gas. Water makes this change from a liquid to a gas at 100°C (212°F). That temperature is the boiling point of water. Other liquids have different boiling points.

Differentiated Instruction

ELL Language Support for Describing Boiling Water

BEGINNING	INTERMEDIATE	ADVANCED
Show students pictures of liquid water freezing to ice, ice melting, and water boiling. Have students identify each process by saying the appropriate word—*boiling, freezing, melting*—as they point to each picture.	Provide Academic Language Frames that students can complete to describe photos of boiling water, such as on page 49. For example: • *The water in this pot is ______.* • *Eventually, all of the water in the pot will ______.*	Have students complete Academic Language Stems as they describe the pictures of boiling water, such as: • *Water will boil when its temperature ...* • *When water boils, ...*

Teach, continued

• Say: **Now sum up how water changes when it evaporates.** Ask: **What is evaporation?** (the change from a liquid to a gas) **What makes water evaporate?** (heat energy)

Describe Boiling Water

• Tell students to reread the caption on page 48. Explain: **Only the water on the surface of this puddle evaporates.**

• Have students read page 49. Ask if they have ever observed a pot of boiling water. Have them describe what they saw. Ask: **Do you think the water evaporated more quickly when it was boiling?** (Possible answer: Yes, faster than if the water wasn't boiling.)

• Have students look at the photo of boiling water and point out the bubbles. Explain: **Evaporation can happen anywhere in boiling water—not only at the surface. That's why the boiling water has rising bubbles but the puddle doesn't.** Ask: **What are the bubbles?** (gas; water vapor)

• Say: **Water boils at a certain temperature.** Direct students to the thermometer: **At what temperature does water boil?** (100°C) **What happens to water when it boils?** (It changes from its liquid state to a gas, water vapor.)

Have students use the Digital Library to find other photos of boiling water.

Integrated Technology

Whiteboard Presentation Students can use the information to make a computer presentation about the process of **evaporation.** Help them show the presentation on a whiteboard.

❯ Monitor and Fix Up

Ask students if anything they have read is confusing. Students may be confused by the distinction between **evaporation** and boiling. Suggest they reread these pages. Then discuss how (surface) water can evaporate at any temperature. As long as there is *some* heat energy, *some* water can evaporate. But when the temperature reaches 100°C, all the water in a pot can evaporate because it is boiling.

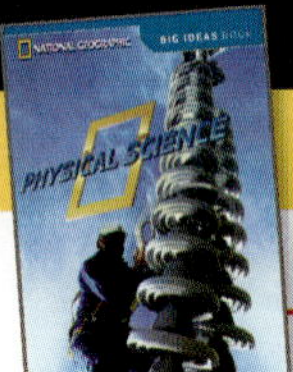

Objectives

Students will be able to:

- Understand that the properties of an object are dependent on the conditions of the present surroundings in which the object exists, such as temperature.

- Recognize that heating and cooling (temperature change) may cause changes in the properties of materials such as water, including phase changes in the state of matter.

Science Academic Vocabulary

condensation

PROGRAM RESOURCES

- Science Inquiry and Writing Book: *Physical Science*
- Science Inquiry and Writing Book **eEdition** at **myNGconnect.com**

Teach, continued

Academic Vocabulary: *condensation*

- Write the words *condense* and **condensation** and sound out the syllables as you pronounce them. Have a volunteer read the definition. Explain that **condensation** occurs when a gas changes state to form a liquid.

Describe **Condensation**

- Have students read pages 50–51.

- Ask students to describe the clouds in the photo.

- Say: **Cool air high in the sky cannot hold as much water vapor as warmer air close to the ground. As air gets cooler, the water vapor in the air changes from a gas to tiny drops of water. This change in state from a gas to a liquid is** **condensation**. **The drops of water stick to pieces of dust in the air and gather to form clouds.**

- Ask: **What causes** **condensation?** (change in temperature) Ask: **What other examples of** **condensation did you read about?** (Water vapor in the air condenses to form liquid water, such as dew and drops of water on the outside of a cold glass.)

Condensation See the white fluffy clouds in the photograph? How did they form? Clouds form because of water. Air contains invisible water in the form of a gas called water vapor. The air high in the sky is cool. As the air cools, tiny drops of water form. **Condensation** is the change from a gas to a liquid. When water condenses high in the air, it sticks to pieces of dust and floats in the air. When those pieces of dust covered with water come together, they make a cloud you can see.

Clouds form when the temperature of the air changes.

50

Raise Your SciQ!

From Gas to Solid Water vapor does not have to move through the liquid state to become a solid. When water vapor touches a surface that is below freezing, it can change directly to a solid. The result is frost.

Condensation doesn't happen only high in the air. On the ground, dew can form on grass or flowers. Dew forms when water vapor in the air condenses. If you have a glass of cold lemonade on a warm day, you may notice drops of water on the glass. The drops of water are condensation. The warmer air causes the water vapor in the air to condense on the cold drinking glass.

Before You Move On

1. Name two actions that cause matter to change from liquid to gas.
2. What changes of state can happen when temperatures get colder?
3. **Apply** After a rainstorm, the sidewalk dries out. What happens? Where does the water go?

51

Science Misconceptions

Sweating Glasses People sometimes refer to glasses as "sweating," as if the moisture on the outside were coming from the liquid on the inside. Glasses and other containers with cold liquids inside "sweat" because heat is transferred from the air around the container to the container and the cold liquid. As a result, the air around the container is cooled. This drop in temperature causes water vapor in the air to condense and collect on the outside of the container.

Classify Solids, Liquids, and Gases 

Science Inquiry and Writing Book: *Physical Science,* page 163

Materials writing utensils, sticky notes, balloon, plastic bottle of seltzer or sparkling water

Read the directions with students. Make sure to have several examples of solids, liquids, and gases available for students to identify. Have a blown-up balloon and a bottle of seltzer or sparkling water in the room. Students should record their observations in their science notebook.

What to Expect Students should place both solid and gas labels on the balloon and solid, liquid, and gas labels on the bottle. Suggest that students hang a sticky note near a fan or vent to label the air in the room.

❸ Assess

» Before You Move On

1. **Name** Name two actions that cause matter to change from liquid to gas. (boiling and evaporation)

2. **Cause and Effect** What changes of state can happen when temperatures get colder? (Freezing and condensation can happen when it gets colder.)

3. **Apply** After a rainstorm, the sidewalk dries out. What happens? Where does the water go? (It evaporates, forming water vapor in the air.)

Objectives

Students will be able to:

- Understand that the properties of an object are dependent on the conditions of the present surroundings in which the object exists, such as temperature.

❶ Introduce

Tap Prior Knowledge

Ask: **How do people measure temperature? Who has seen thermometers? What did they look like?** Have students share their experiences.

Set a Purpose and Read

- Tell students they will read about thermometers and use thermometers to measure temperature.

- Have students read pages 52–53.

❷ Teach

Understand Thermometers

- Explain: **A thermometer is a tool for measuring temperature, or how hot or cold something is.**

- Have students contrast the two thermometers pictured on page 52. Ask: **How do you read the temperature with the thermometer on the stove?** (by lining up the top of the red line with a number that indicates temperature) **How do you read the temperature with the thermometer in the swimming pool?** (by finding the number, or temperature, the arrow is pointing to)

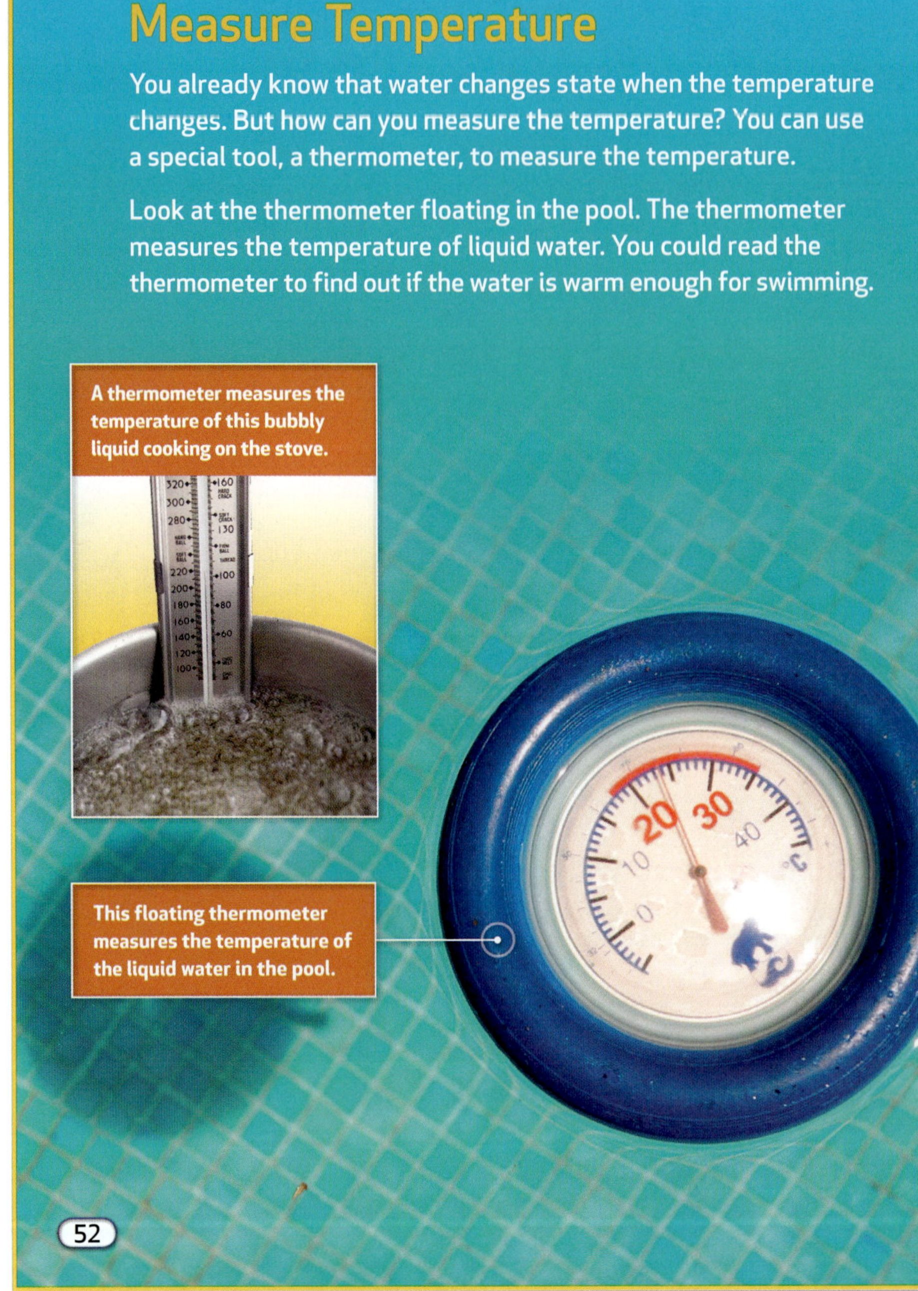

Measure Temperature

You already know that water changes state when the temperature changes. But how can you measure the temperature? You can use a special tool, a thermometer, to measure the temperature.

Look at the thermometer floating in the pool. The thermometer measures the temperature of liquid water. You could read the thermometer to find out if the water is warm enough for swimming.

52

Find the Temperature 

Materials 3 clear containers, vegetable oil, corn syrup, water, 3 thermometers

Allow the vegetable oil, corn syrup, and water to sit until they are at room temperature before this activity. Have students work in pairs. Ask them to read and follow the instructions on page 53 of the Big Ideas Book. Pour an equal amount of each liquid into its own container. Have students observe the consistency and thickness of each liquid. Students should record their observations and measurement results in their science notebook.

You can measure the temperature of solids, too. Meat is a solid. Cooks use thermometers to help them decide when meat is finished cooking. When the meat reaches the correct temperature, the cook knows that it is safe for people to eat.

Look at the oil, corn syrup, and water. Describe them. How are they the same? How are they different?

Put a thermometer in each cup.

What is the temperature of each liquid?

Before You Move On

1. Describe how you would measure the temperature of the soil outside your school.
2. Compare the temperature of water in its liquid state with the temperature of water in its solid state.
3. **Infer** An ice cube falls on the ground outside. A thermometer shows that the outside air temperature is below 0°C. What will happen to the ice cube?

53

Measure and Compare Temperatures

- Say: **You use thermometers to measure the temperature of air. Air is made of gases.** Ask: **You use thermometers to measure the temperature of what other kinds of matter?** (solids such as meat, liquids for cooking, other gases)

- Do the *Science in a Snap!* activity, in which students measure and compare the temperature of different liquids.

❸ Assess

» Before You Move On

1. **Describe** Describe how you would measure the temperature of the soil outside your school. (Place the thermometer in the soil. Wait for about 3 minutes. Then read the temperature at the top of the red line or the number at the end of the arrow.)

2. **Compare** Compare the temperature of water in its liquid state with the temperature of water in its solid state. (The temperature of water in its liquid state is warmer than the temperature of water in its solid state.)

3. **Infer** An ice cube falls on the ground outside. A thermometer shows that the outside air temperature is below 0°C. What will happen to the ice cube? (The ice cube will stay solid because the temperature is below the freezing point.)

What to Expect Students will find no difference in the temperatures of the three liquids, even though they vary in thickness and consistency.

Quick Questions Ask students the following questions:

- **How would the temperatures of the three liquids change if you placed them in a refrigerator?** (Students should infer that the temperatures would decrease.)

- **Would one liquid become colder than the others?** (Students should infer that the temperatures of all three liquids would decrease equally.)

Objectives

Students will be able to:

- Recognize that heating and cooling (temperature change) may cause changes in the properties of materials such as water, including phase changes in the state of matter.

PROGRAM RESOURCES

- Learning Masters Book, page 196, or at
  🌐 **myNGconnect.com**

❶ Introduce

Tap Prior Knowledge

- Ask: **Where do you get water in your home?** (from the faucet or tap; from a well) **How does the water get to the faucets?** (in water pipes)

- Ask: **Who has seen fog? What does fog look like?** Have students share their answers.

Preview and Read

- Read the heading. Then preview the photos on pages 54–55. Tell students they will read about fog and a clever way people have found to use it.

- Have students read pages 54–55.

❷ Teach

Understand the Importance of Liquid Water

- Discuss with students all they ways they have used water today. (to drink, bathe, brush teeth)

- Ask: **What are some other ways people use water?** (cooking, washing and cleaning, swimming or playing, growing crops, providing drinking water for farm animals, making products in factories)

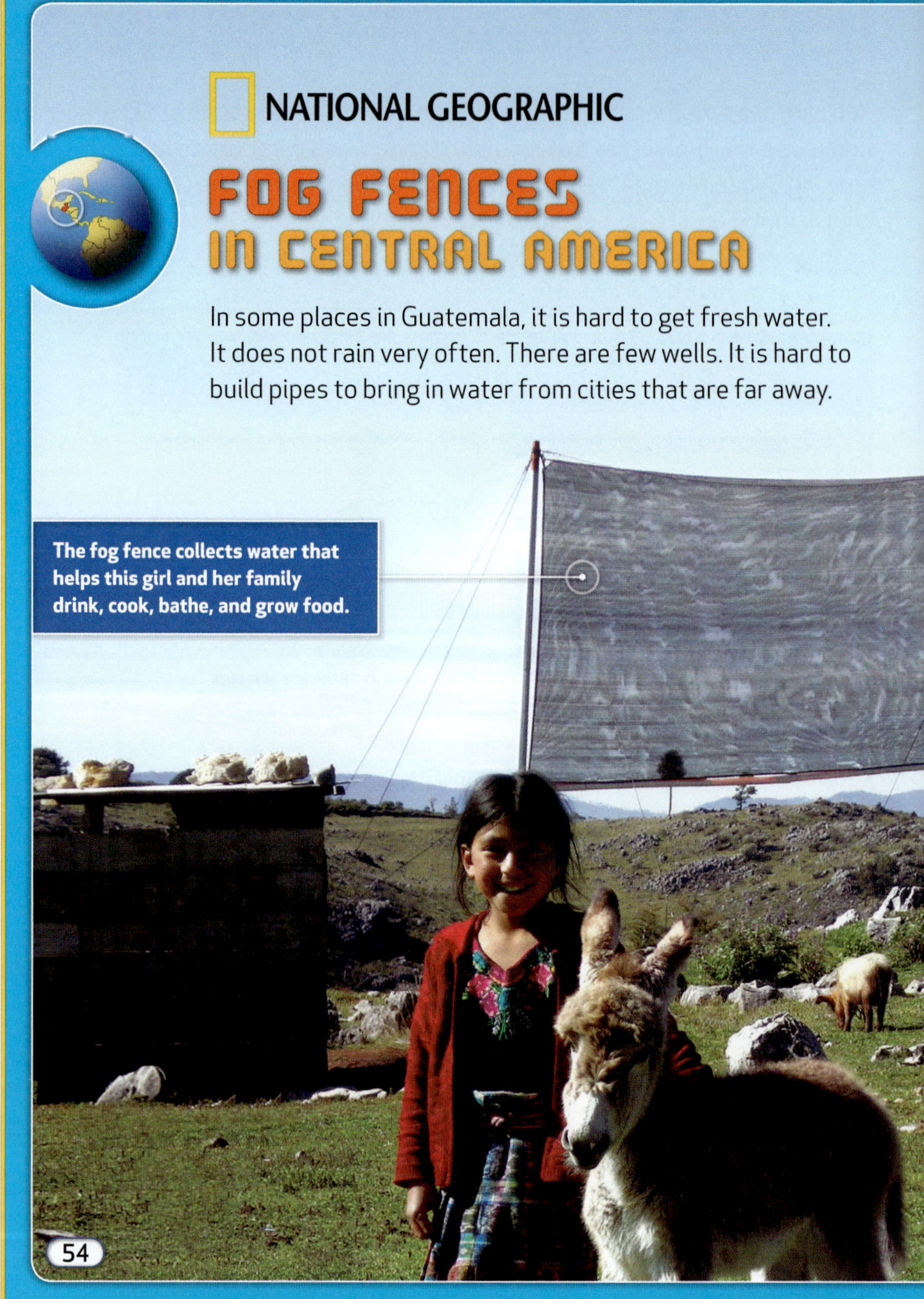

Social Studies in Science

Tojquia, Guatemala Ask students if they have heard of the country of Guatemala, which is in Central America. Have a volunteer locate Guatemala on a map or globe. Explain that this mountainous country has large towns but also many small villages. Tojquia is a village located in Guatemala's western highlands, or mountains, where the climate is cool and dry. The only other source of water for Tojquia is at the bottom of a valley far away.

The people who live in Guatemala still need water. They need it to drink and cook. They need it to feed their animals and grow plants for food. One way to get water is to build a fog fence. A fog fence uses condensation to collect water.

Fog is a cloud of water droplets floating in the air near the ground. A fog fence is a large plastic net. When wind blows fog through the net, the net gets wet. The water rolls or trickles down into pipes. A net that is 8 meters (26.2 feet) high and 50 meters (164 ft) long collects about 200 liters (52.8 gallons) each day. That amount would fill about four bathtubs.

Describe the Condensation of Water Vapor

- Review with students the definition of condensation on page 50.

- Explain: **Fog is a cloud that forms near the ground.** Ask: **Where do fog and other types of clouds come from?** (Water vapor, which is a gas, cools and condenses into drops of liquid water. These water droplets come together to form a cloud.)

- Say: **Remember that for water vapor, which is a gas, to change to liquid water, the temperature must get cooler.**

Text Feature: Photo Flowchart

- Point out the three-photo flow chart. Explain that it shows the steps involved in collecting water with a fog fence.

- Ask: **What must happen first for the fence to work?** (There must be fog.) **Then what happens?** (The droplets of water from the fog collect on the net.) **Where does the water from the fence collect?** (in a large tank)

❸ Assess

1. **Recall** **What is a fog fence?** (a large plastic net that collects water from fog)

2. **Explain** **How does fog form?** (Water vapor in the air condenses into water droplets near the ground, forming a low cloud.)

3. **Evaluate** **Would a fog fence work well in a hot, dry place? Why or why not?** (No. A hot, dry place would not have a lot of water vapor in the air, and the temperature would not get cool enough for fog to form.)

Share and Compare

States, Properties, and Changes of Water Give students the Learning Master. Have them find examples of liquid water, ice, and water vapor in the text. Students should use the examples to help them add additional information to their table, such as properties like "hardness" for Solid. Have partners discuss their findings and add to each other's charts.

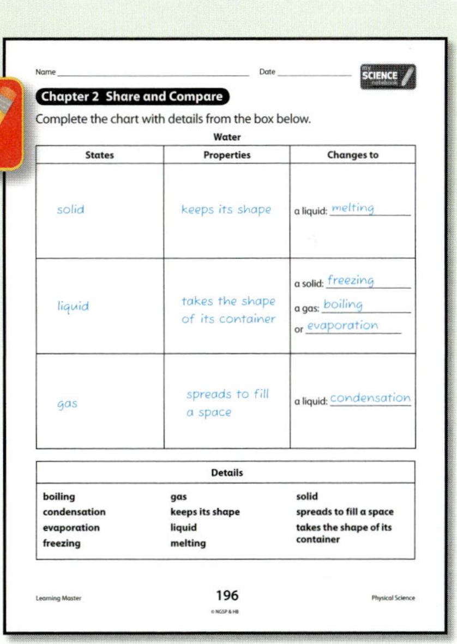

Learning Master 196 at 🌐 **myNGconnect.com**

Physical Science **T55**

Objectives

Students will be able to:

- Develop strategies and skills for information-gathering and problem-solving using appropriate scientific tools and technologies.
- Conduct an investigation in which temperature is measured to the nearest degree Celsius.
- Make observations and collect data using scientific process skills.
- Compare amounts and measurements made by others and seek reasons to explain the differences.
- Express quantities using appropriate number formats.

PROGRAM RESOURCES

- Learning Masters Book, pages 197–198, or at **myNGconnect.com**

❶ Introduce

Tap Prior Knowledge

- Ask: **When do you measure your temperature?** (when you feel sick, when you think you have a fever) **What do you use to measure your temperature?** (a hand on the forehead; a thermometer) **Which gives a more accurate measurement, a hand or a thermometer? Why?** Discuss why a thermometer is more accurate. Have students suggest other ways in which a thermometer might be used. (cooking, measuring air or water temperature)

❷ Teach

Measuring Temperature

- Read page 178 aloud with students and ask them to study the pictures. Say: **These are two different types of thermometers. How are they different? How are they used?** (one is digital and is inserted under the tongue; the other is liquid crystal and is placed on the forehead)

- Explain to students that a thermometer is a tool that gives a number for temperature. Doctors and scientists use tools such as thermometers to make better, more accurate observations.

Math in Science

Measuring Temperature

Scientists use thermometers to measure temperature. Thermometers help scientists make better observations. Scientists can get a more exact temperature with a thermometer than by just feeling how hot or cold something is. Also, scientists can use thermometers to measure the temperatures of objects that are too hot or too cold to touch.

Digital thermometer

Liquid crystal thermometer

178

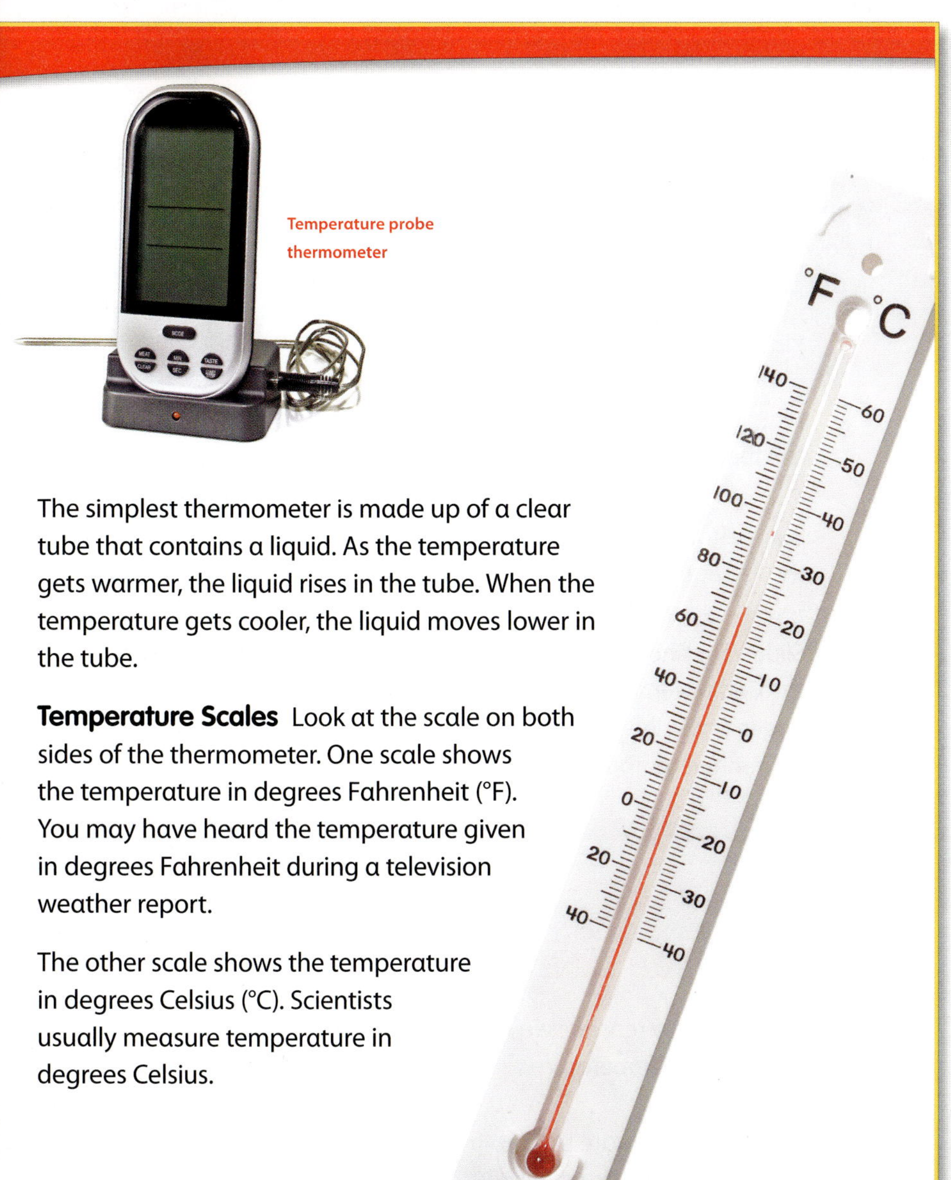

The simplest thermometer is made up of a clear tube that contains a liquid. As the temperature gets warmer, the liquid rises in the tube. When the temperature gets cooler, the liquid moves lower in the tube.

Temperature Scales Look at the scale on both sides of the thermometer. One scale shows the temperature in degrees Fahrenheit (°F). You may have heard the temperature given in degrees Fahrenheit during a television weather report.

The other scale shows the temperature in degrees Celsius (°C). Scientists usually measure temperature in degrees Celsius.

179

NATIONAL GEOGRAPHIC

Raise Your SciQ!

Why Doesn't the Liquid in a Thermometer Freeze?
The first thermometers used water, but because water freezes at 0 degrees Celsius (32 degrees Fahrenheit), there was no way to measure colder temperatures—the thermometer would just get stuck!

Today, most liquid thermometers used in classrooms contain alcohol, which has a much lower freezing point than water. An alcohol thermometer can be used to measure temperatures lower than 0 degrees Celsius without freezing.

Teach, continued

- Say: **Scientists have used thermometers to learn the temperatures at which matter changes its state.** Ask: **What are three states of matter?** (solid, liquid, gas)

- Have students look at the pictures on page 179. Say: **This page shows two other types of thermometers. The thermometer on the left measures temperatures inside refrigerators and ovens, or in foods being cooked. The thermometer on the right is a type of thermometer you might use in science class.**

- Read the first paragraph aloud with students. Explain to students that the liquid inside a thermometer expands and rises when it gets warmer. The liquid contracts and falls when it gets cold.

- Read the last two paragraphs of page 179 aloud with students and have them study the picture of the simple thermometer. Make sure students understand that the thermometer shows two different temperature scales.

- Ask: **What do the letters C and F mean on the thermometer?** (Celsius and Fahrenheit) **Which side of the scale shows the temperature in degrees Fahrenheit?** (the left side) **Which scale will we use in science class?** (Celsius)

- Explain that in the Fahrenheit scale, water freezes at 32° and boils at 212°. In the Celsius scale, water freezes at 0° and boils at 100°.

- Ask: **Why do you think all scientists use the same scale?** (It makes it easier to compare results.)

- Explain that thermometers should be handled carefully, and remind students to notify you immediately if a thermometer breaks. Be sure that the thermometers used by students contain alcohol rather than mercury. Exposure to mercury is a health hazard.

Teach, continued

- Have students read the top half of page 180. Ask: **What would happen if you waited less than 1 minute before looking at the thermometer?** (The liquid in the thermometer may not have enough time to reach its full height. The measurement would be incorrect.)

- Direct students to look at the marks on the thermometer in the picture. Say: **The marks on the thermometer show the number of degrees in each scale. On the Fahrenheit scale, each mark stands for 2 degrees. On the Celsius scale, each mark stands for 1 degree.** Ask: **Why is it important to know the number of degrees each mark shows?** (so you can read the correct measurement)

- Hold up a simple thermometer and say: **A thermometer must be read at eye level.** Ask a volunteer to demonstrate reading the thermometer at eye level. Ask: **What could happen if you didn't read the thermometer at eye level?** (You would read the marks at an angle and get an incorrect measurement.)

- Using the thermometer in the photograph, have students follow steps 2 and 3. Say: **What is the Fahrenheit temperature?** (70 degrees) **What is the Celsius temperature?** (21 degrees) Ask: **Which scale is easier to read? Why?** (Celsius; because there is a mark for every degree)

- Have students work with a partner to answer the questions on page 180.

What Did You Find Out?

1. Scientists can get more exact temperature measurements with a thermometer. They can measure the temperature of things that are too hot or too cold to touch.

2. Fahrenheit and Celsius; scientists usually use the Celsius scale to measure temperature.

Using a Thermometer

When reading a thermometer, follow the steps below. For steps 2 and 3, use the thermometer in the photograph to practice.

1. Place the thermometer on or in the material for which you want to find the temperature. Wait about 1 minute for the thermometer to complete its reading.

2. Put the top of your finger at the top of the red liquid.

3. Slide your finger to the right to read the Celsius temperature. Slide your finger to the left to read the Fahrenheit temperature.

When you measure temperature, make sure you understand the number of degrees each mark shows.

SUMMARIZE

What Did You Find Out?

1. What are two reasons why scientists use thermometers to measure temperature?

2. What two scales are used to measure temperature? Which do scientists usually use?

180

Name _______ Date _______

Think Like a Scientist **Math in Science**

Measuring Temperature
What Did You Find Out?

1. What are two reasons why scientists use thermometers to measure temperature?

   Scientists can get a more exact temperature measurement with a thermometer. They can measure the temperature of things that are too hot or too cold to touch.

2. What two scales are used to measure temperature? Which do scientists usually use?

   Fahrenheit and Celsius; scientists usually use the Celsius scale to measure temperature.

Learning Master   197   Physical Science

Learning Masters 197–198 or at
myNGconnect.com

Measure the Temperature of Water

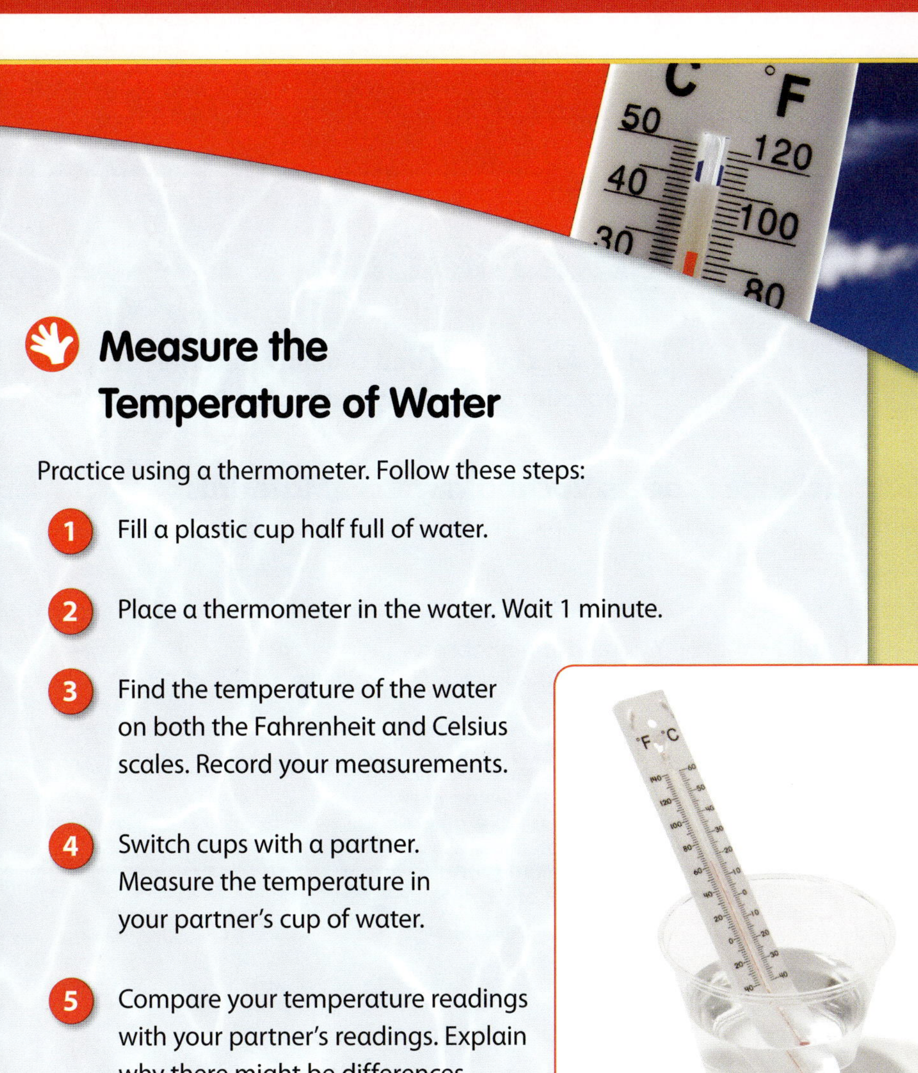

Practice using a thermometer. Follow these steps:

1. Fill a plastic cup half full of water.

2. Place a thermometer in the water. Wait 1 minute.

3. Find the temperature of the water on both the Fahrenheit and Celsius scales. Record your measurements.

4. Switch cups with a partner. Measure the temperature in your partner's cup of water.

5. Compare your temperature readings with your partner's readings. Explain why there might be differences between your measurements and your partner's measurements.

181

Teach, continued

Measure the Temperature of Water

- Explain to students that they will conduct an investigation using a thermometer, which is a tool that measures temperature.

- Read page 181 aloud. Make sure all students understand the procedure. Have students complete the activity in pairs. Give each student a cup of water and a thermometer. If possible, the water should be slightly cooler or slightly warmer than room temperature.

- Tell students to place their thermometers in the water at the same time. Use a clock or stopwatch to time one minute. When the minute is up, tell students to record the temperature readings for both scales.

- Have students remove their thermometers and switch cups with a partner to repeat the activity. Guide students as they compare their temperature readings. Ask: **Are your temperature readings reasonable?** If students are not sure, ask them to feel the cups with their fingers to estimate and compare the temperatures.

- Discuss why partners' readings may vary. Elicit that the water may have gotten closer to room temperature over time, which could affect the readings. Also, students may have made errors as they read the thermometers.

❸ Assess

- Use Learning Master Book pages 197–198. Students should have an understanding of how to use a thermometer and why measurements can differ between groups.

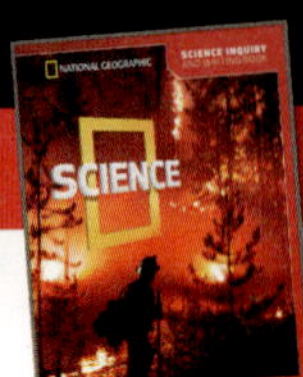

Objectives

Students will be able to:

- Investigate through Guided Inquiry (answer a question; make and compare observations; collect and record data and observations; generate explanations and conclusions based on evidence; share findings; ask questions based on observations to increase understanding; adjust explanations based on findings and new ideas).
- Measure and compare the temperature of liquid and solid water.
- Plan and conduct a fair test.
- Construct simple charts and tables from data and observations and summarize that information to answer scientific questions.

Science Process Vocabulary

experiment, variable

PROGRAM RESOURCES

- Science Inquiry and Writing Book: *Physical Science*
- Science Inquiry and Writing Book **eEdition** at **myNGconnect.com**
- **Inquiry eHelp** at **myNGconnect.com**
- Science Inquiry Kit: *Physical Science*
- Learning Masters Book, pages 199–202, or at **myNGconnect.com**
- Inquiry Rubric: Assessment Handbook, page 212, or at **myNGconnect.com**
- Inquiry Self-Reflection: Assessment Handbook, page 220, or at **myNGconnect.com**

MATERIALS

Kit materials are listed in italics.

2 plastic cups (10 oz); masking tape; *graduated cylinder (100 mL);* 4 plastic cups (9 oz); water; *salt; vinegar; juice; 2 thermometers; stopwatch*

❶ Introduce

Tap Prior Knowledge

- Ask students if they have ever watched an iced fruit treat or plain ice as it melts. Ask: **Which of these melts the fastest? Why do you think so?**

MANAGING THE INVESTIGATION

Time

40 minutes (or until liquid has melted)

Groups

Small groups of 4

Advance Preparation

- Prepare cups of water, salt water, vinegar, and juice for each group. Cups should contain 50 mL of liquid.
- Prepare salt water by adding ½ tsp of salt to 50 mL of water.
- Prepare juice with a powdered mix and water.

Do an Experiment

Write your plan in your science notebook.

Make a Hypothesis

In this experiment, you will investigate how heat energy from sunlight can affect the properties of two frozen materials. You will freeze two liquids. One liquid will be water. You will choose the second liquid. Then you will place the frozen materials in sunlight. What will happen to the temperature and properties of the materials as they sit in sunlight? Write your **hypothesis.**

Identify, Manipulate, and Control Variables

Which variable will you change?
Which variable will you observe or measure?
Which variables will you keep the same?

What to Do

1. Label 1 cup **Water.** Then use the graduated cylinder to **measure** 50 mL of water. Pour the water into the cup.

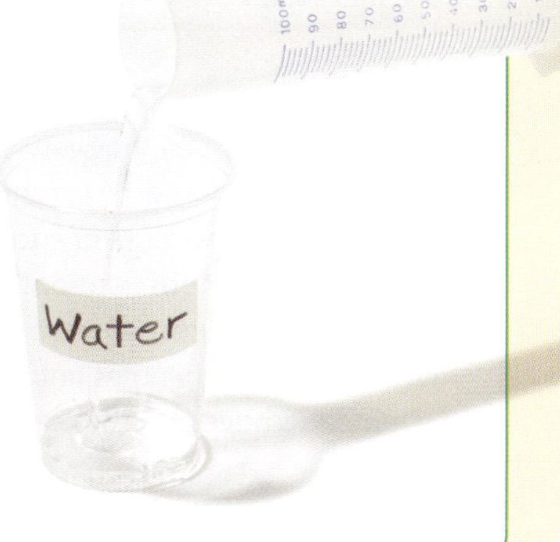

183

Teaching Tips

- The activity can be done without sunlight. However, it will take longer for the frozen liquids to melt. It may take more than 30 minutes for both solids to melt completely.
- Remind students to record their temperatures in degrees Celsius.

What to Expect

- Students will observe and record temperature changes in two liquids as they change from a solid to a liquid. Students may find that salt water melts fastest, vinegar melts slowest, and water and juice take the same amount of time to melt.

Introduce, continued

Connect to the Big Idea

- Review the Big Idea Question, *What are states of matter?* Explain to students that this inquiry will explore the solid and liquid states of matter.
- Have students open their Science Inquiry and Writing Books to page 182. Read the Question and invite students to share ideas about how heating and cooling affect the properties of materials.

❷ Build Vocabulary

Science Process Words: experiment, variable

Use this routine to introduce the words.

1. **Pronounce the Word** Say **variable**. Have students repeat it in syllables.
2. **Explain Its Meaning** Choral read the sentence. Ask students to define **variable**. (something that can change in an investigation)
3. **Encourage Elaboration** Ask: **What is a variable in this activity?** (Possible answers: temperature, type of liquid, amount of heat)

ELL Vocabulary Support

Explain that some **variables** in an experiment will change and some are kept the same. Help students complete the sentence: **One variable that is kept the same in this activity is ______.** (amount of liquid; temperatures in the freezer and the sun; location)

Repeat for the word **experiment**. To encourage elaboration, ask: **How do you know that this is an experiment?** (Variables are identified, manipulated, and controlled.)

❸ Guide the Investigation

- Guide students in writing their hypotheses. Students may write their hypotheses as "If…, then…" statements.
- Guide students in planning an experiment, or fair test. Students may list what they will change, what they will observe or measure, and what they will keep the same. Ask: **What variable will change in this experiment?** (type of liquid)
- Distribute materials. Read the steps on pages 183–184. Clarify steps if necessary.

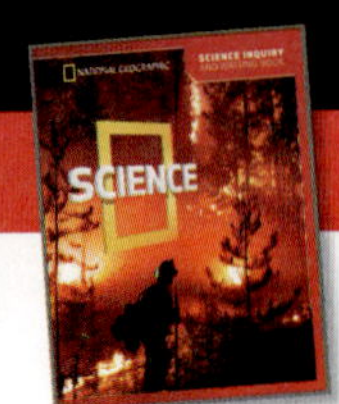

Guide the Investigation, continued

- Ask: **What properties do you observe about the solid water, vinegar, juice, and salt water? How do those properties change when they become liquids?** Encourage students to contrast the substances and states of matter. They should recognize that the sun produces heat energy that melts the substances into liquids.

❹ Explain and Conclude

- Guide students as they record their observations in tables. Point out that tables are a helpful way to organize data and that it is important to be accurate and concise while recording data.

- Have students summarize their data for the class. Students should notice that the liquids have a lower temperature when they are solids and a higher temperature when they are liquids.

- Have students interpret the data and look for relationships. Say: **Compare the temperatures of the water and the other liquid after 10 minutes in the sunlight. How were the temperatures similar or different? Why do you think this happened?**

- Ask: **Why do you think water was used as the control liquid in this experiment?** Guide students to understand that the freezing point and melting point of water is well known.

- Have students make inferences about the melting points of salt water, juice, and vinegar. Students may infer that salt water has the lowest melting point and vinegar has the highest melting point.

Answers

1. The temperature of the materials went down in the freezer. Their temperatures went up in sunlight.

2. Possible answer: As the temperatures of the water and the juice went down, they began to change from liquids to solids. The materials were solids when they were placed in the sunlight, but as their temperatures went up, they began to change to liquids. The results support my hypothesis.

What to Do, continued

2. Choose a different liquid. Label a cup with the name of the liquid. Put 50 mL of the liquid into the cup.

3. Put a thermometer in each cup. **Measure** the temperature of each liquid and **observe** their properties. Record your **data** in your science notebook.

4. Put the cups with liquid in a freezer. The next day, take the cups out of the freezer. Observe the properties and record the temperature of each material.

5. Place the cups in sunlight. Every 10 minutes, observe each cup and measure the temperature of each liquid. Record your observations and the temperatures. Observe the materials every 10 minutes until all the materials have melted.

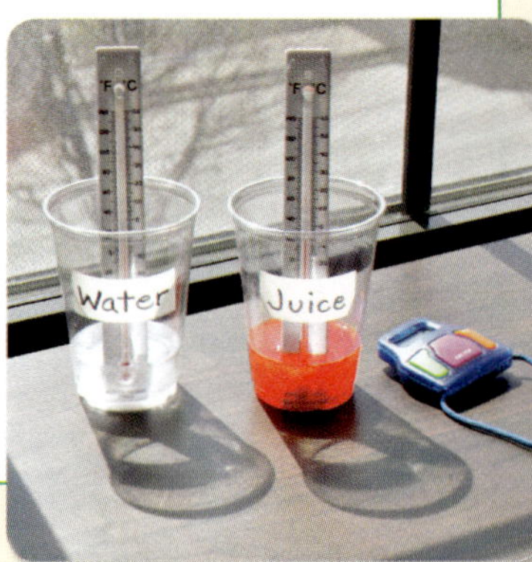

184

Raise Your SciQ!

Why Salt Melts Ice Water freezes at 0°C. When ice is formed, there is a layer of water on the surface that is difficult to see. Water molecules are constantly going back and forth between the liquid phase and the solid phase. If more of these molecules become solid, then the water is freezing and the ice grows. If more of the molecules are going from solid to liquid, then the ice is melting. When salt dissolves in water, it prevents ice crystals from forming. Simply, salt interferes with the process of freezing, but does not interfere with melting. Salt generally lowers the freezing point of water from 0°C to -9°C.

Record

Write in your science notebook.
Use a table like this one.

Liquids in Cups

	Water		My Choice: _______	
	Temperature	Observations	Temperature	Observations
Before freezing				
After freezing				
After 10 min in sunlight				
After 20 min in sunlight				

Explain and Conclude

1. What happened to the temperature of the liquids in the freezer? What happened to the temperature of the materials in sunlight?

2. How did the properties of the materials change as their temperatures changed? Do these results support your **hypothesis**? Explain.

3. **Compare** the results of all groups. What patterns do you **observe?**

Think of Another Question

What else would you like to find out about changes in temperature and properties of materials? How could you find an answer to this new question?

The heat from sunlight melts the ice.

185

Explain and Conclude, continued

3. Possible answer: All the liquids became solid as their temperature went down. All the materials changed back to liquids as the temperature went up in the sunlight. Salt water melted fastest. Vinegar melted slowest. Water and juice took about the same amount of time to melt.

❺ Find Out More

Think of Another Question

- Students should use observations made in this investigation to generate other questions. Discuss what students could do to find answers to the new questions.

❻ Reflect and Assess

- To assess student work with the Inquiry Rubric below, see Assessment Handbook, page 212, or go online at ⊘ **myNGconnect.com**

- Score each item separately and then decide on one overall score.

- Have students use the Inquiry Self-Reflection on Assessment Handbook, page 220, or at ⊘ **myNGconnect.com**

Inquiry Rubric at ⊘ myNGconnect.com	Scale			
The student made a **hypothesis** about how changes in temperature affect the properties of frozen liquids.	4	3	2	1
The student **measured** the change in temperature of two different substances as they melted.	4	3	2	1
The student described how the properties of the substances changed as they changed states.	4	3	2	1
The student found a pattern in the way different liquids changed states.	4	3	2	1
The student **shared** and **compared** results and **conclusions** with others.	4	3	2	1
Overall Score	4	3	2	1

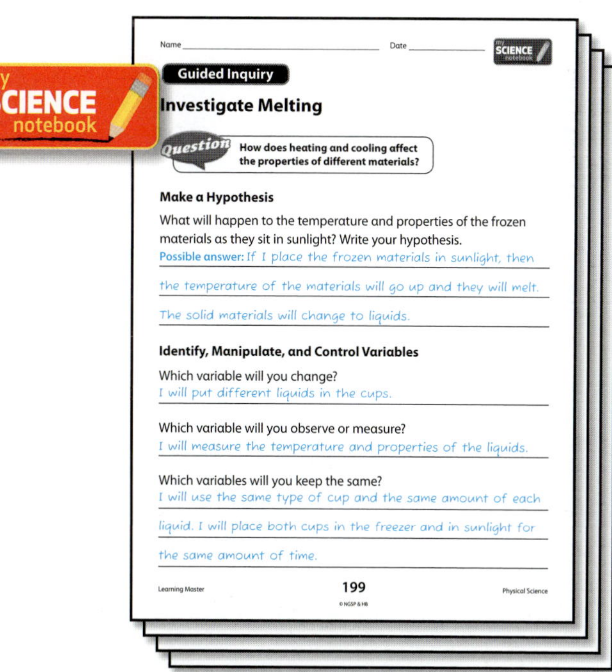

Learning Masters 199–202 or at ⊘ **myNGconnect.com**

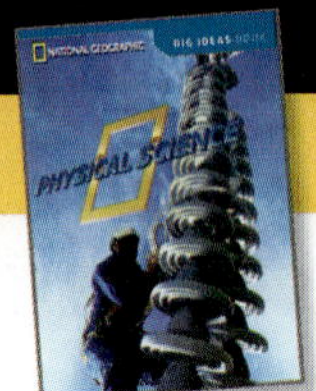

PROGRAM RESOURCES
- Chapter 2 Test, Assessment Handbook, pages 92–95, or at 🌐 **myNGconnect.com**
- NGSP ExamView CD-ROM

❶ Sum Up the Big Idea

- Display the chart from page T40–T41. Read the Big Idea Question. Then add new information to the chart.

- Ask: **What did you find out in each section? Is this what you expected to find?**

What Are States of Matter?

States of Matter

Karyn: There are three states of matter: solid, liquid, and gas.

Matter Changes State

Michael: Liquid can become a gas at a high temperature. It can become a solid at a low temperature.

Measure Temperature

Marcus: We can measure temperature with thermometers.

Conclusion

Matter exists in three states: water, liquid, and gas. Matter changes state because of changes in temperature. Freezing and condensation happen when water is cooled. Melting, boiling, and evaporation happen when water is heated.

Big Idea Matter exists in three states. Changes in state are caused by changes in temperature.

Liquid

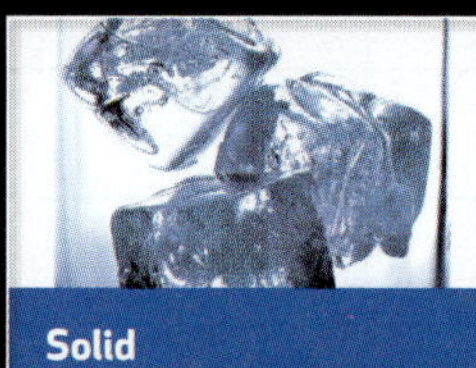

Solid

Gas

Vocabulary Review

Match the following terms with the correct definition.

A. condensation
B. evaporation
C. gas
D. states of matter
E. liquid
F. solid

1. Matter that keeps its own shape
2. Matter that spreads to fill a space
3. Matter that takes the shape of its container
4. The change from a gas to a liquid
5. The change from a liquid to a gas
6. The forms in which a material can exist

56

Review Academic Vocabulary

Academic Vocabulary Tell students that scientists use special terms and words when describing their work to others. For example:

states of matter   liquid   evaporation

solid   gas   condensation

Have students write the list in their science notebook and use each term and word in a sentence that tells about water.

Big Idea Review

1. Define Give definitions for these terms: solid, liquid, gas. Tell what water is like in each of these states.

2. Explain What happens to the shape of a liquid when it is poured into a container with a different shape?

3. Describe How is evaporation part of boiling?

4. Cause and Effect What would happen to a jar of water vapor if you put it in the refrigerator?

5. Evaluate A friend says to you, "Matter is very useful in its liquid and solid states, but not in its gas state." How would you respond?

6. Infer What kinds of places can use fog fences?

Write About Matter

Describe What is happening in this photograph? What states of matter are there? What change is taking place? How is temperature important?

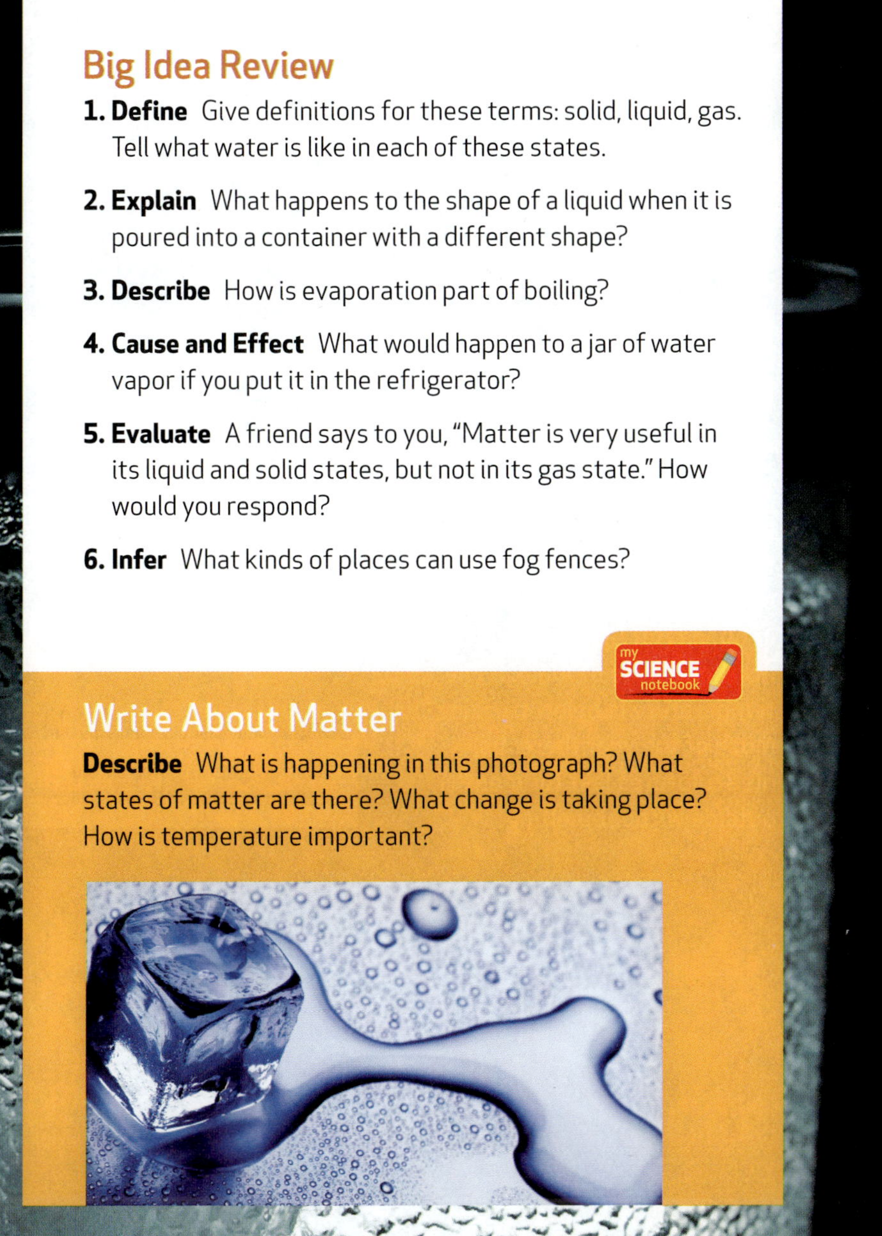

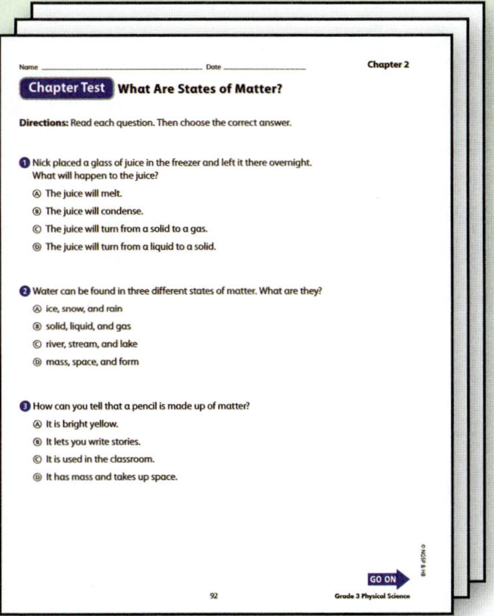

Assess Student Progress Chapter 2 Test

Have students complete their Chapter 2 Test to assess their progress in this chapter.

Chapter 2 Test, Assessment Handbook, pages 92–95, or at 🌐 myNGconnect.com

or NGSP ExamView CD-ROM

❷ Discuss the Big Idea

Notetaking

Have students write in their science notebook to show what they know about the Big Idea. Have them:

1. Write about the three states of matter.

2. Explain how water can change states through evaporation and condensation.

3. Describe how to measure temperature.

❸ Assess the Big Idea

Vocabulary Review

1. F **2.** C **3.** E **4.** A **5.** B **6.** D

Big Idea Review

1. A solid is matter that keeps its own shape. Solid water, or ice, looks sort of like glass. A liquid is matter that takes the shape of its container. Liquid water is clear. A gas is matter that spreads to fill a space. Water vapor is invisible.

2. The shape changes. A liquid takes the shape of its new container.

3. While boiling is happening, evaporation is also happening as the liquid changes into a gas and becomes water vapor.

4. The water vapor would cool and condense to form drops of liquid water in the jar.

5. Sample answer: Water vapor helps make clouds. We need clouds to have rain on Earth. Rain makes it possible to grow crops and other plants for people to eat.

6. Sample answer: Not all places can use fog fences. The air in such a place needs to be cool and moist so that fog can form. There also needs to be wind to blow the fog through the fog fence.

Write About Matter

Ice is melting. Water is changing from its solid to its liquid state. Ice melts at its melting point to become water.

Objectives

Students will be able to:

- Understand that scientists and inventors come from different gender and ethnic backgrounds and describe the work they have contributed to science and technology.

PROGRAM RESOURCES

- Big Ideas Book: *Physical Science*
- Big Ideas Book: *Physical Science* **eEdition** at 🌐 **myNGconnect.com**
- **Digital Library** at 🌐 **myNGconnect.com**

❶ Introduce

Tap Prior Knowledge

- Ask students if they have ever seen something made out of ice, such as icicles, ice cubes, or even a frozen pond.

- Ask: **What does a sculptor do?** (A sculptor carves or shapes statues or other art objects.) **What do you think an *ice* sculptor does?**

Set a Purpose and Read

- Read the headings. Say: **You will read about an ice sculptor and the kind of work he does.**

- Have students read pages 58–59.

❷ Teach

Describe Ice Sculpting and Ice

- Ask: **What does an ice sculptor like Steve Brice do?** (carves shapes from ice) **Why must ice sculptors work in the cold?** (It has to be below freezing for the blocks of ice that they carve to remain solid.) **What happens when the temperature rises above 0°C?** (The sculptures melt.)

- Ask: **Why do you think ice is good for making big sculptures, but snow is not?** (Ice is harder and holds its shape better than snow.)

Social Studies in Science

Winter in Alaska Have a volunteer locate Alaska on a globe or world map. Chena Hot Springs is located about 96 kilometers (60 miles) northeast of Fairbanks. Explain that a place so far north has very cold, long winters. Freezing temperatures, snow, and ice can occur from late September to early May. Tell students that this winter weather spans much of their school year.

Steve Brice invented this special tool for carving details in ice. He uses it to put writing on his sculptures.

Ice is hard like stone and wood. An ice sculptor uses many of the same tools to carve ice that other sculptors use to shape other solids. Brice uses chain saws, sanders, knives, and drills to shape the ice.

As long as it stays cold, the finished sculptures keep their shape. Every year, the sculptures melt in the spring. Brice makes new ones when it freezes again in winter.

Steve Brice uses a special tool to smooth the face of one of his ice sculptures.

59

Differentiated Instruction

ELL Language Support for Describing Ice Sculpting and Ice

BEGINNING	INTERMEDIATE	ADVANCED
Help students learn words to describe ice sculpting. Give them choices. For example: **Is this artist a sculptor or a sculpture? Is an ice sculpture freezing cold or warm? Does an ice sculpture freeze or melt in the spring?**	Help students describe ice sculpting using Academic Language Frames: • *This man is an ice _____.* • *His _____ are made of ice.* • *He uses _____ to shape the ice.*	Help students elaborate on ice sculpting using Academic Language Stems: • *An ice sculptor is an artist who . . .* • *An ice sculptor uses tools such as . . .* • *Ice sculptors work with ice because . . . .*

Teach, continued

Technology

- Explain: **Steve Brice uses many different kinds of tools; some are power tools that need electricity, but others are not** Ask: **What kinds of tools does Steve Brice use to shape ice?** (chain saws, sanders, knives, drills)

- Have students study the photos and captions. Ask: **What other tools does he use?** (a special tool he made to carve details and write in ice) **Why do you think Steve invented his own tool?** (As he was shaping the ice, he may have thought of a tool that could make his work easier.)

Digital Library

myNGconnect.com

Have students use the Digital Library to find more ice sculptures and photos of Steve Brice at work.

Integrated Technology

Digital Booklet Students can use the photos to make a digital booklet describing the tools and techniques used by ice sculptors to create beautiful works of art.

Find Out More

- Ask: **Is this a career you would find interesting? Why or why not?** (Accept all answers, positive or negative.)

- Encourage interested students to research other solid and liquid media that artists use to express their thoughts and feelings.

❸ Assess

1. **Recall** **What happens to Steve Brice's sculptures in the spring?** (They melt.)

2. **Compare and Contrast** **How is ice like other solids such as wood or stone?** (It is hard and keeps its shape; it can be sculpted.) **How is it different?** (It melts when the temperature gets above 0°C.)

3. **Predict** **Predict which ice sculptures shown on pages 58–59 would melt first. Explain your reasoning.** (Sample answer: The smallest ones would melt first because they contain less ice.)

NATIONAL GEOGRAPHIC
BECOME AN EXPERT

TECHTREK
myNGconnect.com
Student eEdition
Digital Library

Sweden: Ice Hotel Construction

Buildings have to keep their shapes. They cannot bend or flow. This is why they have to be made out of **solid** things. Did you know that a building can be made out of water? A hotel in the far northern part of the country of Sweden is made out of water in its solid state—ice!

The temperature has to be well below 0°C (32°F) to keep the ice—the hotel—solidly frozen.

Freezing Ice Blocks

In some places, it gets very cold in the winter. When it gets below 0°C (32°F), water freezes. In cold places such as northern Sweden, the **liquid** water in lakes and rivers can freeze. When a river freezes, the water at the top turns solid. The people who build the ice hotel use large blocks of this frozen river water.

Workers haul away large blocks of ice. They move ice from the river to the place where they will build the hotel.

solid
A **solid** is matter that keeps its own shape.

60

liquid
A **liquid** is matter that takes the shape of its container.

61

PROGRAM RESOURCES

- Big Ideas Book: *Physical Science*
- Big Ideas Book: *Physical Science* **eEdition** at ⊚ **myNGconnect.com**
- **Digital Library** at ⊚ **myNGconnect.com**

Access Science Content

Describe How Ice Can Be Used in Building

- Preview the pictures and headings on pages 60–61. Ask: **What do you think you will read about here?**

- Have students read pages 60–61. Ask: **Why must buildings be made of solids?** (to keep their shape) **What kinds of solids are most buildings made from?** (metal, wood, stone or rock, glass) **How is this building made in a different way?** (It's made of ice, or solid water.)

- Explain: **Sweden and other places that are very far north have very long, cold winters. Liquid water in lakes and rivers freezes and becomes solid.** Ask: **Why is it possible to build a hotel out of ice in Sweden?** (because the ice stays frozen solid during the long winter) Have a volunteer locate Sweden on a map or globe.

Notetaking

Have students begin a list of the different things used to make this special hotel in Sweden. Encourage them to describe how each item is used.

Access Science Content

Sequence the Steps of Building with Snow and Ice

- Have students read page 62 and study both process diagrams.

- Explain: **Solid water in the form of snow and ice is used to build the hotel.** Ask: **What is the first step in building with snow?** (Forms are set in place to give the snow the right shape.) **What is the second step?** (A snow cannon shoots snow onto the forms, and the snow is packed down.) **What is the third step?** (The form is taken out.)

- Remind students that ice is water in its frozen state. Water freezes when the temperature drops below 0°C. Ask: **Where does the ice come from that is used for building?** (Blocks of ice are cut from a frozen river.) **What do the builders do with the ice?** (They shape it to make pillars, beds, and other parts of the building.)

Make Inferences

Have students review how snow and ice are used for building. Ask: **Why do you think that snow, rather than ice, is used to make the hotel's walls and ceilings?** (Snow can be packed onto forms to give it the needed shape. It might be hard to cut curved shapes out of ice.)

Assess

1. **Recall** Why is it possible to build with ice and snow? (They are solids that keep their shape.)

2. **Cause and Effect** What must happen in order to build with ice and snow? Why? (The temperature must stay below 0°C, the point at which liquid water freezes to form these solids. If the temperature rises above 0°C, a building made from ice and snow will melt.)

BECOME AN EXPERT

Living in Ice

To keep the ice frozen, it must stay cold inside the hotel. The rooms are usually five degrees colder than 0°C. Outside, it can be 20 or 30 degrees colder than that temperature.

Seeing Your Breath A guest at the ice hotel might be able to see her breath. When it is cold, sometimes it looks like clouds are coming out of your mouth when you breathe. The puffs are not a **gas**. They are tiny drops of water. They formed when warm air that you breathed out cooled. In other words, **condensation** occurred.

gas
A **gas** is matter that spreads to fill a space.

condensation
Condensation is the change from a gas to a liquid.

64

65

PROGRAM RESOURCES

• **Digital Library** at ⊘ **myNGconnect.com**

Access Science Content

Describe Condensation

• Have students read pages 64–65. Say: **Your warm breath contains water vapor, an invisible gas. Condensation occurs when the air that you breathe out cools, and tiny drops of water form.**

Differentiated Instruction

ELL Language Support for Describing Condensation

BEGINNING	INTERMEDIATE	ADVANCED
Have students hold a mirror close to their mouth and breathe out. Point to the cloud on the mirror and have them say *condensation*.	Have students use an Academic Language Frame to describe condensation. *When condensation occurs, water changes from a gas to a _____.*	Have students use Academic Language Stems to describe what happens: *When I breathe on the mirror, condensation forms because . . . .*

Digital Library

⊘ **myNGconnect.com**

Have students use the **Digital Library** to find other images from the ice hotel.

Integrated Technology

Make Sound Files Have pairs of students use the photos to create ads for the ice hotel, recording their narration as a sound file.

Summarizing and Evaluating

Have students summarize how water in its solid state is used to build the hotel in Sweden. Then have them write about whether they would like to stay in the ice hotel, giving reasons to support their response.

Access Science Content

Explain Melting and Evaporation of Ice and Snow

- Have students read and look at the diagram on page 66. Explain: **The ice hotel cannot stay solid all year long.** Ask: **What happens to the hotel in spring?** (The ice and snow begin to melt, and the hotel disappears.) **Why does this happen?** (The temperature rises above 0°C, the point at which ice changes to liquid water.)

- Have students read page 67 and point out the term **states of matter.** Say: **The ice hotel was built of ice and snow, which are both forms of water in its solid state. How does the water's state of matter change as the weather gets warmer?** (The ice melts; water changes from a solid to a liquid.)

- Explain: **When snow and ice melt in the spring, some of the water soaks into the ground. Some runs off into rivers and lakes. But evaporation also occurs. What happens to the water that evaporates?** (It becomes water vapor, a gas in the air.) Point out how the diagram visually shows the hotel's water in its three states of matter.

Assess

1. **Identify** Give an example of each state of water you can find inside the ice hotel as the temperature warms above 0°C. (solid in the form of ice; liquid in the form of water from melting ice; gas in the form of water vapor in the air)

2. **Interpret Diagrams** Look at the diagram. What happens when a puddle of water that came from the melting ice hotel is warmed? (It evaporates.)

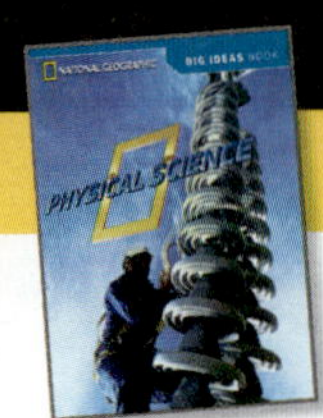

Share and Compare

Turn and Talk

Ask students to turn to partners and talk about how the three states of matter affect the ice hotel. Prompt students by asking:

1. **Recall** **What state of water was used to build the hotel in Sweden? Why?** (solids, or ice and snow, because solids keep their shape)

2. **Describe** **How were these materials used?** (Snow was packed onto forms and used to build walls and ceilings. Ice was cut in blocks from the frozen river and used for furniture and columns.)

3. **Explain** **What happened to the hotel in the spring? Why?** (It melted because the temperature rose above freezing.)

Read

Ask students to choose the two pages they think are most interesting. Ask them to read the text several times, including any captions or labels. Then have them read the pages aloud to a partner and talk about why the pages are interesting to them.

Write  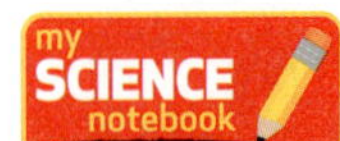

Have students review the Become an Expert section and then write a concluding paragraph for the section as well as a statement of the Big Idea. Then ask them to compare what they wrote with a classmate. Ask if partners included the same ideas in their conclusions.

Draw  

Encourage students to draw stages, or phases, such that the stages are represented equally. Their drawings should include the appropriate states of matter and labels such as *ice, water, water vapor, freezing at 0°C, melting ice,* and *evaporating water.* Have students combine their drawings with those of their classmates who drew other stages.

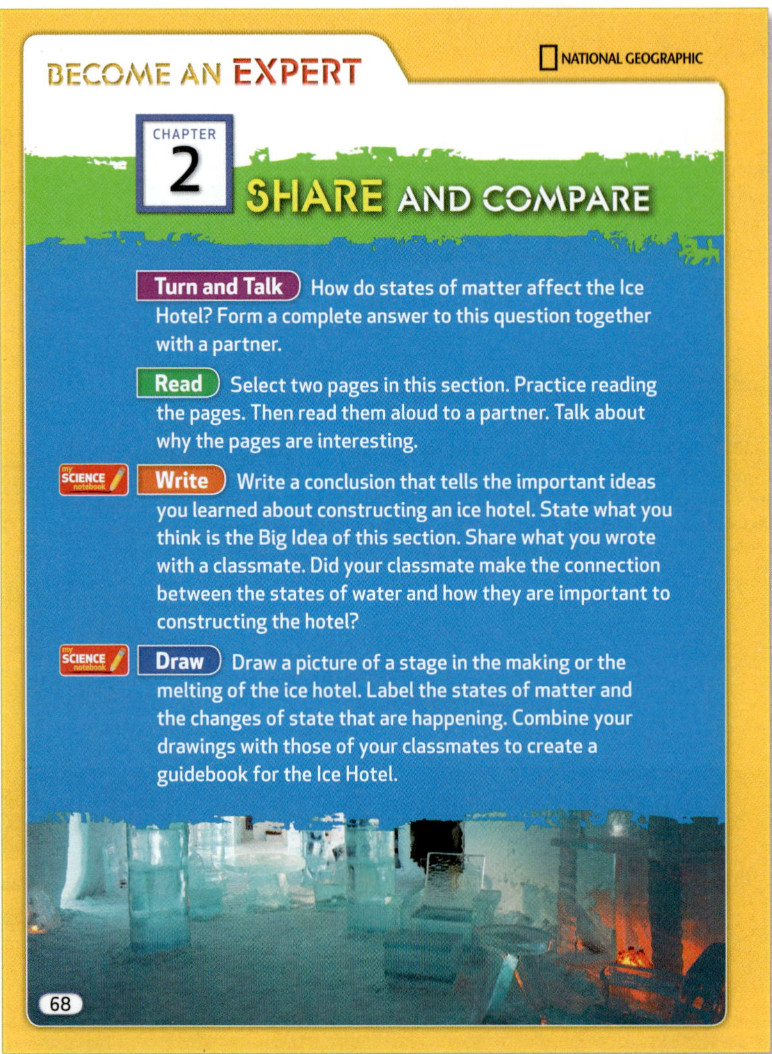

❯ Sum Up

Tell students that to sum up a text helps them pull together the ideas of the text. Remind students that they summed up the Become an Expert lesson in the Write section of Share and Compare. Have students take turns reading the conclusions they wrote to a partner. Invite them to compare and contrast their conclusions.

Read Informational Text

Explore on Your Own books provide additional opportunities for your students to:

• deepen their science content knowledge even more as they focus on one Big Idea.

• independently apply multiple reading comprehension strategies as they read.

After applying the strategies throughout the chapter, students will independently read Explore on Your Own books. Facsimiles of the Explore on Your Own pages are shown on pages T68a–T68h.

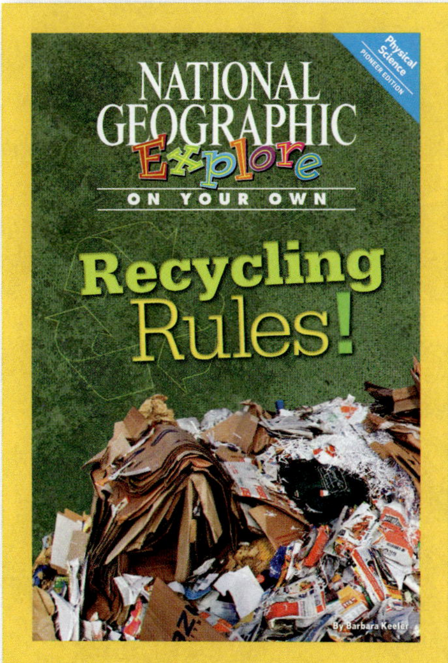

Pioneer   **Pathfinder**

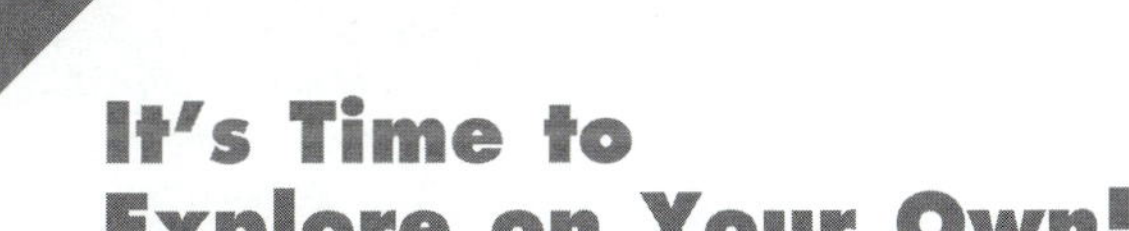

It's Time to Explore on Your Own!

Good readers use multiple strategies as they read on their own. Use the four key reading comprehension strategies below:

1 PREVIEW AND PREDICT
• Look over the text.
• Form ideas about how the text is organized and what it says.
• Confirm ideas about how the text is organized and what it says.

2 MONITOR AND FIX UP
• Think about whether the text is making sense and how it relates to what you know.
• Identify comprehension problems and clear up the problems.

3 MAKE INFERENCES
• Use what you know to figure out what is not said or shown directly.

4 SUM UP
• Pull together the text's big ideas.

Remember that you can choose different strategies at different times to help you understand what you are reading.

NATIONAL GEOGRAPHIC
School Publishing

Recycling Rules!

PIONEER EDITION

By Barbara Keeler

CONTENTS

Recycling Rules!

By Barbara Keeler

In the U.S., a family of four can throw out 2,907 kilograms (6,409 pounds) of trash each year. That's about as much as 115 third graders weigh! Trash can pollute the planet and hurt wildlife. The good news is that some trash can be recycled. Recycling means less trash for the Earth.

Recycling Resources

Why should we **recycle**? It takes lots of resources to make new things. When we recycle, Earth is not harmed to get new resources. Making new things also takes lots of energy. Recycling saves energy.

Trashing the Earth

Trash is usually made up of solids, liquids, and gases. A lot of trash goes to **landfills**. Landfills can take up a lot of space. Think about an empty juice can. The metal can, a solid, takes up space even if it is squished. Any leftover liquid in the can takes up space, too. Even the air in the can, which is a gas, takes up space.

Stripped Bare. Miners have stripped this forest land to mine **bauxite** ore.

Trash Heap. Recycling can help keep some things out of landfills.

Recycling Glass

Glass is often recycled. Glass is sorted by color. Then it is crushed and melted to a liquid in a hot furnace.

The melted glass is blown into molds. It hardens back into a solid. It is used to make a bottle or jar.

Recycling one glass jar saves enough energy to light a 100-watt lightbulb for four hours.

Recycling Metals

What happens to a can when it's recycled? First, it goes to a recycling plant. Then, giant blades shred it. Next a huge furnace melts it from a solid to a liquid. This uses a lot of energy. But it does not use as much energy as melting new metal.

The liquid metal cools to a solid. The solid metal is used to make new cans or other metal products.

5

Recycling Plastic

Plastic can last for hundreds of years. So it's a good idea to reuse it! Recycled plastic can be used for a lot of things. Some carpet fiber is made from it. So are some clothes.

Plastic is hard to recycle. Why? Different plastics contain different materials. They have different properties. They melt at different temperatures. This makes it hard to recycle them.

When it is recycled, most trash changes from solid to liquid and then back to solid. Sometimes gases are added during the recycling process. When gas is added to liquid plastic, it makes plastic foam. Plastic foam is used for insulated cups.

We can recycle trash by changing its state. Recycling helps the environment.

Tree Musketeers

The city of El Segundo, California, picks up and recycles plastic. Brook Church remembers when the city had no recycling program. That's why he started one at age 12. He, his ten-year-old sister, and some friends were in a group. It was called Tree Musketeers.

Making a Difference. Tree Musketeers helped residents recycle until the city set up a program.

The Tree Musketeers set up huge bins. People dropped off items to recycle. Parents drove the solids to a recycling center.

Years passed. People wanted recycling picked up from their homes. Tree Musketeers asked the city to do this, but the city said no. So Tree Musketeers found another way.

Success!

Tree Musketeers called a waste hauler to pick up **recyclables**. But using the hauler cost money. The program cost $6.00 a month. People signed up for it.

For years, Tree Musketeers collected money for the waste hauler. Finally, the city took over. It started picking up recyclables at people's homes for free.

Recycling Everywhere

Now many cities have curbside recycling. Kids are making recycling part of their lives.

The kids in Santa Rosa Beach, Florida helped their town. They worked to get more blue bags for the blue bag program. People put recyclables into blue bags. The bags are picked up with the trash. At the landfill, workers take the blue bags out. They recycle what is in them.

Gwen Wright's 1st-grade class at Butler Elementary helped out. Her class had a petition asking stores to sell blue bags. They got 239 students to sign. Some stores began to sell bags. The county then gave out some free bags.

Ella Robinson lives in Santa Rosa Beach, Florida. **She is learning not to make solid trash.** This way, no energy is used to melt solids for recycling. Ella takes her own cloth napkin to school. She washes her own dishes there. When she shops with her grandmother, they take their own shopping bags.

These kids and many others are making recycling a bigger part of their communities.

Watching Your Waste: The 4 Rs

Here are a few ways you can help reduce waste.

REDUCE
Buy things with no or little packaging.
Take your own bags to the store.
Take your own containers for leftovers.
Buy reusable, not disposable (dishes, napkins, towels, dishcloths).

REUSE
Reuse containers.
Write notes on used paper.
Buy used things. Repair, sell, or give away old things.

RECYCLE
Look for recycling bins when you are out.
Write to companies who make plastic containers. Ask them to put symbols on lids so recycling centers will take them.

RECLAIM
Buy products made from recycled materials.
Buy products with recycled packaging and containers.

Wordwise

bauxite: the ore from which aluminum is made

landfill: an area of land where trash is stored

recycle: make into new products

recyclables: things that can be made into new products

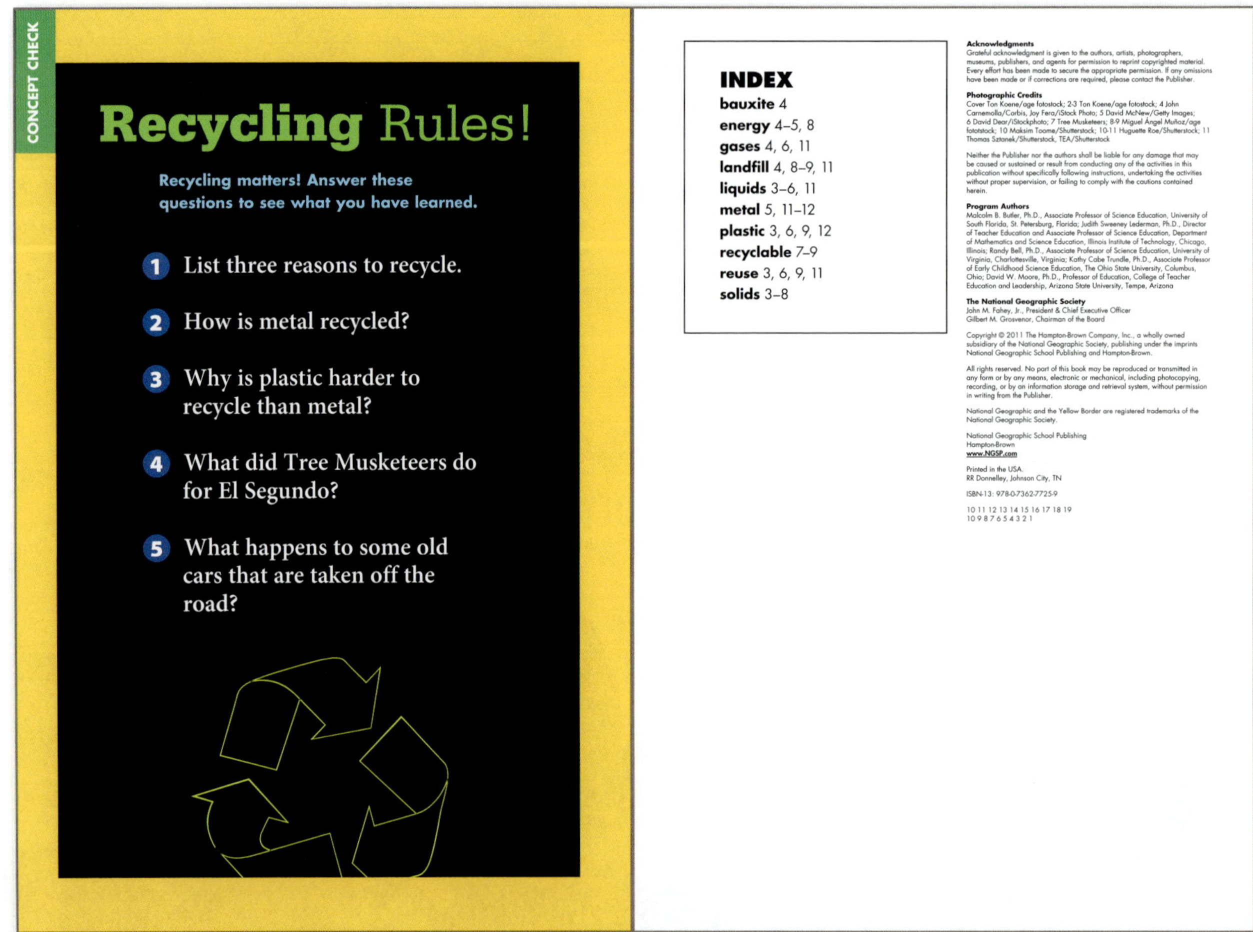

Acknowledgments
Grateful acknowledgment is given to the authors, artists, photographers, museums, publishers, and agents for permission to reprint copyrighted material. Every effort has been made to secure the appropriate permission. If any omissions have been made or if corrections are required, please contact the Publisher.

Photographic Credits
Cover Ton Koene/age fotostock; 2–3 Ton Koene/age fotostock; 4 John Carnemolla/Corbis, Joy Fera/Stock Photo; 5 David McNew/Getty Images; 6 David Dean/iStockphoto; 7 Tree Musketeers; 8–9 Miguel Angel Muñoz/age fotostock; 10 Maksim Toome/Shutterstock; 10–11 Huguette Roe/Shutterstock; 11 Thomas Sztanek/Shutterstock, TEA/Shutterstock.

Neither the Publisher nor the authors shall be liable for any damage that may be caused or sustained or result from conducting any of the activities in this publication without specifically following instructions, undertaking the activities without proper supervision, or failing to comply with the cautions contained herein.

Program Authors
Malcolm B. Butler, Ph.D., Associate Professor of Science Education, University of South Florida, St. Petersburg, Florida; Judith Sweeney Lederman, Ph.D., Director of Teacher Education and Associate Professor of Science Education, Department of Mathematics and Science Education, Illinois Institute of Technology, Chicago, Illinois; Randy Bell, Ph.D., Associate Professor of Science Education, University of Virginia, Charlottesville, Virginia; Kathy Cabe Trundle, Ph.D., Associate Professor of Early Childhood Science Education, The Ohio State University, Columbus, Ohio; David W. Moore, Ph.D., Professor of Education, College of Teacher Education and Leadership, Arizona State University, Tempe, Arizona

The National Geographic Society
John M. Fahey, Jr., President & Chief Executive Officer
Gilbert M. Grosvenor, Chairman of the Board

Copyright © 2011 The Hampton-Brown Company, Inc., a wholly owned subsidiary of the National Geographic Society, publishing under the imprints National Geographic School Publishing and Hampton-Brown.

All rights reserved. No part of this book may be reproduced or transmitted in any form or by any means, electronic or mechanical, including photocopying, recording, or by an information storage and retrieval system, without permission in writing from the Publisher.

National Geographic and the Yellow Border are registered trademarks of the National Geographic Society.

National Geographic School Publishing
Hampton-Brown
www.NGSP.com

Printed in the USA.
RR Donnelley, Johnson City, TN
ISBN-13: 978-0-7362-7725-9

10 11 12 13 14 15 16 17 18 19
10 9 8 7 6 5 4 3 2 1

It's Time to Explore on Your Own!

Good readers use multiple strategies as they read on their own. Use the four key reading comprehension strategies below:

1 PREVIEW AND PREDICT
- Look over the text.
- Form ideas about how the text is organized and what it says.
- Confirm ideas about how the text is organized and what it says.

2 MONITOR AND FIX UP
- Think about whether the text is making sense and how it relates to what you know.
- Identify comprehension problems and clear up the problems.

3 MAKE INFERENCES
- Use what you know to figure out what is not said or shown directly.

4 SUM UP
- Pull together the text's big ideas.

Remember that you can choose different strategies at different times to help you understand what you are reading.

Recycling Rules!

PATHFINDER EDITION

By Barbara Keeler

CONTENTS

In the U.S., the average family of four throws out 2,907 kilograms (6,409 pounds) of paper, glass, plastic, food scraps, and other trash each year. That's about as much as 115 third graders weigh! Those pounds of trash add up. It can pollute the planet and hurt wildlife.

The good news is that people can recycle some of this trash. You've probably dropped containers into a recycle bin. What happens after that? A truck hauls them to a recycling plant. There, the solid containers are melted into liquids. Then the liquids are made into new solids. People can then use the new solids. Using solids again means less trash for the Earth.

Recycling Resources

Why is recycling important? People are quickly using up Earth's resources to make new things. But when people recycle, Earth is not harmed to get new resources. Recycling also saves energy because making new products uses a lot of energy.

Stripped Bare. Miners have stripped this forest land to mine **bauxite** ore. This is used to make aluminum products.

Trashing the Earth

Trash is made up of solids, liquids, and gases. A lot of trash ends up in a **landfill**. Solids have a definite shape, but liquids and gases take the shape of their containers. All these kinds of matter take up space.

That means landfills can take up a lot of room. Think about a half-empty juice can. The solid metal takes up space whether or not it's flattened. Any leftover juice, or liquid, in the can also takes up space. Even the air in the can, which is a gas, takes up space.

Trash Heap. Some cities run out of space for landfills. Recycling helps to keep some items out of landfills.

Recycling Glass

Glass is an item that is often recycled. Glass containers are sorted by color. Next, the glass is crushed and melted in a furnace.

Molten glass pours out of the furnace. The molten glass is blown into molds. It cools, hardens back into a solid, and is taken out of the mold. Then it is cut into pieces to make a bottle or a jar.

Recycling one glass jar saves enough energy to light a 100-watt lightbulb for four hours.

Recycling Metals

Have you ever recycled a can? What happens to it? First, it goes to a recycling plant. There, giant blades shred it up. Next, the coatings are removed. Then a huge furnace changes it from solid to liquid. Melting an old can into liquid metal doesn't use nearly as much energy as it takes to make new liquid metal.

The molten, or melted, metal may be formed into bars and cooled into solids. Later, the bars are squeezed flat between huge rollers. Then they are rolled into sheets of metal to make new cans.

Recycling Plastic

Plastic can last for hundreds of years. So we might as well reuse it! Some plastics are melted and recycled much like glass. Recycled plastic gets used for a lot of things. You probably walk on recycled plastic every day. Some carpet fiber is made using it. You probably also wear recycled plastic. Polyester clothing is often made from it.

Plastic is harder to recycle than glass. Why? Not all plastic is the same. Each type contains different materials with different properties. Some plastics melt at different temperatures than others. That makes it hard for recycling plants to recycle them.

A lot of trash changes from solids to liquids and back to solids during the recycling process. Solids can be melted and formed into new products. Sometimes gases are added. Adding gas into some kinds of liquid plastic makes plastic foam for insulated cups.

We can recycle a lot of trash by changing its state.

Tree Musketeers

The city of El Segundo, California, picks up and recycles plastic. Brook Church remembers when El Segundo had no recycling program. That's why he started one at age 12. He, his ten-year-old sister, and some friends were in a group called Tree Musketeers.

Making a Difference. Tree Musketeers helped residents recycle until the city set up a program. By 2009, Tree Musketeers had won four Presidential awards.

Tree Musketeers began by writing columns about recycling in the newspaper. Then they set up some huge bins. People dropped off recyclables. Parents drove the solids to a recycling center.

After a few years, people in El Segundo wanted their recycling to be picked up from their homes. This is called curbside pickup. Tree Musketeers asked the city council to start picking up **recyclables** at the curb. The council said no. So Tree Musketeers found another way.

Success!

Tree Musketeers called a waste hauler. The waste hauler agreed to pick up recyclables for a fee. The program cost people $6.00 a month. Tree Musketeers collected payments to pay for the waste hauler.

Finally, after 13 years, the city took over. The city started picking up recyclables from every home. Curbside pickup was free.

Recycling Everywhere

Now many cities have curbside recycling. Kids all over the country are making recycling a part of their lives.

The kids in Santa Rosa Beach, Florida, helped to make it easier to recycle in their community. The kids worked to get more blue bags for the blue bag program. With this program, people put their recyclables into blue bags. The bags are picked up with the rest of the trash. At the landfill, workers separate the blue bags from the rest of the trash. Then the recyclables are taken to recycling plants! Some places do the same thing, just with different color bags.

Gwen Wright's 1st-grade class at Butler Elementary in Florida led the way. The class collected 239 student signatures for a petition. The petition asked stores to sell blue bags. Some stores began to sell them. As the bags became popular, the county gave out some free bags.

It takes energy to melt solids for recycling. In Santa Rosa Beach, Florida, **Ella Robinson is learning not to make solid trash in the first place.** Ella takes her own cloth napkin and washes her own dishes at school. When Ella shops with her grandmother, they take their own shopping bags.

These kids and many others are working hard to make recycling a bigger part of their communities.

Watching Your Waste: The 4 Rs

Here are a few ways you can help reduce waste.

REDUCE

Buy things with no or little packaging.
Take your own bags to the store.
Take your own cups when you buy drinks.
Buy reusable, not disposable (dishes, napkins, towels, dishcloths).
Compost, or make into fertilizer, yard waste and some food scraps.

REUSE

Reuse containers (boxes, bags, bottles, jars, plastic tubs).
Write notes on used paper.
Buy used things. Repair, sell, or give away old things.

RECYCLE

Look for recycling bins when you are out.
Write to companies that make plastic containers. Ask them to put symbols on lids so recycling centers will take them.

RECLAIM

Buy products made from recycled materials.
Buy products with recycled packaging and containers.

Wordwise

bauxite: the ore from which aluminum is made

landfill: an area of land where trash is stored

recyclables: things that can be made into new products

Clunkers
From the Road to the Recycler

As cars get old, sometimes they are no longer good for driving. People sometimes call these cars clunkers. Some of them are sold for scrap. Many of their parts are recycled.

What happens when a clunker is sold for scrap? First, the liquids, such as oil, are removed. Some are recycled. Then some of the parts of the car are sold and reused.

Next, the car is often flattened and crushed. The parts are shredded. Huge magnets separate the metals from the rest of the matter. Then the metals are melted down and formed into other products.

What happens to the tires? Many landfills will not accept tires. They take up too much space and don't break down very easily. Good tires are used on other cars. Some are used for other purposes, such as building walls to stop flooding. Others are recycled without being melted. They may be cut up and used for other products, such as sandals. Some tires are ground and shredded. They are used in carpet padding or as filler for roads.

CONCEPT CHECK

Recycling Rules!

Recycling matters! Answer these questions to see what you've learned.

1. List three reasons to recycle.

2. How is metal recycled?

3. Why is plastic harder to recycle than metal?

4. How did Tree Musketeers change recycling in their community?

5. What are old tires used for after they are recycled?

Acknowledgments
Grateful acknowledgment is given to the authors, artists, photographers, museums, publishers, and agents for permission to reprint copyrighted material. Every effort has been made to secure the appropriate permission. If any omissions have been made or if corrections are required, please contact the Publisher.

Photographic Credits
Cover Ton Koene/age fotostock; 2-3 Ton Koene/age fotostock; 4 John Carnemolla/Corbis, Joy Fera/iStock Photo; 5 David McNew/Getty Images; 6 David Dear/iStockphoto; 7 Tree Musketeers; 8-9 Miguel Ángel Muñoz/age fototstock; 10 Maksim Toome/Shutterstock; 10-11 Huguette Roe/Shutterstock; 11 Thomas Sztanek/Shutterstock, TEA/Shutterstock

Neither the Publisher nor the authors shall be liable for any damage that may be caused or sustained or result from conducting any of the activities in this publication without specifically following instructions, undertaking the activities without proper supervision, or failing to comply with the cautions contained herein.

Program Authors
Malcolm B. Butler, Ph.D., Associate Professor of Science Education, University of South Florida, St. Petersburg, Florida; Judith Sweeney Lederman, Ph.D., Director of Teacher Education and Associate Professor of Science Education, Department of Mathematics and Science Education, Illinois Institute of Technology, Chicago, Illinois; Randy Bell, Ph.D., Associate Professor of Science Education, University of Virginia, Charlottesville, Virginia; Kathy Cabe Trundle, Ph.D., Associate Professor of Early Childhood Science Education, The Ohio State University, Columbus, Ohio; David W. Moore, Ph.D., Professor of Education, College of Teacher Education and Leadership, Arizona State University, Tempe, Arizona

John M. Fahey, Jr., President & Chief Executive Officer
Gilbert M. Grosvenor, Chairman of the Board

Copyright © 2011 The Hampton-Brown Company, Inc., a wholly owned subsidiary of the National Geographic Society, publishing under the imprints National Geographic School Publishing and Hampton-Brown.

All rights reserved. No part of this book may be reproduced or transmitted in any form or by any means, electronic or mechanical, including photocopying, recording, or by an information storage and retrieval system, without permission in writing from the Publisher.

National Geographic and the Yellow Border are registered trademarks of the National Geographic Society.

National Geographic School Publishing
Hampton-Brown
www.NGSP.com

Printed in the USA.
RR Donnelley, Johnson City, TN

ISBN-13: 978-0-7362-7728-0

10 11 12 13 14 15 16 17 18 19
10 9 8 7 6 5 4 3 2 1

PHYSICAL SCIENCE

Chapter 3 Contents

How Does Force Change Motion?

After reading Chapter 3, you will be able to:

- Describe a force as a push or a pull, and recognize that some forces will only make objects move if they are touched. **MOTION, GRAVITY, MAGNETISM**

- Describe the position, direction, and motion of objects, and how the motion of objects changes by speeding up or slowing down. **POSITION, MOTION**

- Recognize that an object's speed depends on the time it takes to go a certain distance. **MOTION**

- Explain how the mass of objects and the strength of forces affect motion. **MOTION, GRAVITY, MAGNETISM**

- Explain that friction slows the motion of objects. **MOTION**

- Explain that gravity is a force that pulls objects toward Earth. **GRAVITY**

- Explain that magnetism is a force that pulls on some objects without touching them. **MAGNETISM**

- **Snap!** Recognize that an object's speed depends on the time it takes to go a certain distance. **MOTION**

69

❭ Preview and Predict

Tell students that they can preview a text by looking over it to get an idea of what the text is about and how it is organized. Tell students that predicting is forming ideas about what they will read. Students should confirm their ideas with others.

Have students preview and make predictions about the chapter. Remind them to use section heads, vocabulary words, pictures, and their background knowledge to make predictions.

CHAPTER 3 □ How Does Force Change Motion?

LESSON	PACING	OBJECTIVES
Directed Inquiry **1** *Investigate Forces and Motion* pages T69e–T69h **Science Inquiry and Writing Book** pages 186–189	**30** minutes	Investigate through Directed Inquiry (answer a question; make and compare observations; collect and record data and observations; generate explanations and conclusions based on evidence; share findings; ask questions based on observations to increase understanding). Relate a change in motion to the force that causes it. Identify contact and noncontact forces that affect the motion of an object, including gravity. Demonstrate that when an object does not move in response to a force, it is because another force is acting on it.
2 Big Idea Question and Vocabulary pages T70-T71–T72-T73 **Position** pages T74–T75	**30** minutes	Describe how pushes and pulls make objects move. Describe the position of objects.
3 Motion pages T76–T79	**25** minutes	Describe the motion of objects. Describe the direction of objects. Recognize that an object's speed depends on the time it takes to go a certain distance. Describe how the motion of objects changes by speeding up or slowing down.
4 Force pages T80–T85	**30** minutes	Describe a force as a push or a pull. Explain how the mass of objects and the strength of forces affects motion. Explain how the strength of forces affects motion. Explain that when a force does not move an object, it is because an equal force is acting on it. Explain that friction slows the motion of objects.
5 Gravity pages T86–T89	**25** minutes	Explain that gravity is a force that pulls objects toward Earth.
6 Magnetism pages T90–T93	**25** minutes	Understand that magnetism is a force that pulls on some objects without touching them.

VOCABULARY	RESOURCES	ASSESSMENT
measure **compare**	Science Inquiry and Writing Book: *Physical Science* Science Inquiry Kit: *Physical Science* Directed Inquiry: Learning Masters 203–205	Inquiry Rubric: Assessment Handbook, page 213 Inquiry Self-Reflection: Assessment Handbook, page 221 Reflect and Assess, page T69h
	Vocabulary: Learning Master 206 Big Ideas Book: *Physical Science*	Assess, page T75
motion **speed**	Science Inquiry and Writing Book: *Physical Science* Extend Learning: Learning Master 207	Assess, page T79
force **friction**		Assess, page T85
gravity		Assess, page T89
magnetism		Assess, page T93

TECHNOLOGY RESOURCES

STUDENT RESOURCES

myNGconnect.com

- Student eEdition
- Big Ideas Book
- Science Inquiry and Writing Book
- Read with Me
- Vocabulary Games
- Digital Library
- Enrichment Activities

National Geographic Kids

National Geographic Explorer!

TEACHER RESOURCES

myNGconnect.com

- Teacher eEdition
- Teacher's Edition
- Science Inquiry and Writing Book
- Online Lesson Planner
- National Geographic Unit Launch Videos
- Assessment Handbook
- Presentation Tool
- Digital Library

NGSP ExamView CD-ROM

IF TIME IS SHORT...
FAST FORWARD.

CHAPTER 3 □ How Does Force Change Motion?

LESSON	PACING	OBJECTIVES
7 NATIONAL GEOGRAPHIC **Parachutes: Floating Through the Air** pages T94–T95	**20** minutes	Explain that friction slows the motion of objects. Explain that gravity is a force that pulls objects toward Earth.
Guided Inquiry **8** *Investigate Motion and Position* pages T95a–T95d **Science Inquiry and Writing Book** pages 190–193	**30** minutes	Investigate through Guided Inquiry (answer a question; make and compare observations; collect and record data and observations; generate explanations and conclusions based on evidence; share findings; ask questions based on observations to increase understanding; adjust explanations based on findings and new ideas). Describe an object's position by relating it to other objects. Describe an object's motion by measuring its position over time. Recognize that when solving a problem, it is important to plan and get ideas and help from others.
9 Conclusion and Review pages T96–T97	**15** minutes	
10 NATIONAL GEOGRAPHIC **PHYSICAL SCIENCE EXPERT** *Materials Scientist* pages T98–T99 NATIONAL GEOGRAPHIC **BECOME AN EXPERT** *High-Speed Trains* pages T100-T101–T108	**35** minutes	Recognize how different scientists' work has contributed to general scientific understanding. Describe the motion of objects. Understand that magnetism is a force that pulls on some objects without touching them.

FAST FORWARD ▶▶▶
ACCELERATED PACING GUIDE

DAY 1 🌗 **30** minutes

Directed Inquiry

Investigate Forces and Motion, page T69e

DAY 2 🌗 **35** minutes

NATIONAL GEOGRAPHIC **PHYSICAL SCIENCE EXPERT** *Materials Scientist,* page T98

NATIONAL GEOGRAPHIC **BECOME AN EXPERT** *High-Speed Trains,* page T100–T101

VOCABULARY	RESOURCES	ASSESSMENT
	Share and Compare: Learning Master 208	Assess, page T95
plan	Science Inquiry and Writing Book: *Physical Science* Science Inquiry Kit: *Physical Science* Guided Inquiry: Learning Masters 209–212	Inquiry Rubric: Assessment Handbook, page 213 Inquiry Self-Reflection: Assessment Handbook, page 222 Reflect and Assess, page T95d
		Assess the Big Idea, page T97 Chapter 3 Test: Assessment Handbook, pages 97–100 NGSP ExamView CD-ROM
		Assess, pages T99, T102–T103, T106–T107

TECHNOLOGY RESOURCES

STUDENT RESOURCES

myNGconnect.com

- Student eEdition
- Big Ideas Book
- Science Inquiry and Writing Book
- Read with Me
- Vocabulary Games
- Digital Library
- Enrichment Activities

National Geographic Kids

National Geographic Explorer!

TEACHER RESOURCES

myNGconnect.com

- Teacher eEdition
- Teacher's Edition
- Science Inquiry and Writing Book
- Online Lesson Planner
- National Geographic Unit Launch Videos
- Assessment Handbook
- Presentation Tool
- Digital Library

NGSP ExamView CD-ROM

DAY 3 · 30 minutes

Guided Inquiry

Investigate Motion and Position, page T95a

Objectives

Students will be able to:

- Investigate through Directed Inquiry (answer a question; make and compare observations; collect and record data and observations; generate explanations and conclusions based on evidence; share findings; ask questions based on observations to increase understanding).
- Relate a change in motion to the force that causes it.
- Identify contact and noncontact forces that affect the motion of an object, including gravity.
- Demonstrate that when an object does not move in response to a force, it is because another force is acting on it.

Science Process Vocabulary

measure, compare

PROGRAM RESOURCES

- Science Inquiry and Writing Book: *Physical Science*
- Science Inquiry and Writing Book **eEdition** at ⊘ **myNGconnect.com**
- **Inquiry eHelp** at ⊘ **myNGconnect.com**
- Science Inquiry Kit: *Physical Science*
- Learning Masters Book, pages 203–205, or at ⊘ **myNGconnect.com**
- Inquiry Rubric: Assessment Handbook, page 213, or at ⊘ **myNGconnect.com**
- Inquiry Self-Reflection: Assessment Handbook, page 221, or at ⊘ **myNGconnect.com**

MATERIALS

Kit materials are listed in italics.

3 books; *table tennis ball; meterstick;* masking tape; *fleece fabric*

❶ Introduce

Tap Prior Knowledge

- Encourage students to share their experiences sledding or coasting downhill on bicycles. Ask: **Do you go farther and faster down a big hill or a small hill? a rough hill or a smooth hill? Why do you think that's so?**

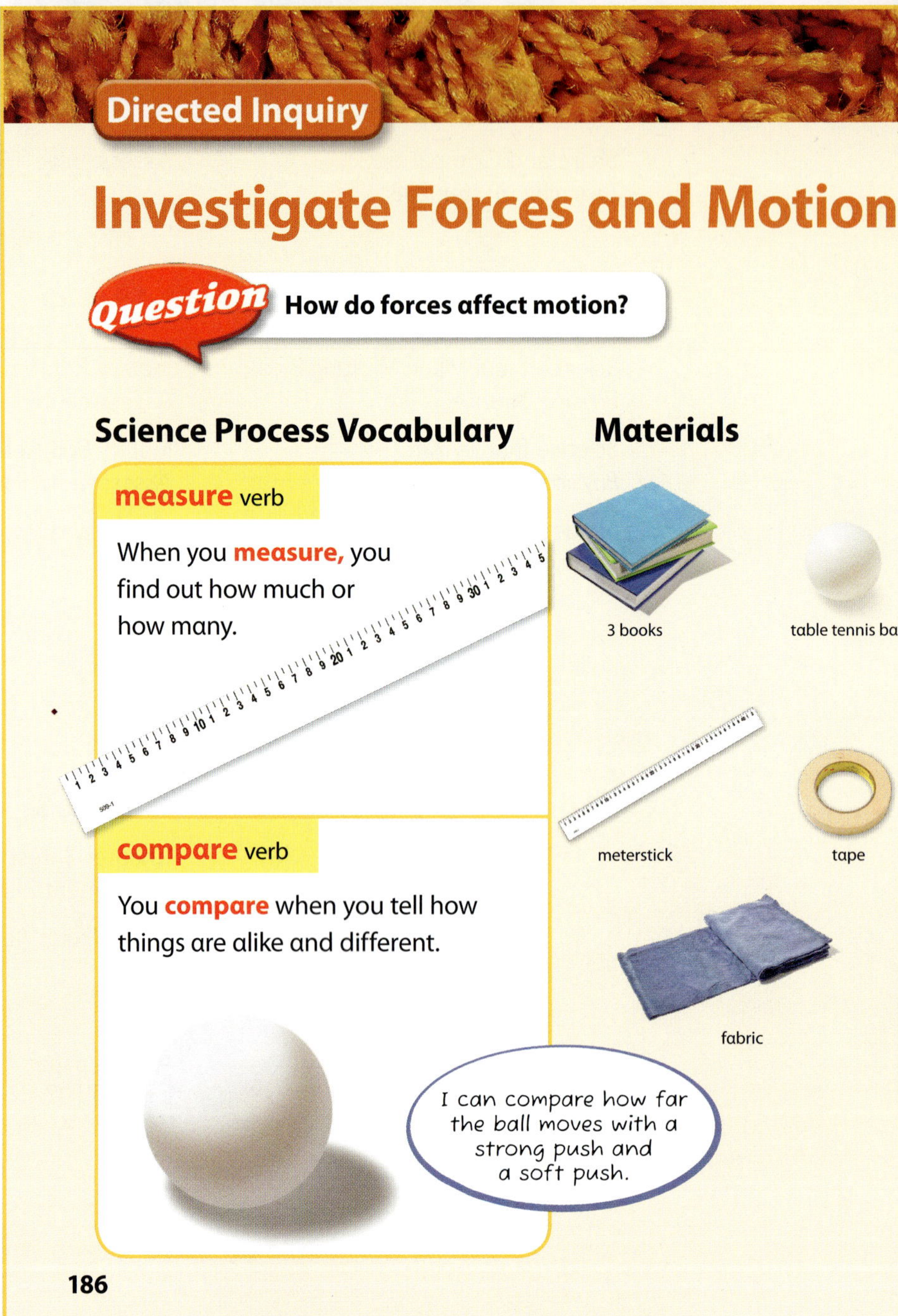

MANAGING THE INVESTIGATION

Time

 30 minutes

Groups

Small groups of 4

Advance Preparation

- Cut the fleece fabric into strips 1 ft wide by 6 ft long.
- Gather two thick and one thin hardcover book for each group. The two thick books should be about the same size, if possible.
- Clear a smooth surface, such as a wood or linoleum floor, for students to set up the ramps.

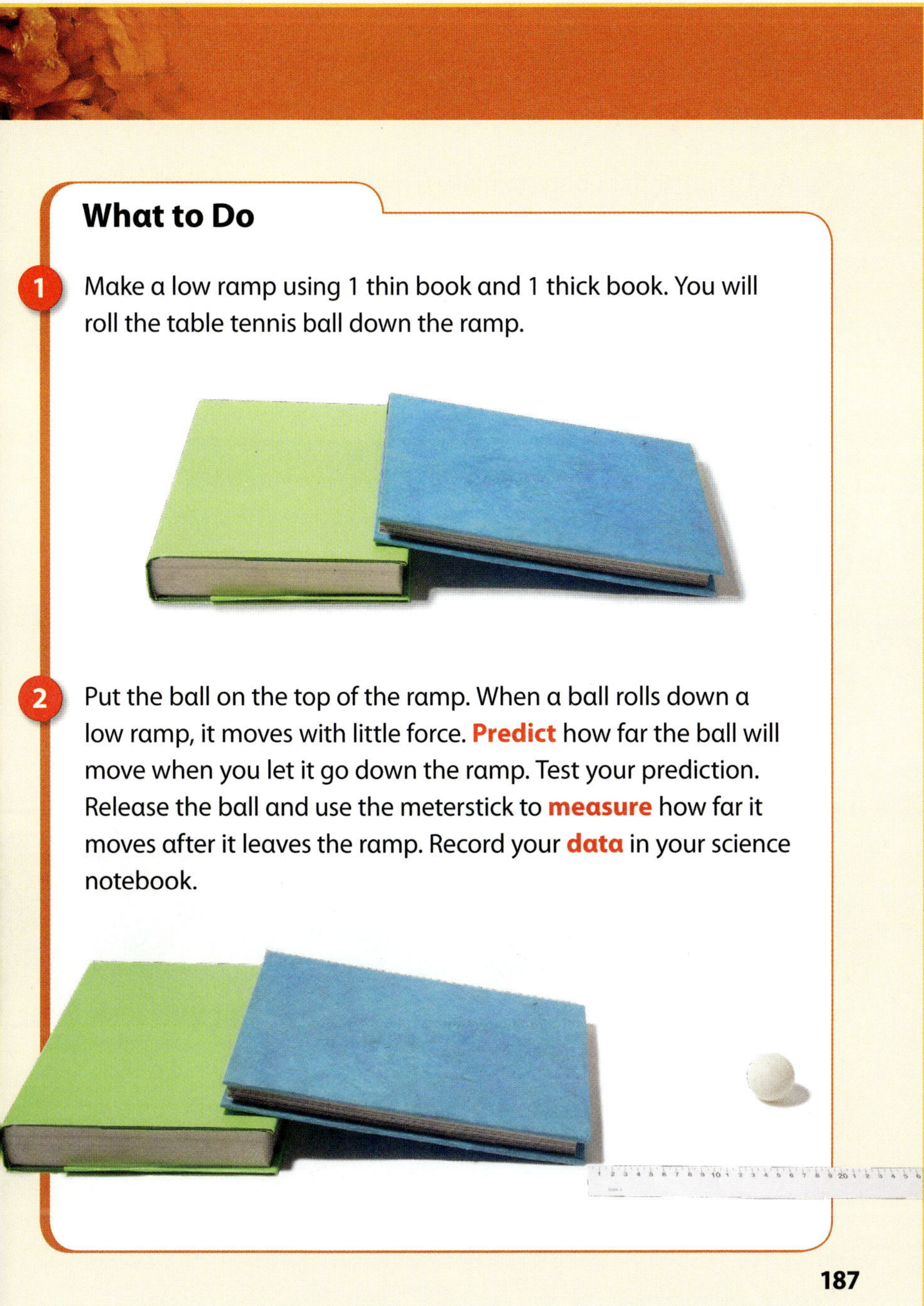

What to Do

1 Make a low ramp using 1 thin book and 1 thick book. You will roll the table tennis ball down the ramp.

2 Put the ball on the top of the ramp. When a ball rolls down a low ramp, it moves with little force. **Predict** how far the ball will move when you let it go down the ramp. Test your prediction. Release the ball and use the meterstick to **measure** how far it moves after it leaves the ramp. Record your **data** in your science notebook.

187

Teaching Tips

- Before step 4, have students rub their hands over the floor and then over the piece of fleece and describe the difference.
- Tell students to position the ball at the top of the ramp and then just let go. They should not push the ball.

What to Expect

- Students should observe that the ball moved farther when rolled down the high ramp than the low ramp. They should also observe that the ball did not roll as far across the fabric as it did across the smooth floor.

Introduce, continued

Connect to the Big Idea

- Review the Big Idea Question, *How does force change motion?* Explain to students that this inquiry focuses on how two forces—gravity and friction—change the motion of a ball.

- Have students open their Science Inquiry and Writing Books to page 186. Read the Question and invite students to share ideas about how a force will affect the motion of a ball.

❷ Build Vocabulary

Science Process Words: measure , compare

Use this routine to introduce the words.

1. **Pronounce the Word** Say compare. Have students repeat it in syllables.

2. **Explain Its Meaning** Choral read the sentence. Ask students for another way to say compare. (explain how one thing is like or different from another)

3. **Encourage Elaboration** Ask: **How would you compare a crayon and a pencil?** (You can write or draw with both. A crayon is made of wax; a pencil is made of wood and graphite.)

   ELL **Vocabulary Support**

   Show a thick book and a thin book collected for this activity. Say: I compare these two things. I see they are *alike* because they are both books. I see they are *different* because one is bigger than the other.

Repeat for the word measure. To encourage elaboration, ask: **What tool could you use to measure your height? (meterstick, tape measure)**

❸ Guide the Investigation

- Distribute materials. Read the inquiry steps on pages 187–188. Move from group to group and clarify steps if necessary.

- Have students write down their predictions about how far the ball will move on the low ramp. Have them discuss their predictions. Ask: **Why do you think the ball will move that far?**

- As students measure the distance the ball moved, explain that gravity is the force that pulls the ball down the ramp toward Earth's center.

Guide the Investigation, continued

- As students make predictions, ask: **Do you think the ball will move farther? Why or why not?** Students may predict that the ball will move farther.

- Remind students that all changes in motion are caused by forces. As students predict how far the ball will move on the fabric, explain that friction is a force that acts when two surfaces rub together. Friction can slow down a moving object. Say: **The force of friction can stop a moving object. Rough, bumpy surfaces cause more friction than smooth surfaces.**

❹ Explain and Conclude

- Guide students as they record their data in the table. Ask: **How do your results compare with your predictions?** Have students share their predictions and results with the class.

- Ask: **What patterns did you observe?** (The ball rolled farther from the high ramp than from the low ramp. It moved farther on the floor than on the fabric.)

- Ask: **Why do you think the ball moved farther from the high ramp?** (because it was moving with more force) **What force was pulling it?** (gravity) **When you pull something, you touch it, or make contact with it. Did gravity touch the ball?** (no) Say: **We call gravity a noncontact force.**

- Ask: **Why do you think the ball moved farther on the floor than on the fabric?** (There was more friction between the ball and the fabric so the ball slowed down more.) **Is friction a contact or a noncontact force?** (contact)

Answers

1. Answers will vary but should include mention of the predictions, the results, and whether the predictions were supported.

2. Possible answer: The ball did not roll as far on the low ramp as it did on the high ramp. The ball rolled with more force on the high ramp.

3. Possible answer: The ball moved farther on the floor. Friction from the fabric made the ball move more slowly.

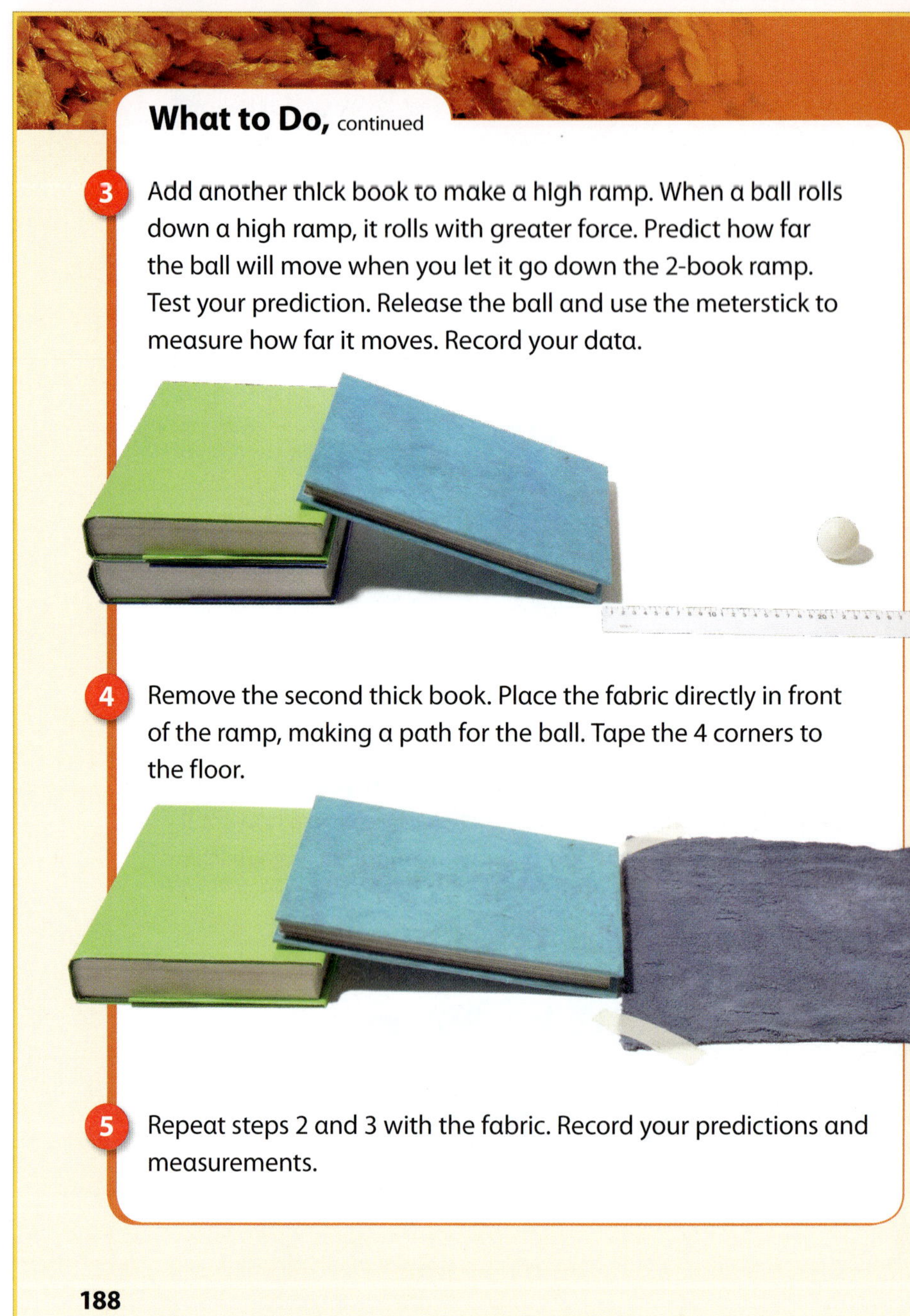

What to Do, continued

3 Add another thick book to make a high ramp. When a ball rolls down a high ramp, it rolls with greater force. Predict how far the ball will move when you let it go down the 2-book ramp. Test your prediction. Release the ball and use the meterstick to measure how far it moves. Record your data.

4 Remove the second thick book. Place the fabric directly in front of the ramp, making a path for the ball. Tape the 4 corners to the floor.

5 Repeat steps 2 and 3 with the fabric. Record your predictions and measurements.

188

NATIONAL GEOGRAPHIC Raise Your SciQ!

Inertia and Force An object at rest has the tendency to stay at rest unless a force acts on it. This tendency is called inertia. The mass of the object affects the amount of force needed to overcome inertia and move it. The mass of the heavy object in the Extend activity requires more force to move it than students could apply. The forces of friction and gravity also act on the object and prevent it from being moved.

Record

Write in your science notebook.
Use a table like this one.

Motion of Ball		
Trial	Prediction (cm)	How Far Ball Moves (cm)
Low ramp on floor		
High ramp on floor		
Low ramp on fabric		
High ramp on fabric		

Explain and Conclude

1. Did your results support your **predictions?** Explain.
2. **Compare** how the ball moved when you rolled it down the low and the high ramps. What do you think caused the difference?
3. Compare how the ball moved on the floor and on the fabric. What do you think caused the difference?

Think of Another Question

What else would you like to find out about forces and motion? How could you find an answer to this new question?

The roughness of the sand slows down the ball.

189

❺ Find Out More

Think of Another Question

- Students should use their observations to generate new questions. For example, students may ask: *Is more force necessary to move a heavy object down a ramp than a light object?*

❻ Reflect and Assess

- To assess student work with the Inquiry Rubric shown below, see Assessment Handbook, page 213, or go online at ✈ **myNGconnect.com**
- Use the Inquiry Self-Reflection on Assessment Handbook, page 221, or at ✈ **myNGconnect.com**

❼ Extend

- Have students apply a weak force to an object that will not move, such as a heavy book or a desk. Ask: **Why do you think your force does not cause the object to move?** (Possible answer: The force I applied is not greater than the other forces acting on the object to keep it in place.) Explain that, if a force is applied to an object and it does not move, it is because other forces are acting on it.

Inquiry Rubric at ✈ myNGconnect.com	Scale			
The student **predicted** how far a ball would move down a ramp, changing the height and amount of friction.	4	3	2	1
The student **measured** the distances the ball moved and recorded the **data** in a table.	4	3	2	1
The student **compared** predictions with results.	4	3	2	1
The student made a **conclusion** about how the forces of gravity and friction affect how far a ball moves.	4	3	2	1
The student **shared** and compared data and conclusions.	4	3	2	1
Overall Score	4	3	2	1

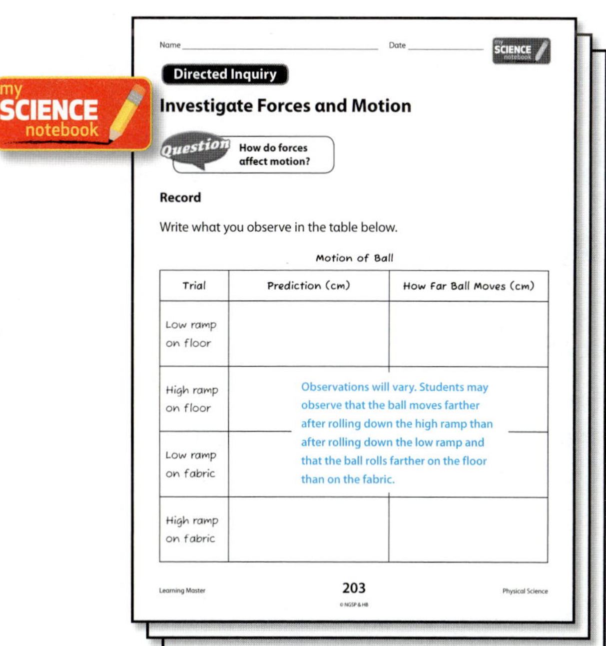

Learning Masters 203–205 or at ✈ myNGconnect.com

Objectives

Students will be able to:

• Describe how pushes and pulls make objects move.

Science Academic Vocabulary

motion, speed, force, friction, gravity, magnetism

PROGRAM RESOURCES

• Big Ideas Book: *Physical Science*
• Big Ideas Book: *Physical Science* **eEdition** at **myNGconnect.com**
• **Vocabulary Games** at **myNGconnect.com**
• **Digital Library** at **myNGconnect.com**
• **Enrichment Activities** at **myNGconnect.com**
• **Read with Me** at **myNGconnect.com**
• Learning Masters Book, page 206, or at **myNGconnect.com**

❶ Introduce

Tap Prior Knowledge

• Ask students what happens when they push or pull an object.

• Then have students study the photo on pages 70–71. Explain that the scarab beetle is pushing a ball of animal dung, which is animal waste. The beetle will bury the dung and use it for food. Ask: **What happens when the beetle pushes the ball?** (It rolls forward.)

❷ Focus on the Big Idea

Big Idea Question

• Read the Big Idea Question aloud and have students echo it.

• Preview pages 74–75, 76–79, 80–85, 86–89, and 90–93, linking the headings with the Big Idea Question.

Differentiated Instruction
ELL Vocabulary Support

BEGINNING	INTERMEDIATE	ADVANCED
After students have been introduced to the science vocabulary, ask yes/no questions to check for understanding. For example: **Is a push a force?** **Is only one surface needed for friction?**	Help students use these Academic Language Frames to reinforce their understanding of the vocabulary: • *A force is a _____ or pull.* • *Two surfaces rubbing together cause _____.*	Have students describe the vocabulary words with Academic Language Stems. For example: • *Speed depends on …* • *A force is …*

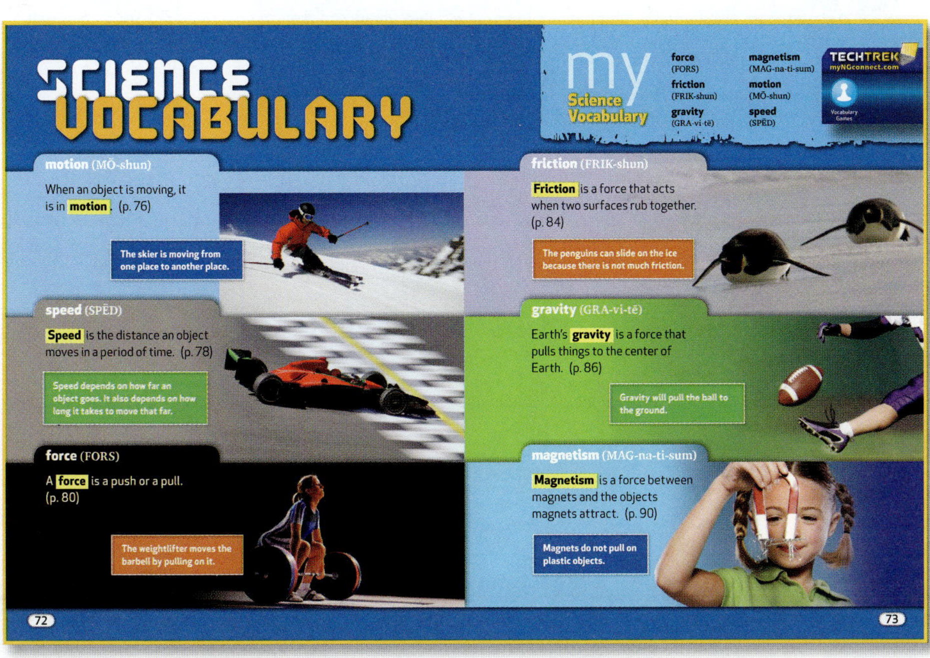

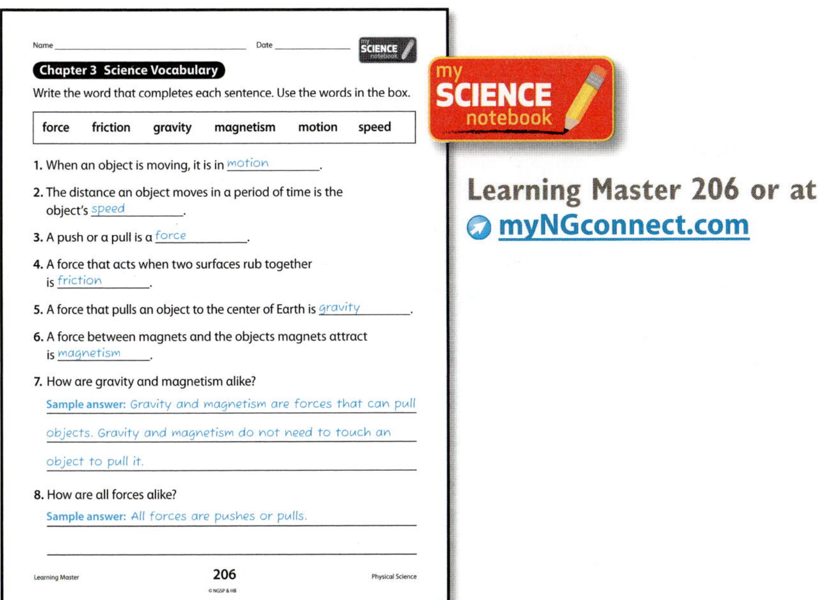

Learning Master 206 or at
🌐 myNGconnect.com

With student input, post a chart that displays the chapter headings. Have students orally share what they expect to find in each section. Then read page 70 aloud.

❸ Teach Vocabulary

Have students look at pages 72–73, and use this routine to teach each word. For example:

1. **Pronounce the Word** Say **motion** and have students repeat it.

2. **Explain Its Meaning** Read the word, definition, and sample sentence, and use the photo to explain the word's meaning. Point out the skier: **The skier is in motion.**

3. **Encourage Elaboration** Have students look at the photo again. Ask: **How might you describe the skier's motion?** (downhill, fast, smooth, turning)

Repeat for the words **speed**, **force**, **friction**, **gravity**, and **magnetism** using the following Elaboration Prompts:

- **What two things are needed to describe speed?** (distance an object moves and period of time)

- **When will the weightlifter use a pulling force?** (when she lifts the weight off the ground)

- **What causes friction?** (two surfaces rubbing together)

- **Is the force of gravity a push or a pull?** (pull)

- **Does magnetism push or pull paper clips?** (pull)

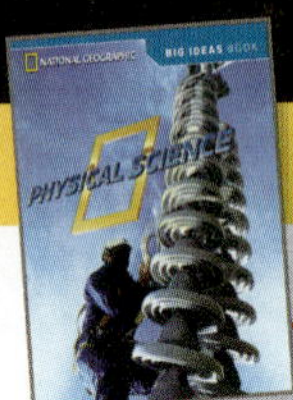

Objectives

Students will be able to:

• Describe the position of objects.

❶ Introduce

Tap Prior Knowledge

• **Ask students to choose an object near them and describe its position. Record the words and phrases students used in their descriptions.** (Sample answers: on top of, below, above, to the right of, in front of)

Set a Purpose and Read

• Read the heading. Tell students they will read about how to describe position.

• Have students read pages 74–75.

❷ Teach

Text Feature: Photo and Caption

• Have students study the photo and caption on page 74. Say: **Position describes where an object is. To describe an object's position, the object needs to be compared to another object. You can describe the position of the acrobats by comparing the acrobats to the balls.**

• Ask: **What is the position of the acrobats compared to the balls?** (above the balls)

Position

Look at the photo. Where are the acrobats? You can use the balls to describe the positions of the acrobats. The position of the acrobats is above the balls. Position describes where an object is. You always compare one object to another object when you describe position.

74

Differentiated Instruction

Extra Support

To help students understand different ways to describe position, name and point to an object in the room. Ask: **What words can you use to describe the position of this object?** (Descriptions should name at least two objects and include words such as between, in front of, behind, to the right of, to the left of, on top of, above, below, beside, and over.)

Challenge

Ask students to write a paragraph describing their position compared to at least four objects.

There are many ways you can describe position. Look at the photo of the three friends. The boy is in the middle. He is holding a giant pumpkin in front of his body. One girl is wearing a hat on top of her head. She is holding a pumpkin beside her head. The other girl is holding a pumpkin over her head. You compare the location of objects using phrases that describe position.

Before You Move On

1. What is position?
2. What are some words you can use to describe the position of objects?
3. **Apply** A bird is flying over a cat. Describe the position of the cat by comparing its position with that of the bird.

75

Teach, continued

Describe the Position of Objects

- Have students reread the paragraph on page 75 and study the photo. Ask: **If you wanted to describe the position of the biggest pumpkin, what objects can it be compared to?** (the boy, the two girls)

- Say: **You can use words to describe position in many ways.** Ask: **What different words can you use to describe the boy's position compared to the other objects in the photo?** (Sample answers: behind the big pumpkin, between the two girls, between the two smaller pumpkins, in the middle)

❸ Assess

» Before You Move On

1. **Define** **What is position?** (Position is where an object is located.)

2. **Classify** **What are some words you can use to describe the position of objects?** (Answers will vary. Possible answers: above, after, behind, before, below, beside, close, distant, far, in front of, near, next to, over, under)

3. **Apply** **A bird is flying over a cat. Describe the position of the cat by comparing its position with that of the bird.** (The cat's position is below or under the bird.)

Think Like a Scientist Math in Science

- Tell students that scientists also use distance to describe position. Distance is a measurement of how far two objects are from each other. An object might be one (1) meter (about 1 yard) above another object.

- Provide sample problems for describing position using distance. Say: **The book is one (1) meter to the right of the lamp. A pencil is twice as far from the lamp.** Ask: **How far is the pencil from the lamp?** (2 meters)

LESSON 3 □ Motion

Objectives

Students will be able to:
- Describe the motion of objects.
- Describe the direction of objects.

Science Academic Vocabulary

motion

PROGRAM RESOURCES
- Learning Masters Book, page 207, or at
  myNGconnect.com

❶ Introduce

Tap Prior Knowledge
- Have students describe the route they take when they walk or ride to school, using words such as *left, right,* and *straight.*
- As you move around the classroom, have students describe how your position changes.

Preview and Read
- Read the heading. Then preview the pictures on pages 76–79. Tell students they will read about the motion of objects and what causes objects to move. Have students read pages 76–78.

❷ Teach

Academic Vocabulary: *motion*
- Write the words *move, moves,* and *moving.* Then pronounce and write **motion.**

Extend Learning

MANAGING THE INVESTIGATION

Time

20 minutes

Groups
Small groups of 4

PROGRAM RESOURCES
- Learning Master 207

MATERIALS
- toy car with moving wheels
- tennis ball
- a long board or other surface that can be tilted

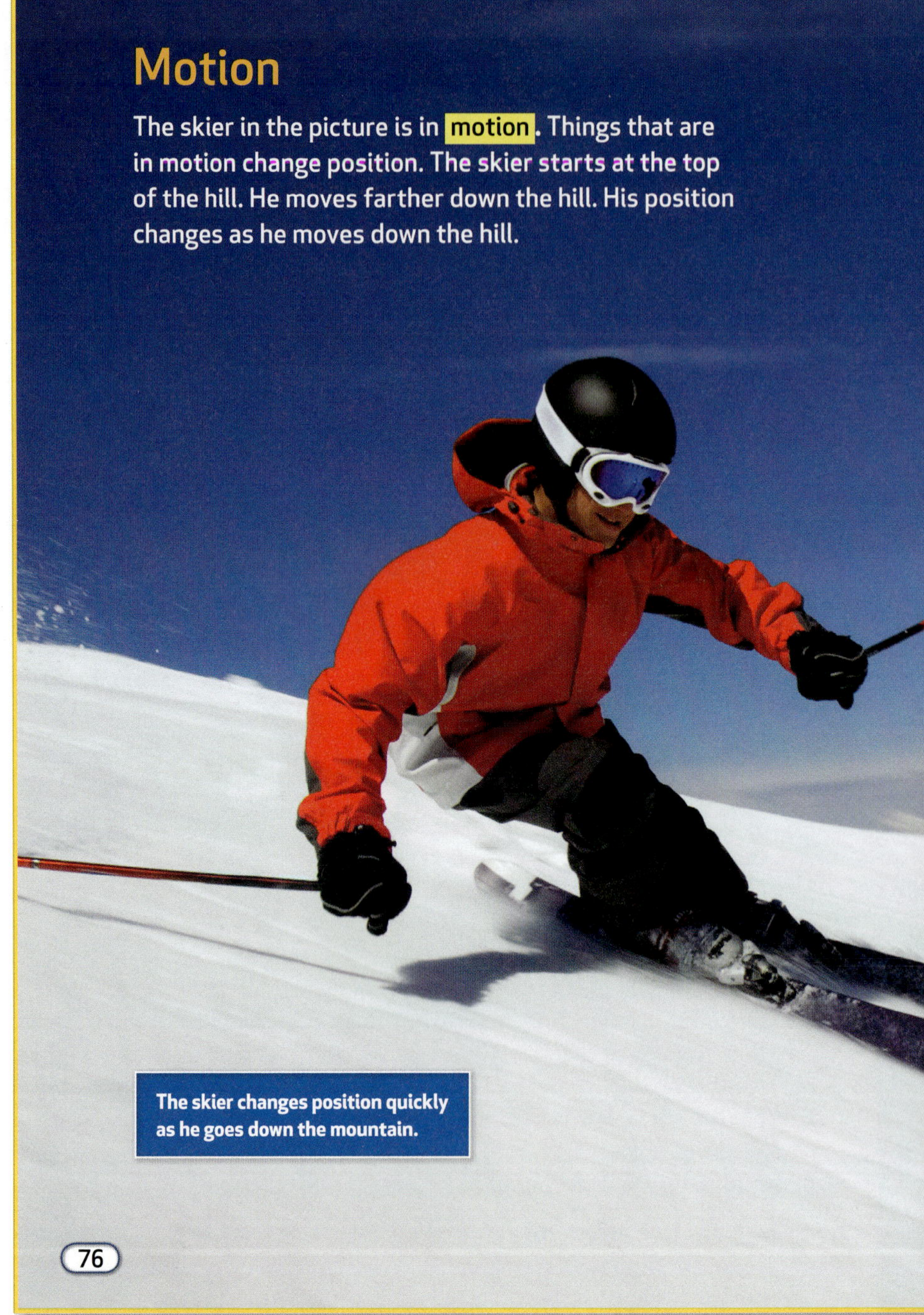

Motion

The skier in the picture is in **motion**. Things that are in motion change position. The skier starts at the top of the hill. He moves farther down the hill. His position changes as he moves down the hill.

76

Investigate Changes in Motion

Preview	What To Do
 **Question** How can an object's motion change? Students will think about all the ways an object's motion can change. They will use a toy car and a tennis ball to demonstrate each change in motion. Then they will identify and explain each change of motion they observed.	1. Use a toy car to demonstrate ways a moving object's motion can change. The car can be rolled up or down a sloped surface so that its speed increases or decreases. 2. Use a tennis ball to demonstrate each change. The ball can be rolled toward a wall to demonstrate a change in direction after it strikes the wall.

One way you describe motion is by describing direction. Direction is the path an object takes. Look at the picture. Some of these cars are going straight. Others might turn left or right. *Left*, *right*, and *straight* are words that describe direction. *North*, *south*, *east*, *west*, *up*, and *down* are also direction words.

77

Teach, continued

- Say: **An object changes position when it moves. When an object is moving, it is in motion.**

- Have students study the photo of the skier. Ask: **How does the skier's position change when he is in motion?** (Possible answer: He moves farther down the hill and shifts to the left or right.)

Describe the Motion of Objects

- Have students reread the paragraph on page 77. Say: **Motion, or movement, can be described in many different ways. Describing direction is one way to describe motion.** Point out the sentence that tells what direction is. Ask: **What is direction?** (the path an object takes)

- Ask: **What are some words that describe direction?** (left, right, straight, north, south, east, west, up, down)

- Have students stand and change their direction as you say the words *left*, *right*, and *straight*. Then show them the directions of north, south, east, and west in the classroom and repeat the activity.

- Have students study the photo and read the caption. Ask: **What can you tell about the motion of people and objects in the photo? Use words that describe their direction.** (Answers may vary. Possible answers: People and objects are moving in different directions. A person is running to the left. A taxi is moving to the right.)

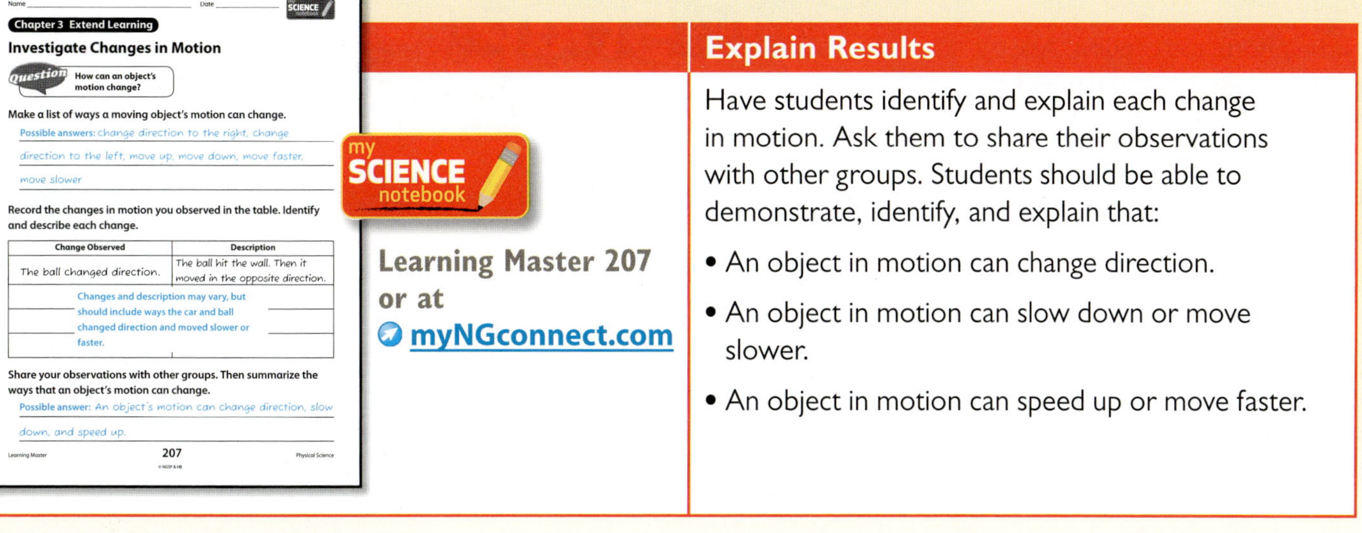

Explain Results

Have students identify and explain each change in motion. Ask them to share their observations with other groups. Students should be able to demonstrate, identify, and explain that:

- An object in motion can change direction.

- An object in motion can slow down or move slower.

- An object in motion can speed up or move faster.

Learning Master 207
or at
🌐 **myNGconnect.com**

LESSON 3 □ Motion

Objectives

Students will be able to:

- Recognize that an object's speed depends on the time it takes to go a certain distance.
- Describe how the motion of objects changes by speeding up or slowing down.

Science Academic Vocabulary

speed

PROGRAM RESOURCES

- Science Inquiry and Writing Book: *Physical Science*
- Science Inquiry and Writing Book **eEdition** at ⊘ **myNGconnect.com**
- **Digital Library** at ⊘ **myNGconnect.com**

Teach, continued

Academic Vocabulary: *speed*

- Point out the word **speed** and write it. Read the sentence that defines **speed**.

- Ask: **What does speed tell you about an object's motion?** (how fast an object is moving) **When might knowing speed be important?** (Possible answers: when driving a car, running a race, or trying to get somewhere at a certain time)

Recognize What Speed Depends On

- Say: **An object's speed depends on the time it takes to change position.**

- Ask: **Can you find an object's speed if you only know how far it moved?** (no) **What information do you need to know?** (You have to know both how far it moved and how long it took to go that far.)

- Have students study the photo and caption. Ask: **What is the race car's speed?** (280 km or 175 miles per hour) **What would its speed be if it went a distance of 100 km in one hour?** (100 km per hour) **Is that a faster or slower speed than the car in the photo?** (slower)

- Say: **A race car does not travel the same speed for the whole race. When we say what the car's speed is at a certain moment, we are saying how many kilometers it would travel if it kept going at that speed for a whole hour.**

Speed Motion can be fast or slow. **Speed** is the rate at which an object changes position. It tells you how fast an object is moving. A race car moving at a high speed changes position faster than a race car moving at a slower speed. The car that completes the race in the least amount of time has the greatest speed. The car that completes the race in the greatest amount of time has the slowest speed.

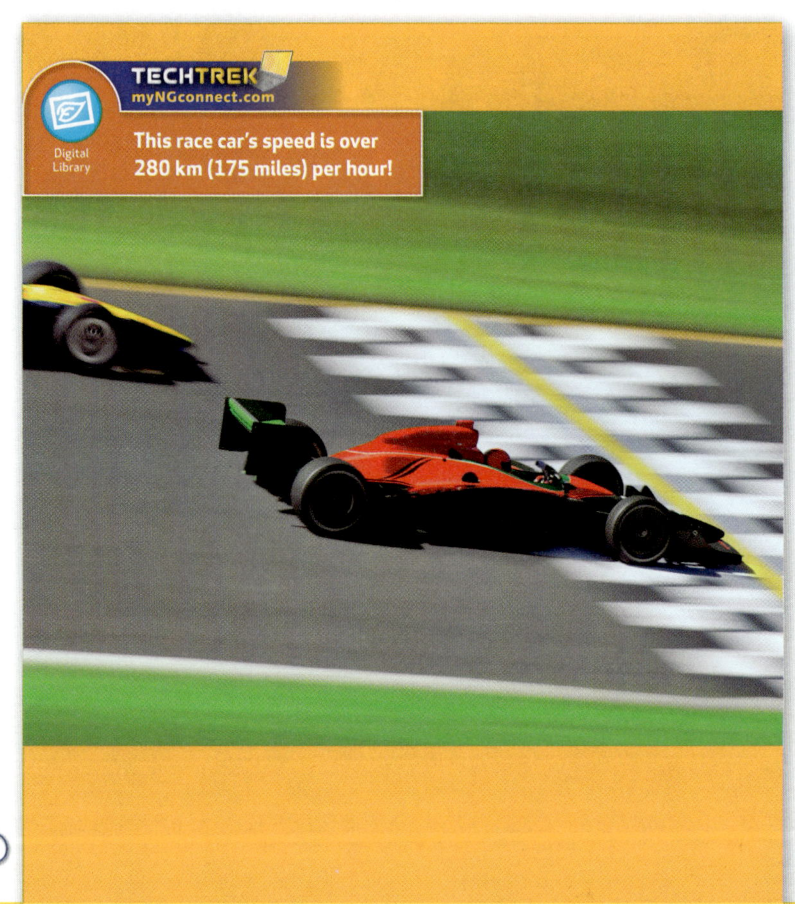

Science in a Snap!

Measure Your Speed 

Materials meter stick, masking tape, clock or watch with second hand, Big Ideas Book

Describe the activity to students. Discuss with students what measuring tool would be appropriate for measuring distance in this activity. Have students work in pairs. Distribute the materials to each pair of students. Have students read and follow the instructions on page 79 of the Big Ideas Book. Students should record their observations in their science notebook.

Science in a Snap! Measure Your Speed

Put a piece of tape at your starting position. Starting at the piece of tape, measure ten meters with a partner. Put another piece of tape at the end of ten meters.

Walk ten meters. Time your partner and have your partner time you. Observe your speed and your partner's speed closely.

How were your speeds different? Who had the greater speed?

Before You Move On

1. What is motion?
2. How are speed and motion related?
3. **Apply** You are watching two people running in a race. How do you know which person is moving at a greater speed?

79

What to Expect By dividing the distance in meters by the time in seconds, they can calculate their speed. Speed should be expressed in meters per second. Students should compare their speeds with words, such as *faster* and *slower*. They can also subtract to find out how much faster or slower their speeds were.

Quick Questions Ask students the following questions:

- **What measuring tools did you use to find your speed?** (meter stick and clock)
- **How did you use each tool?** (I used the meter stick to measure distance and the clock to measure time.)

- Say: **Motion, or movement, can be described in many ways.** Ask: **What are two ways you learned to describe motion?** (direction, speed)

 Digital Library

 myNGconnect.com

Have students use the **Digital Library** to find photos of race cars.

Science in a Snap!

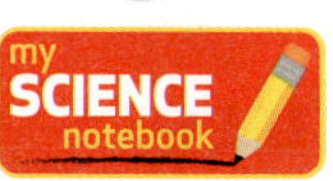

Compare Motion

Science Inquiry and Writing Book: *Physical Science,* page 163

Materials book, ball

Read and follow the instructions together with students. Challenge students to think of all the ways the motion can be described and compared and discuss it with their group. Students should record their observations and ideas in their science notebook.

What to Expect Students will push a book slowly across a table and describe its motion. Then students will roll a ball quickly across the floor and explain all the ways the motion of the ball differed from that of the book. Students should describe and compare how far, how fast, and in what direction the book and the ball moved. The ball will move faster and farther than the book.

❸ Assess

» Before You Move On

1. **Define** **What is motion?** (Motion is when an object changes position.)

2. **Compare** **How are speed and motion related?** (To find the speed of an object, you need to know the distance an object moved and the amount of time it took to move there.)

3. **Apply** **You are watching two people running in a race. How do you know which person is moving at a greater speed?** (The person who travels a greater distance in the same amount of time is moving at a greater speed.)

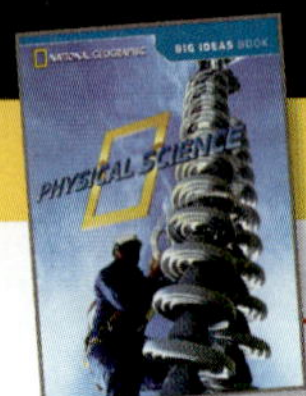

LESSON 4 ▫ Force

Objectives

Students will be able to:

- Describe a force as a push or a pull.
- Explain how the mass of objects and the strength of forces affect motion.

Science Academic Vocabulary

force

PROGRAM RESOURCES

- **Digital Library** at ⊘ **myNGconnect.com**

❶ Introduce

Tap Prior Knowledge

- Ask: **Do you use more effort to lift a box full of books or a box full of pillows?** (books) **Which box has more mass?** (the box of books)

- Have students talk about objects that take a large amount of effort to move and other objects that take a smaller amount of effort.

Set a Purpose

- Tell students they will be reading about different kinds of **force** and how different amounts of **force** affect the motion of objects.

❷ Teach

Academic Vocabulary: *force*

- Write the word **force**. Say: **A force is a push or a pull.**

- Push an object away from you. Ask: **What force am I using?** (a push) Pull an object toward you. Ask: **What force am I using?** (a pull)

- Ask: **What do forces do to an object?** (They make an object move, or change its position.)

Describe a Force

- Have students read and study the main photo on pages 80–81. Ask: **What force is the woman using?** (a pull) Have students predict what will happen. (The barbell will move off the floor; the woman will lift the barbell.) Say: **The barbell will stay in place until a force makes it move.**

Force

Look at the photos. What do you observe about the woman and the barbell? How does she get the barbell to move?

The woman is pulling on the barbell. She must use **force** to move it. A force is a push or a pull. Forces can change the position of objects. You pull objects toward you. The weightlifter will move the barbell as she pulls it.

80

NATIONAL GEOGRAPHIC Raise Your SciQ!

Inertia According to Newton's first law, an object that is not moving will remain at rest until there is a force acting on it. Once a force has set that object in motion, it will continue moving forever, in the same direction and at the same speed, until another force acts on it to change its motion. An object's inertia is how much its position wants to stay the same. If it's not moving, then its inertia keeps it from moving; or if it's moving in one direction, inertia keeps it moving in that same direction until another force makes it change direction. The heavier an object is, the more inertia it has. In other words, the heavier the object, the more force is needed to move it.

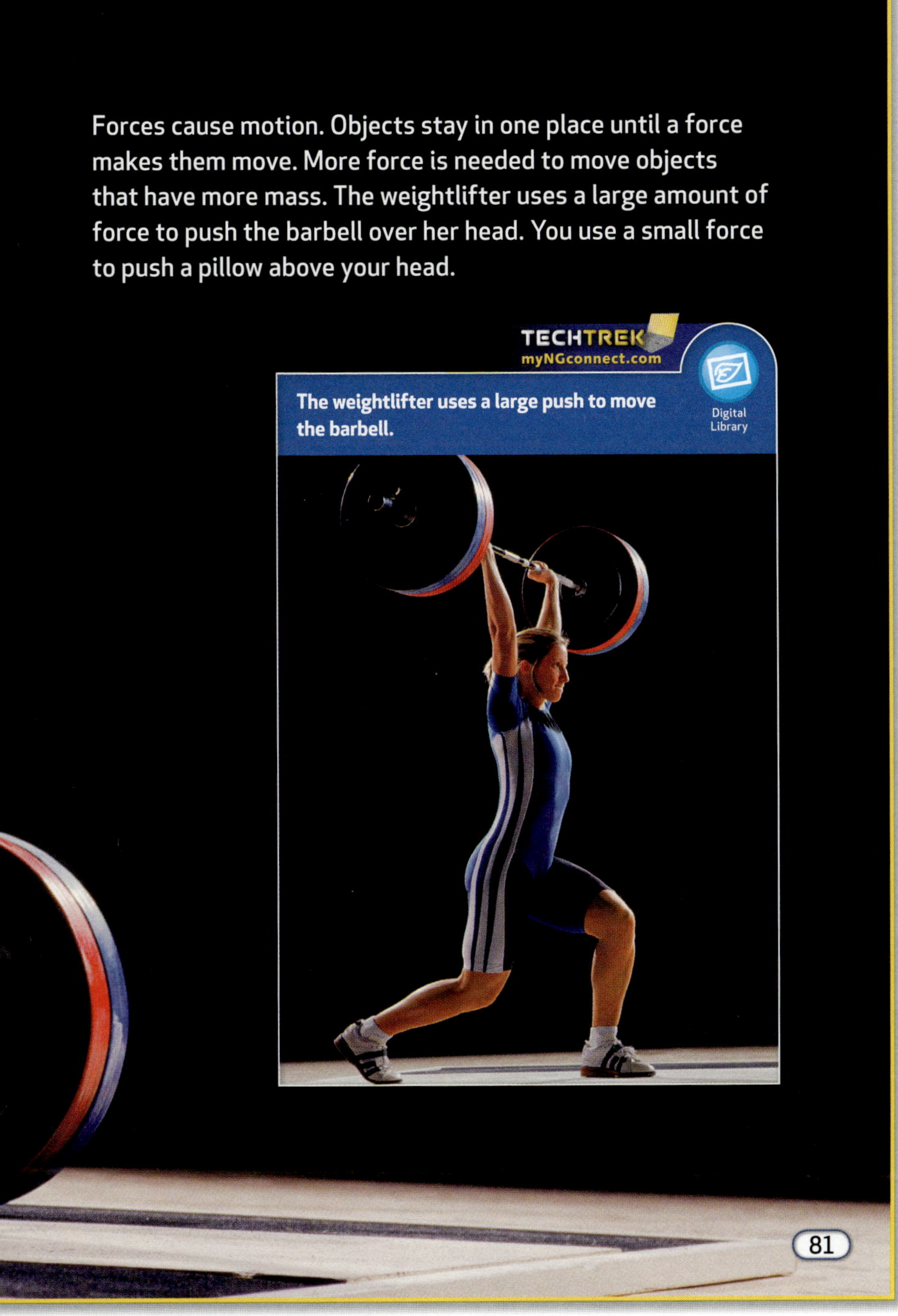

Forces cause motion. Objects stay in one place until a force makes them move. More force is needed to move objects that have more mass. The weightlifter uses a large amount of force to push the barbell over her head. You use a small force to push a pillow above your head.

The weightlifter uses a large push to move the barbell.

81

- Have students study the photo and caption on page 81. Ask: **What force did the woman use to move the barbell to this position?** (a push)

Explain How Force and Mass Affect Motion

- Ask: **How would you describe the amount of force needed to move the barbell?** (Possible answer: It would take a lot of force.)

- Say: **The barbell has a large mass, or amount of matter. A pillow has less mass. It takes more force to move an object with more mass. It takes less force to move an object with less mass. Ask: Would you need to use more force to move a bowling ball or a volleyball? Explain why.** (More force would be needed to move a bowling ball because it has more mass than a volleyball.)

 **Digital Library**

myNGconnect.com

Have students use the **Digital Library** to find photos of professional weightlifters.

 **Integrated Technology**

Electronic News Report Students can use the photos to create an electronic news report of a weightlifting competition. The report can include descriptions of the performance of each weightlifter. Help students use the verbs *push* and *pull* to describe how the lifters pushed or pulled each weight. Have students describe the mass of each weight lifted.

Extend Learning

Demonstrate How Motion Is Related to Force and Mass

Find Out	Think and Do	Describe and Compare
Provide students with two identical hard cover books. Have them think about these questions as they investigate how the motion of an object is related to its mass: "Does the object being pushed or pulled have a large mass or a small mass?" "Is a large amount of force being used or a small amount of force?"	**my SCIENCE notebook** Have students push one book and observe how far and how fast it moves. Then have students place the second book on top of the first and use the same amount of force to push them.	**my SCIENCE notebook** Have students observe and compare the motion of one book and two books and describe their results in their science notebook: • How much farther and faster did one book move compared to two books? • What is the relationship between the motion of the books and their mass?

Objectives

Students will be able to:

- Explain how the strength of forces affects motion.

- Explain that when a force does not move an object, it is because an equal force is acting on it.

Teach, continued

Recognize That Objects Must Touch

- Have students read pages 82–83 and study the photo.

- Say: **There are different kinds of forces. In this lesson you are learning about forces that need contact between objects to cause motion. In the next two lessons you will learn about forces that do not need contact with an object to cause motion.**

- Ask: **What are all the children touching?** (the rope) **Can they pull on the rope without touching it in some way?** (no) Say: **The children cannot cause motion with the rope without touching it.**

- Say: **Sometimes the type of force that needs touch between objects is called a collision. This contact between objects is a quick force, which happens when two or more objects crash into each other. A collision, or crash, between a baseball bat and a ball is one example of a quick force that causes motion.**

- Ask: **What are some other collisions that cause motion?** (Possible answers: a bowling ball and pins, kicking a soccer ball, two bumper cars)

Different amounts of force can change how much an object moves. Greater forces change motion more than smaller forces do. If the children on one side of the rope pull harder than the children on the other, the entire group will move in the direction of the pull.

82

Differentiated Instruction

ELL ### Language Support for Explaining How Forces and Mass Affect Motion

BEGINNING	INTERMEDIATE	ADVANCED
Help students pantomime greater or smaller forces needed to move classroom objects with different masses. Then help students pantomime greater or smaller forces causing different motion of classroom objects.	Help students use Academic Language Frames to explain how forces and mass affect motion: • *More force is needed to move an object with more ______.* • *A greater ______ moves an object more.*	Help students elaborate on how forces and mass affect motion with these Academic Language Stems: • *The strength of forces affects motion by …* • *The mass of objects affects motion by …*

However, even though children are using force to pull the rope, it does not mean that the rope will move. If the children on each side of the rope pull with equal force, neither side will move. The pulling forces cancel each other out.

Explain How the Strength of Forces Affects Motion

- Say: **Greater forces affect motion more than smaller forces do.** Ask: **If the children on the right side in the photo pull with a greater force, what will happen?** (The rope will move to the right.) **If the children on the left side pull with a greater force, what will happen?** (The rope will move to the left.) **What will happen to the rope if both sides pull with the same force?** (The rope won't move.)

Explain How Equal Forces Affect Motion

- Say: **Greater forces affect motion more than smaller forces do.** Then ask: **What happens when the forces acting on an object are equal, or the same?** (The object does not move; they cancel each other out.) Say: **When an object does not move, it is because another force is acting on the object with an equal amount of force.**

Make a Cause and Effect Chart

Have students make a cause-and-effect chart that identifies how the mass of objects and the strength of forces affect motion.

Differentiated Instruction

Extra Support

Create a three-column classroom chart to list *push*, *pull*, and *did not move*. Over the course of a school day, have students make a record of every time they use a push or a pull to move an object during normal classroom activities. Also have them record instances when they push or pull and no movement occurs because forces cancel each other out.

Challenge

Have students observe the photo on pages 82–83. Then have students write a paragraph describing the forces used by the children and how the forces changed the motion of the rope.

Objectives

Students will be able to:

- Explain that friction slows the motion of objects.

Science Academic Vocabulary

friction

PROGRAM RESOURCES

- **Digital Library** at ⊙ **myNGconnect.com**

Teach, continued

Academic Vocabulary: *friction*

- Point out the word **friction** and write it. Read aloud the sentence that defines **friction**.

- Say: **Friction is a force between objects that works against motion. Two objects must be touching for friction to occur between them.**

Explain That Friction Slows Motion

- Have students read pages 84–85 and study the photos.

- Ask: **In the bicycle photo, what two objects rub together to cause friction?** (the bicycle tires and the ground) **How does friction help the bike's rider?** (It helps the bicycle slow down and stop.) Say: **Friction is a helpful force when you want something to slow down.**

- Ask: **In the penguin photo, what two objects rub together to cause friction?** (the penguins' bellies and the ice)

- Say: **The ice is very smooth. There is only a little friction between the penguins and the ice. This friction will slow the penguins' motion, but only a little. Friction will eventually stop their motion. A smooth surface that makes only a little friction is helpful if you want something to slide.**

Friction Suppose you are riding your bike. You stop pedaling. Your bike will start to slow down. Eventually it will stop. It stops because of the friction between the tires and ground. Friction is a force between two objects that rub together. It acts against motion. It slows moving objects. Your bike tires and the ground rub together. This helps the bike stop.

NATIONAL GEOGRAPHIC Raise Your SciQ!

Friction in Liquids and Gases Friction not only occurs when an object contacts solid material. Friction happens in liquids and gases as well. As a ship moves through water, the friction where water contacts the hull slows the ship. When the space shuttle re-enters Earth's atmosphere, friction from the fuselage contacting the air slows the craft and creates extreme heat, requiring special heat shielding to protect it.

Look at the picture of the penguins. What do you observe about their motion? What does this tell you about the force between the penguins and the ice?

The penguins are sliding on the ice. There is only a little friction between their bellies and the ice. They would not be able to slide on a road. The road has a rougher surface, so there is more friction.

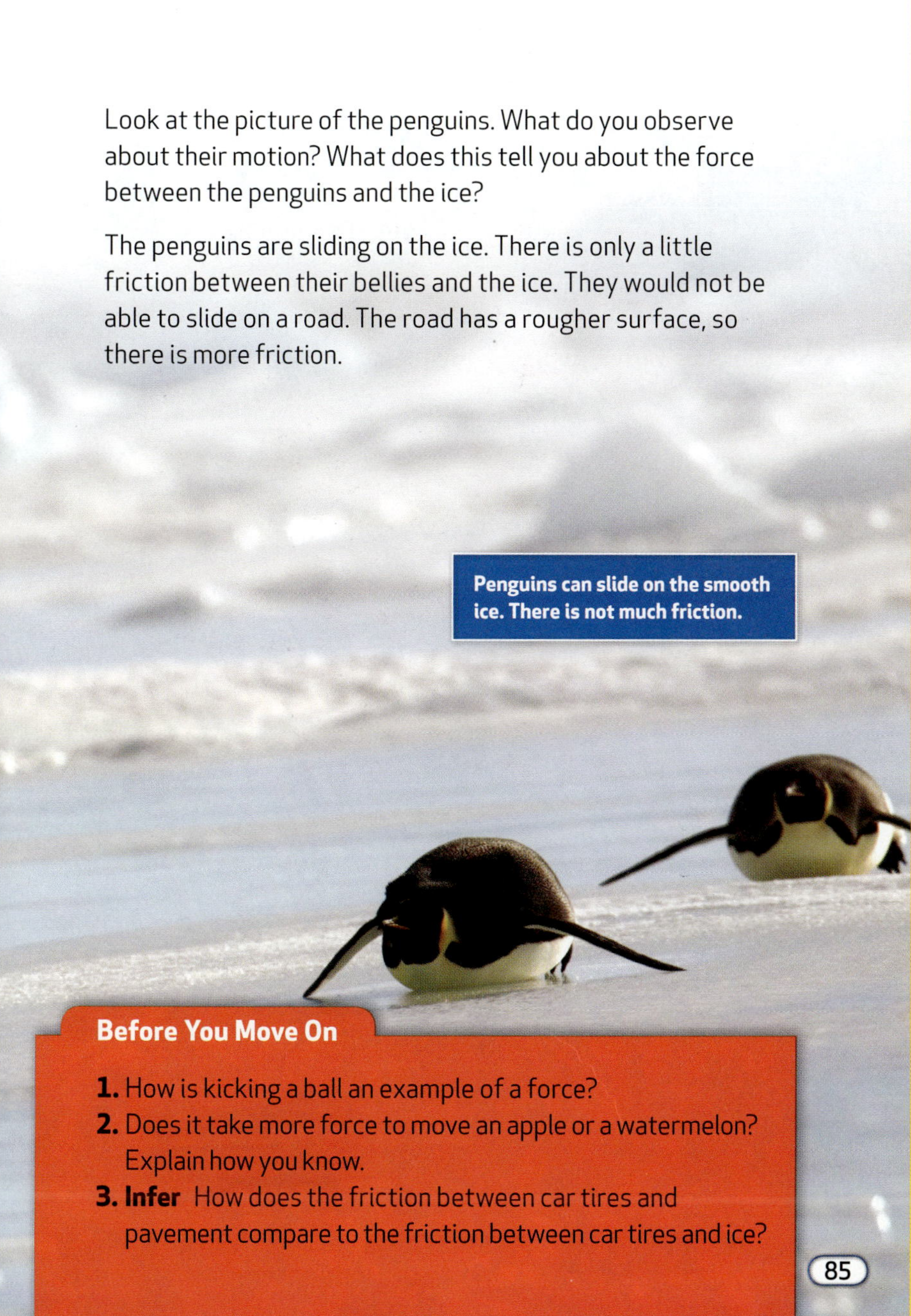

Before You Move On

1. How is kicking a ball an example of a force?
2. Does it take more force to move an apple or a watermelon? Explain how you know.
3. **Infer** How does the friction between car tires and pavement compare to the friction between car tires and ice?

85

Science Misconceptions

Useful Friction Some students may think that, because friction always hinders motion, you always want to eliminate friction. Friction is necessary for walking, riding bicycles, driving a car, and other daily tasks. Tires are designed to create friction between wheels and pavement. Brakes slow cars by creating friction between the brake pads and discs. Icy or wet surfaces can be slippery, so builders often install textured surfaces on sidewalks and building entryways to create more friction when people walk on them.

myNGconnect.com

Have students use the **Digital Library** to find photos of bicycles.

Integrated Technology

Digital Advertisement Students can use the photos to create a digital advertisement that describes bicycles sold by a bike company. The advertisement should describe how the brakes and bike tires use **friction** to make the bikes slow down or stop quickly. Have students share their work.

Predict Changes in Motion

- Have students study the photos again. Ask: **What do you predict would happen if the bicycle were on ice?** (Sample answer: There would be little friction between the tires and ice. It would be hard to slow down and stop.) **What might happen if you were walking on ice?** (Possible answers: I would have fun sliding. I might slip and fall down.)

- Ask: **What do you predict would happen if the penguins slid off the ice onto ground with no ice? Explain your prediction.** (Sample answer: They would suddenly slow down and stop because of the increased friction between their bellies and the new surface.)

❸ Assess

» Before You Move On

1. **Identify** **How is kicking a ball an example of a force?** (Sample answer: A force is a push or a pull. Kicking a ball is an example of a push.)

2. **Contrast** **Does it take more force to move an apple or a watermelon? Explain how you know.** (It takes more force to move a watermelon. A watermelon has more mass than an apple. It takes more force to move an object with more mass.)

3. **Infer** **How does the friction between car tires and pavement compare to the friction between car tires and ice?** (Friction is greater between car tires and pavement than it is between car tires and ice.)

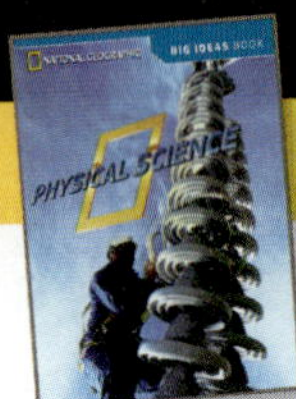

Objectives

Students will be able to:

- Explain that gravity is a force that pulls objects toward Earth.

Science Academic Vocabulary

gravity

❶ Introduce

Tap Prior Knowledge

- Hold a ball in the air and ask students to predict what will happen when you let go of the ball.

- Let the ball drop. Ask: **Why did the ball fall downward?** (Possible answer: A force pulled it downward. Gravity pulled it downward.) Discuss what other objects fall when you let go of them.

Preview and Read

- Read the heading and preview the photographs on pages 86–89 with students. Tell students they will read about the force of **gravity**.

- Have students read pages 86–89.

❷ Teach

Academic Vocabulary: *gravity*

- Pronounce **gravity** and write it.

- Say: **Earth's gravity is a force that pulls objects toward the center of Earth. Objects don't reach the center of Earth because they are stopped by the ground.**

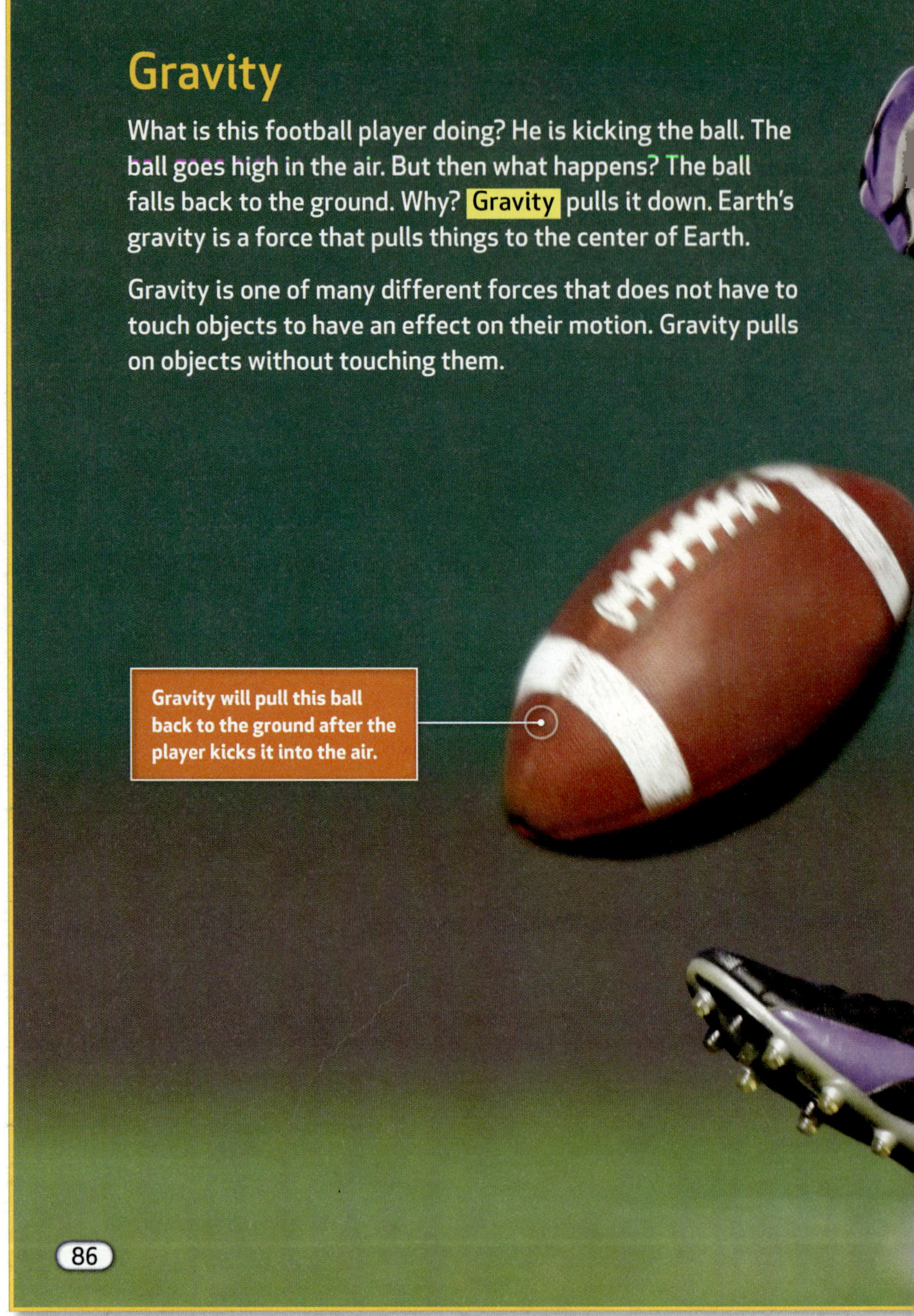

Gravity

What is this football player doing? He is kicking the ball. The ball goes high in the air. But then what happens? The ball falls back to the ground. Why? Gravity pulls it down. Earth's gravity is a force that pulls things to the center of Earth.

Gravity is one of many different forces that does not have to touch objects to have an effect on their motion. Gravity pulls on objects without touching them.

86

NATIONAL GEOGRAPHIC **Raise Your SciQ!**

Gravity Gravity is a force of attraction, or pulling, between any two objects. It is one kind of force that pulls without touching or contact. The strength of gravity depends partly on the mass of the two objects. Objects are pulled toward the center of Earth because Earth's large mass causes it to have a strong force of gravity. An object's weight is a measure of how much Earth's gravity pulls the object. Objects with more mass are pulled more and have more weight. Without gravity, all objects on Earth would float away.

Teach, continued

Text Feature: Photo and Caption

- Have students study the photo. Say: **The football has been kicked.** Ask: **What force causes the football to move up through the air?** (the pushing force of the foot kicking it) Say: **The foot needs to touch the ball in order to push it.**

- Ask: **What force will cause the ball to fall to the ground after it's kicked?** (gravity) **Is gravity a pushing force or a pulling force?** (a pulling force) Say: **Unlike the force of the kick, gravity does not touch or contact the ball. It pulls without touching.**

Recording Observations

Have students think about some of the ways in which they are noticeably affected by **gravity**, such as when they are doing certain activities. Have them record their observations in their science notebook.

Differentiated Instruction

ELL Language Support for Explaining Gravity

BEGINNING	INTERMEDIATE	ADVANCED
Help students understand gravity by asking yes/no questions. For example:	Help students explain gravity using Academic Language Frames. For example:	Help students elaborate on gravity using Academic Language Stems. For example:
Does gravity pull everything towards the center of Earth? Does gravity stop pulling when objcts are touching Earth's surface?.	• *Gravity ______ everything towards Earth's center..* • *Gravity is a ______ that acts on objects without ______ them.*	*Gravity is . . .*

LESSON 5 ▫ Gravity

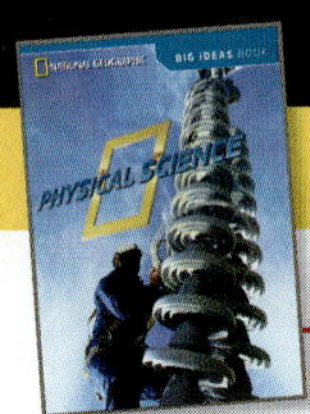

Objectives

Students will be able to:

- Explain that gravity is a force that pulls objects toward Earth.

Teach, continued

Explain the Force of Gravity

- Say: **When you hold your arms above your head, you are using a force to hold them up. Your muscles are pulling them up.**

- Ask: **What other force affects your arms?** (gravity) **Why does gravity make your arms feel tired?** (It pulls my arms toward the center of Earth.) **What else does gravity pull toward the center of Earth?** (everything on Earth)

- Have students study the photo on page 88. Say: **The girl is using a jump rope. She has jumped off the ground. Ask: What force did the girl use to get off the ground?** (a push) **What do you think will happen next?** (She will come back down to the ground; gravity will pull her back to the ground.)

- Say: **Gravity pulls everything toward the center of Earth, even when it's not on the ground.**

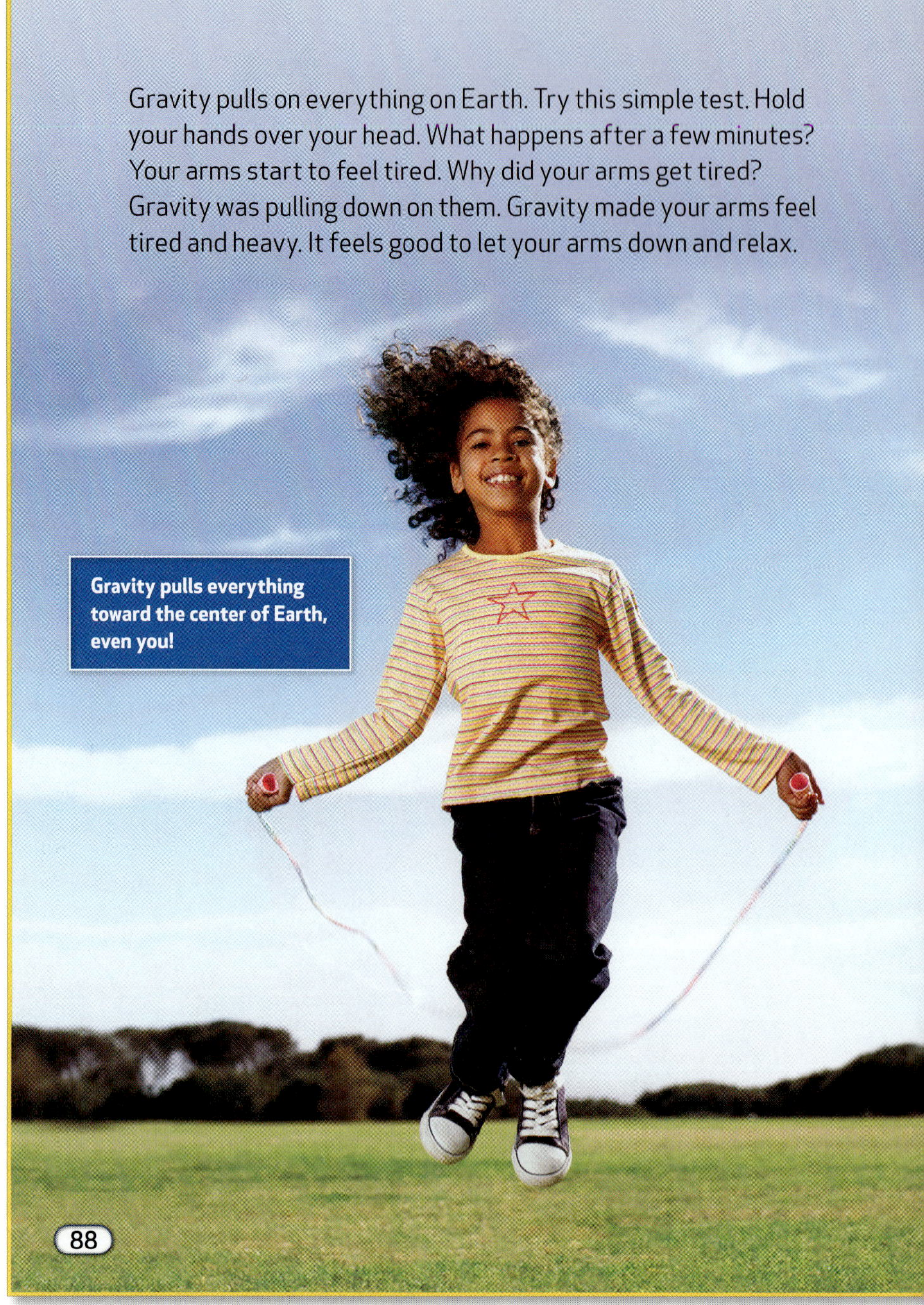

Gravity pulls on everything on Earth. Try this simple test. Hold your hands over your head. What happens after a few minutes? Your arms start to feel tired. Why did your arms get tired? Gravity was pulling down on them. Gravity made your arms feel tired and heavy. It feels good to let your arms down and relax.

Extend Learning

Identify Forces in Nature

Find Out	Think and Do	Describe and Compare
Provide students with a variety of balls, such as a table tennis ball, tennis ball, baseball, blow-up beach ball, basketball, and medicine ball. Have students think about this question as they investigate how gravity affects balls of different size and mass: "Does the size and mass of the ball affect how fast it falls?"	**my SCIENCE notebook** Have students drop two balls, with distinctly different sizes and masses, simultaneously from the same height and observe which ball falls faster. Have them repeat the experiment with different combinations of balls.	**my SCIENCE notebook** Have students observe and compare the effect of gravity on the various combinations of balls and describe the results in their science notebook. (Note: All balls should land at the same time.) • Which ball fell fastest? Slowest? • What do you think is the relationship between gravity and the size and mass of the balls?

Has a doctor ever measured your weight? Weight depends on gravity. It also depends on the mass of objects. The pull of gravity is stronger on objects with more mass. This means objects with more mass weigh more. A stack of books feels heavier than one book. The pull of gravity is greater on the stack of books. The weight of the books is the measure of the pull of Earth's gravity on the books.

Before You Move On

1. What is gravity?
2. What does weight measure?
3. **Apply** One grocery bag is filled with jars of peanut butter and jelly. Another grocery bag is filled with bread. Why does the second grocery bag weigh less than the first one?

89

Science Misconceptions

Mass and Weight Students may think that mass and weight are the same and they are equal at all times. Mass is the amount of matter in an object. An object's mass is the same wherever it is. An object's weight is a measure of the pull of gravity on an object. If the object were on the moon, which has a much smaller pull of gravity than Earth, its mass would be the same but its weight would be less. If the object were on a planet, such as Jupiter, with a stronger force of gravity than Earth, its weight would be more.

Teach, continued

Explain Gravity and Weight

• Have students reread page 89.

• Say: **Weight is a measure of how much Earth's gravity is pulling on something. For example, a big dictionary is heavier than a little notebook. They are both made out of similar material—paper—but one is heavier than the other.**

• Ask: **Why does gravity pull more on the dictionary and make it heavier than the notebook?** (The dictionary has more mass than the notebook.)

• Ask: **What two things does an object's weight depend on?** (gravity and the mass of the object) **How does an object's mass affect its weight?** (The more mass an object has, the stronger the pull of gravity on the object, and the more it weighs.)

• Say: **Every object's weight is a measure of the pull of gravity on it.**

❸ Assess

» Before You Move On

1. **Recall** **What is gravity?** (Gravity is a force that can pull on objects without touching them.)

2. **Explain** **What does weight measure?** (Weight measures the pull of Earth's gravity on objects.)

3. **Apply** **One grocery bag is filled with jars of peanut butter and jelly. Another grocery bag is filled with bread. Why does the second grocery bag weigh less than the first one?** (The grocery bag filled with peanut butter and jelly has more mass than the bag filled with bread. The pull of gravity is greater on objects with more mass, and so it weighs more.)

LESSON 6 □ Magnetism

Objectives

Students will be able to:

- Understand that magnetism is a force that pulls on some objects without touching them.

Science Academic Vocabulary

magnetism

PROGRAM RESOURCES

- **Enrichment Activities** at ⊘ **myNGconnect.com**

❶ Introduce

Tap Prior Knowledge

- Ask students to share what they know about magnets.
- Have them study the photos and describe what the magnets are doing (e.g., pulling, attracting, holding, picking up). Record their descriptions.

Set a Purpose and Read

- Read the heading. Tell students they will read about the force of magnetism.
- Have students read pages 90–93.

❷ Teach

Academic Vocabulary: *magnetism*

- Pronounce **magnetism** and write it. Say: **Magnetism is a force between magnets and the objects they attract.**
- Use a magnet to pick up some paper clips. Ask: **What force causes the paper clips to be picked up?** (magnetism)

Enrichment Activities

⊘ **myNGconnect.com**

Have students use the **Enrichment Activities** to view animations of magnets in action.

Integrated Technology

Computer Presentation Students can use the digital animations to demonstrate how **magnetism** works. Help them create captions that describe different features of a magnet, such as the north and south poles, and how these features interact to attract or repel objects or other magnets.

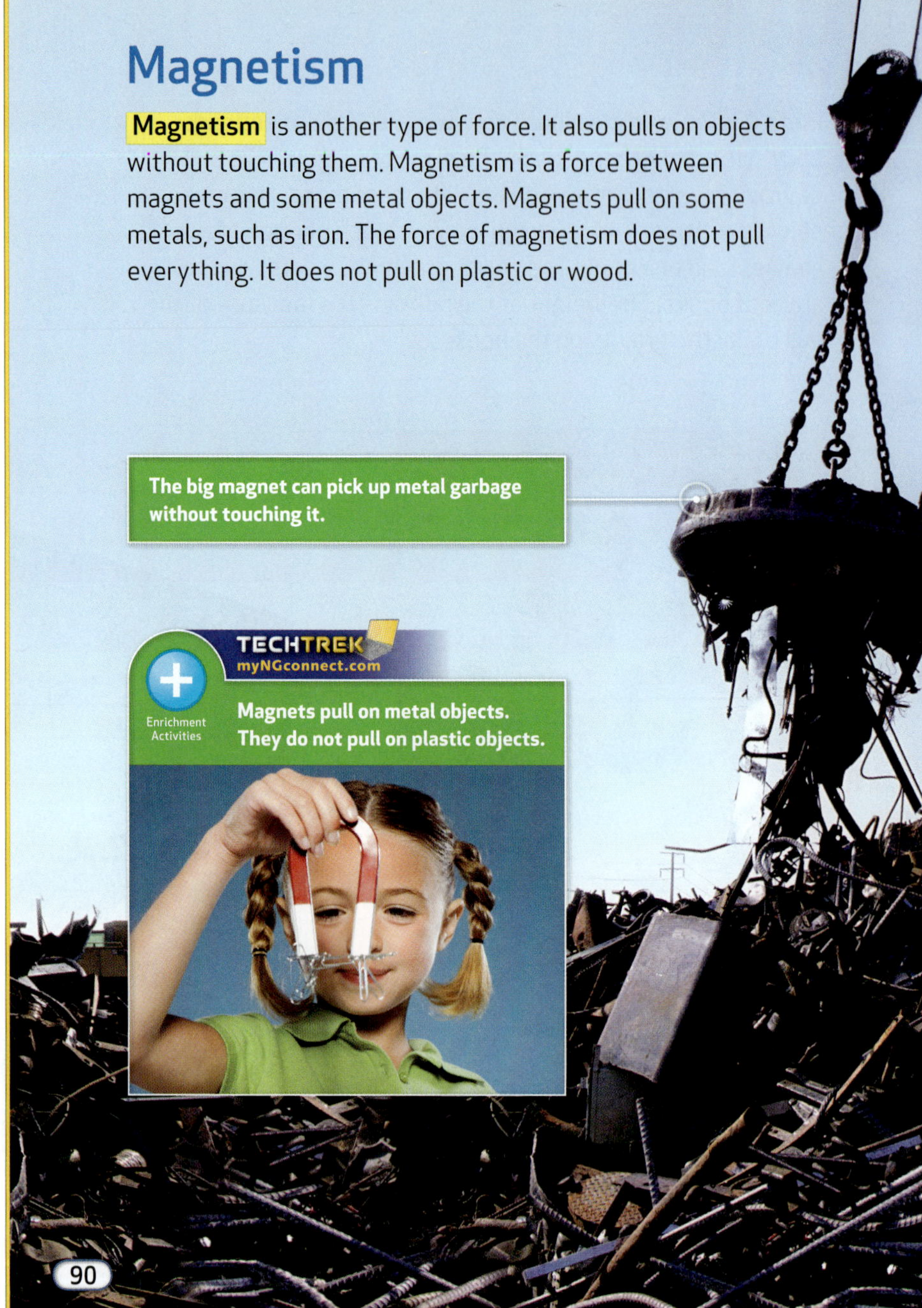

Magnetism

Magnetism is another type of force. It also pulls on objects without touching them. Magnetism is a force between magnets and some metal objects. Magnets pull on some metals, such as iron. The force of magnetism does not pull everything. It does not pull on plastic or wood.

90

Science Misconceptions

Objects Magnets Attract Some students may think that all metals are strongly attracted to a magnet. Only certain metals, such as iron, are strongly attracted to a magnet. To reinforce this concept, have students try to pick up an aluminum can. Aluminum is only weakly attracted to a magnet, and cannot be picked up by most magnets.

Magnets can pull on objects through other materials. Small magnets can pull through paper. That is how they can hold a piece of paper on the refrigerator. The magnets are not attracted to the paper. They pull on the door through the paper. The force of magnetism also depends on how far the magnet is from the object it pulls on. The force gets weaker as the magnet gets farther away.

Magnets pull on the refrigerator door, not the paper.

91

Understand the Force of Magnetism

- Say: **Magnetism affects the motion of some objects by pulling on them. Magnetism can pull paper clips.**

- Ask: **What do objects that are pulled by magnetism contain?** (metal, such as iron) Say: **Magnetism does not pull as strongly on some kinds of metal. For example, a magnet will not pull very strongly on aluminum.**

- Use a magnet to try to pick up a pile of toothpicks. Ask: **Is there magnetism between the magnet and the toothpicks?** (no) **How do you know?** (The toothpicks were not pulled by the magnet and did not move.) Say: **Magnetism does not pull all objects.**

- Have students study the photo on page 91 and the caption. Say: **Like gravity, magnetism is a force that does not have to touch an object to affect its motion.** Ask: **What is being pulled by the magnets?** (refrigerator door) **What does this tell you about the door?** (It has metal in it.) **How can the magnets hold the paper to the door?** (Magnetism can pull through the paper.)

- Ask: **Why do you think a magnet needs to be close to an object for magnetism to pull it?** (Possible answer: The force of magnetism isn't strong enough until it gets close to the object.)

Differentiated Instruction

Extra Support

To help students understand that magnetism is a force that pulls without touching, have one student hold a stiff piece of paper flat. Have another student place a paper clip on top of the paper and use a magnet under the paper to move the paper clip. Ask: **Did the magnet move the paper clip without touching it?** (yes)

Challenge

Provide groups of students with a magnet, a paper clip, and a metric ruler. Allow students to investigate how close a magnet needs to be to a paper clip to pull it without the two touching.

Objectives

Students will be able to:
- Understand that magnetism is a force that pulls on some objects without touching them.

Teach, continued

Describe a Magnet's Poles

- Have students study the photo. Point out the letters on the magnet. Ask: **What do these letters identify?** (The magnet's north pole and south pole.) Say: **All magnets have a north pole and a south pole.**

- Say: **The dark objects in the photo are iron filings or tiny pieces of iron.** Ask: **What can you observe about the iron filings?** (Sample answers: They are pulled by the magnet; they make a pattern; most are at two opposite parts or ends of the magnet.)

❯ Make Inferences

Ask students what inference they can make about this magnet by observing the filings. Students should infer that the magnetism is strongest at the magnet's poles and that the strength of the magnetism at both poles is about the same. Say: **Magnetism is strongest at a magnet's poles.**

Look at the picture of the magnet. Why are the iron filings moving toward certain places on the magnet more than others? The iron filings are attracted to the poles of the magnet. All magnets have two poles. One end of the magnet is the north pole. The other end is the south pole. The force of the magnet is strongest at the poles. It gets weaker farther away from the poles.

Differentiated Instruction

ELL Language Support for Describing Magnetic Poles

BEGINNING	INTERMEDIATE	ADVANCED
Help students describe a magnet's poles by asking either/or questions. **Does a magnet have one pole or two poles? Is a magnet strongest or weakest at its poles?**	Help students describe a magnet's poles using Academic Language Frames. • *Every magnet has two _____.* • *A magnet's force is _____ at its poles.*	Help students elaborate on a magnet's poles using Academic Language Stems. • *A magnet has . . .* • *Poles are . . .*

Poles that are the same repel, or push away from each other.
A north pole of one magnet pushes away from the north pole of
another magnet. Two south poles also push away from each other.
Opposite poles attract, or pull toward each other. A north pole
and a south pole pull together.

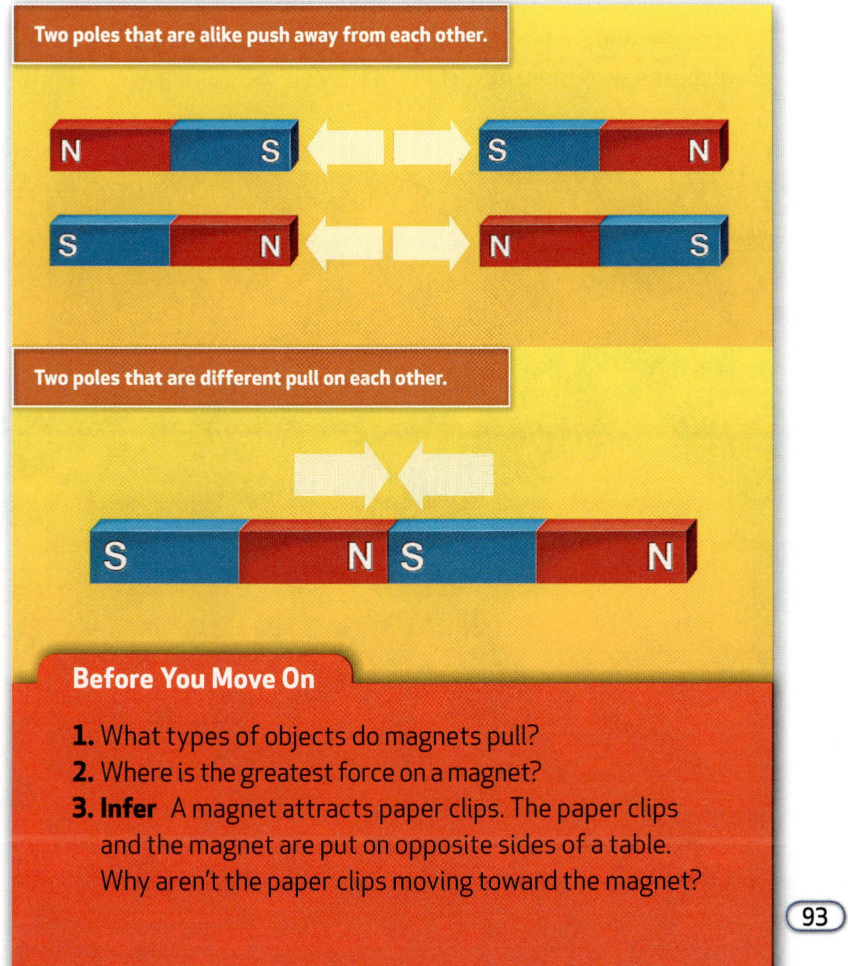

Before You Move On

1. What types of objects do magnets pull?
2. Where is the greatest force on a magnet?
3. **Infer** A magnet attracts paper clips. The paper clips
   and the magnet are put on opposite sides of a table.
   Why aren't the paper clips moving toward the magnet?

93

Social Studies in Science

Earth as a Magnet Earth is an enormous magnet with a north
pole and a south pole. Earth's magnetic poles are used to determine
direction. Show students a globe and ask them to identify the direction
north. Ask them to find south and help them find east and west. Bring
a compass to class and demonstrate how it is used to find direction.
Compare the compass to a map of the school or neighborhood, and
ask students to find the direction of various locations in relation to the
classroom.

Teach, continued

Text Feature: Diagrams

- Have students study the diagrams on page 93. Ask
  a volunteer to read each caption.

- Say: **The diagrams show you what happens
  at two magnets' poles when they are
  brought near each other.** Ask: **What
  happens when two different poles are
  brought near each other?** (They pull on each
  other, or attract.)

- Ask: **What happens when two poles that are
  alike are brought near each other?** (They
  repel, or push away from, each other.)

- Say: **Magnetism is a *pulling* force between
  two poles that are different. Magnetism is
  a *pushing* force between two poles that are
  the same.**

❸ Assess

❯❯ Before You Move On

1. **Identify** **What types of objects do
   magnets pull?** (Magnets pull on some metal
   objects, such as iron. They also pull on other
   magnets.)

2. **Explain** **Where is the greatest force on a
   magnet?** (at the north pole and the south pole)

3. **Infer** **A magnet attracts paper clips.
   The paper clips and magnet are put on
   opposite sides of a table. Why aren't the
   paper clips moving toward the magnet?**
   (Sample answer: The paper clips are too far away.
   Magnetism gets weaker as objects get farther
   away from the magnet.)

Objectives

Students will be able to:

- Explain that friction slows the motion of objects.
- Explain that gravity is a force that pulls objects toward Earth.

PROGRAM RESOURCES

- Learning Masters Book, page 208, or at 🌐 **myNGconnect.com**

❶ Introduce

Tap Prior Knowledge

- Ask students what they know about parachutes.
- Ask students if they have any ideas about what force slows the motion of a parachute.

Preview and Read

- Read the heading. Then look at the pictures on pages 94–95. Tell students that they will read about how parachutes work and how they use forces to change the motion of objects.
- Have students read pages 94–95.

❷ Teach

Text Feature: Photos and Captions

- Have students study the photos and read the captions. Ask: **How can you tell that forces affect parachutes?** (Possible answer: They are open and raised off the ground, not wilted or laying on the ground; you can see that the rainbow parachute is puffed up with air between its layers.)
- Say: **Air is matter and has mass. Friction is the force that occurs when air touches and pushes the parachute's surface.**

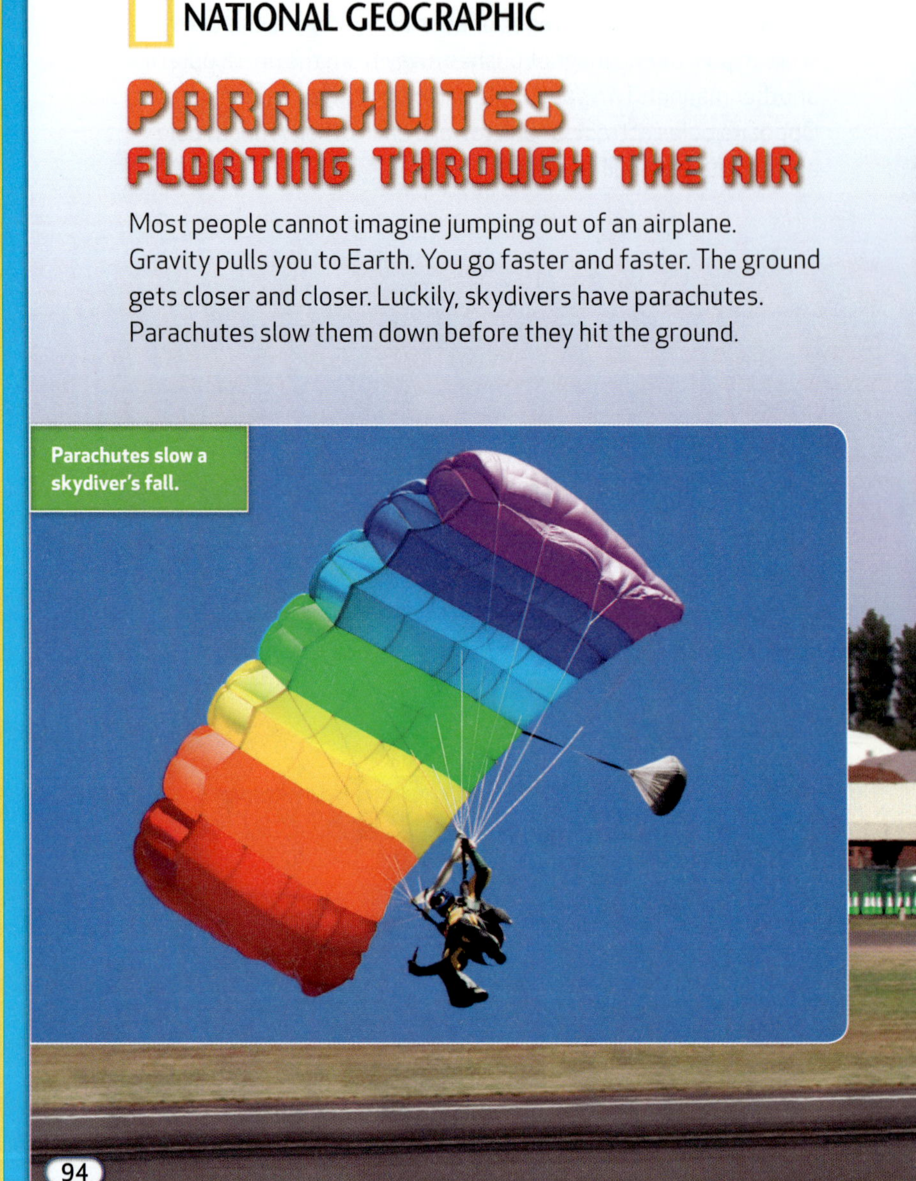

Most people cannot imagine jumping out of an airplane. Gravity pulls you to Earth. You go faster and faster. The ground gets closer and closer. Luckily, skydivers have parachutes. Parachutes slow them down before they hit the ground.

NATIONAL GEOGRAPHIC Raise Your SciQ!

Drag The kind of friction that slows the motion of parachutes is known as air resistance, or drag. Particles of air push against the surface of a parachute as it is pulled down by gravity or as it is pulled quickly forward on the ground. Drag also resists motion as air pushes against a person riding a bicycle or a fast-moving car.

Modern parachutes are made of nylon. Nylon is a strong material. It is also thin and light. Parachutes have a large surface. Air pushes on this surface when they open. This pushing is a kind of friction. It resists the force of gravity. The parachute slows the motion toward the ground.

Parachutes have many uses other than skydiving. Some planes use them to slow down when they land. Parachutes are also used to drop supplies from planes.

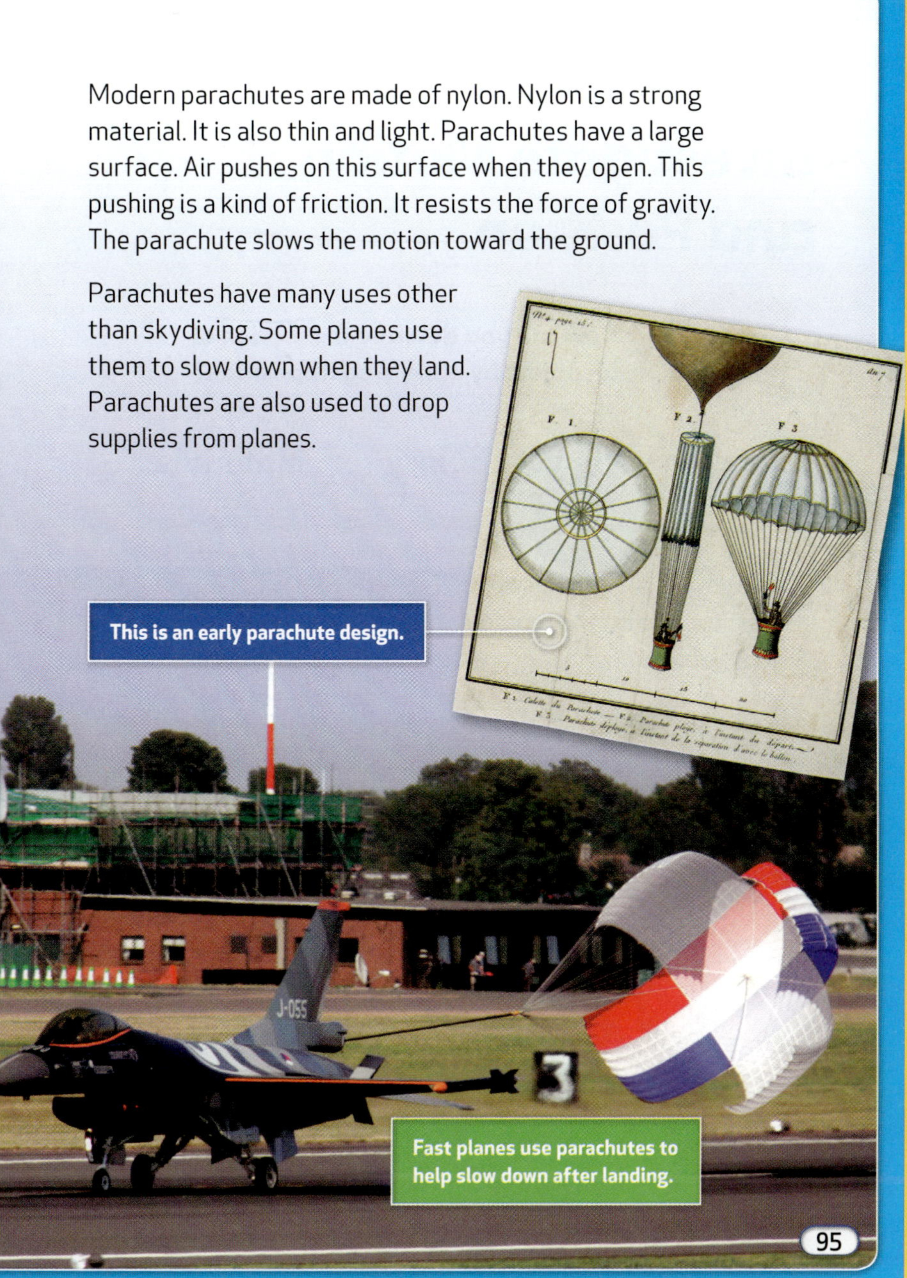

Share and Compare

Give students the Learning Master. Have them find words in Chapter 3 that describe forces and changes in motion. Help them write the words in the correct category on the T-Chart. Check their work. Then have them choose two forces from the chart that are alike and describe how they are alike. Repeat for two forces that are different. Have them share their comparisons with a partner and explain how forces change motion.

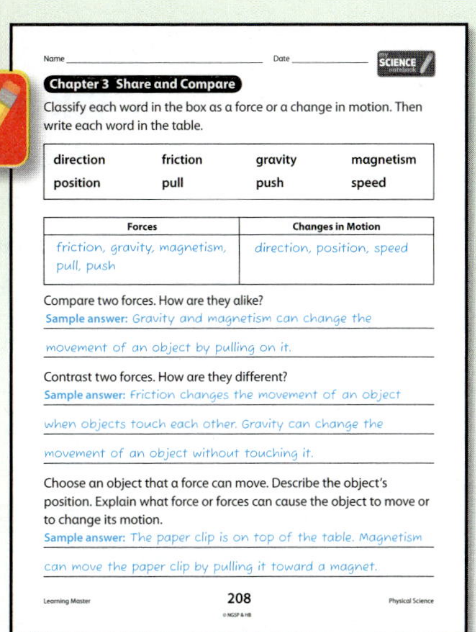

Name _______ Date _______
Chapter 3 Share and Compare
Classify each word in the box as a force or a change in motion. Then write each word in the table.

| direction | friction | gravity | magnetism |
| position | pull | push | speed |

Forces	Changes in Motion
friction, gravity, magnetism, pull, push	direction, position, speed

Compare two forces. How are they alike?
Sample answer: Gravity and magnetism can change the movement of an object by pulling on it.

Contrast two forces. How are they different?
Sample answer: friction changes the movement of an object when objects touch each other. Gravity can change the movement of an object without touching it.

Choose an object that a force can move. Describe the object's position. Explain what force or forces can cause the object to move or to change its motion.
Sample answer: The paper clip is on top of the table. Magnetism can move the paper clip by pulling it toward a magnet.

Learning Master   208   Physical Science

Learning Master 208 or at 🌐 **myNGconnect.com**

Explain How Friction and Gravity Affect Parachutes

- Ask: **What two forces affect the skydiver in the photo?** (gravity and friction) Ask: **How does each force affect the skydiver's motion?** (Gravity pulls the skydiver toward Earth faster and faster. Friction slows the skydiver's motion.)

- Say: **Remember that gravity is a force that pulls an object without touching it. Gravity does not touch the skydiver or parachute to pull them toward Earth. But friction slows their motion when air touches and pushes on the parachute's surface.**

- Ask: **How is the parachute changing the motion of the plane on page 95?** (Air is filling the parachute. Friction causes the airplane to slow down.)

❯ Monitor and Fix Up

Ask students if anything they read was confusing. Students may be confused by more than one force affecting the motion of a parachute. Discuss possible fix-up strategies. For example, students may write the definitions of *gravity* and *friction* in their own words. Students can also use a plastic grocery bag to simulate a parachute. By pulling the bag through the air, they can see and feel how air and friction affect the motion of the bag.

❸ Assess

1. **Recall** What two forces affect the motion of a skydiver using a parachute? (gravity and friction)

2. **Explain** How does a parachute slow motion? (Possible answer: Air pushes against the surface of a parachute. This causes friction, which resists the force of gravity and slows motion.)

3. **Infer** Why do parachutes have a large surface area? (Possible answer: The larger the surface area, the more friction occurs and the more motion is slowed. If the surface area were not large, a person would fall too fast.)

Objectives

Students will be able to:

- Investigate through Guided Inquiry (answer a question; make and compare observations; collect and record data and observations; generate explanations and conclusions based on evidence; share findings; ask questions based on observations to increase understanding; adjust explanations based on findings and new ideas).
- Describe an object's position by relating it to other objects or the background.
- Describe an object's motion by measuring its position over time.
- Recognize that when solving a problem, it is important to plan and get ideas and help from others.

Science Process Vocabulary

plan

PROGRAM RESOURCES

- Science Inquiry and Writing Book: *Physical Science*
- Science Inquiry and Writing Book **eEdition** at **myNGconnect.com**
- **Inquiry eHelp** at **myNGconnect.com**
- Science Inquiry Kit: *Physical Science*
- Learning Masters Book, pages 209–212, or at **myNGconnect.com**
- Inquiry Rubric: Assessment Handbook, page 213, or at **myNGconnect.com**
- Inquiry Self-Reflection: Assessment Handbook, page 222, or at **myNGconnect.com**

MATERIALS

Kit materials are listed in italics.

poster board; masking tape; 3 markers; *ball; 2 straws; plastic spoon; string; ruler*

❶ Introduce

Tap Prior Knowledge

- Ask students to think of a time they were on a bus, looking out a window. Ask: **How could you tell when the bus was moving?** (The objects outside appeared to move.) Explain that motion can be described by explaining how an object's position changes compared to other objects.

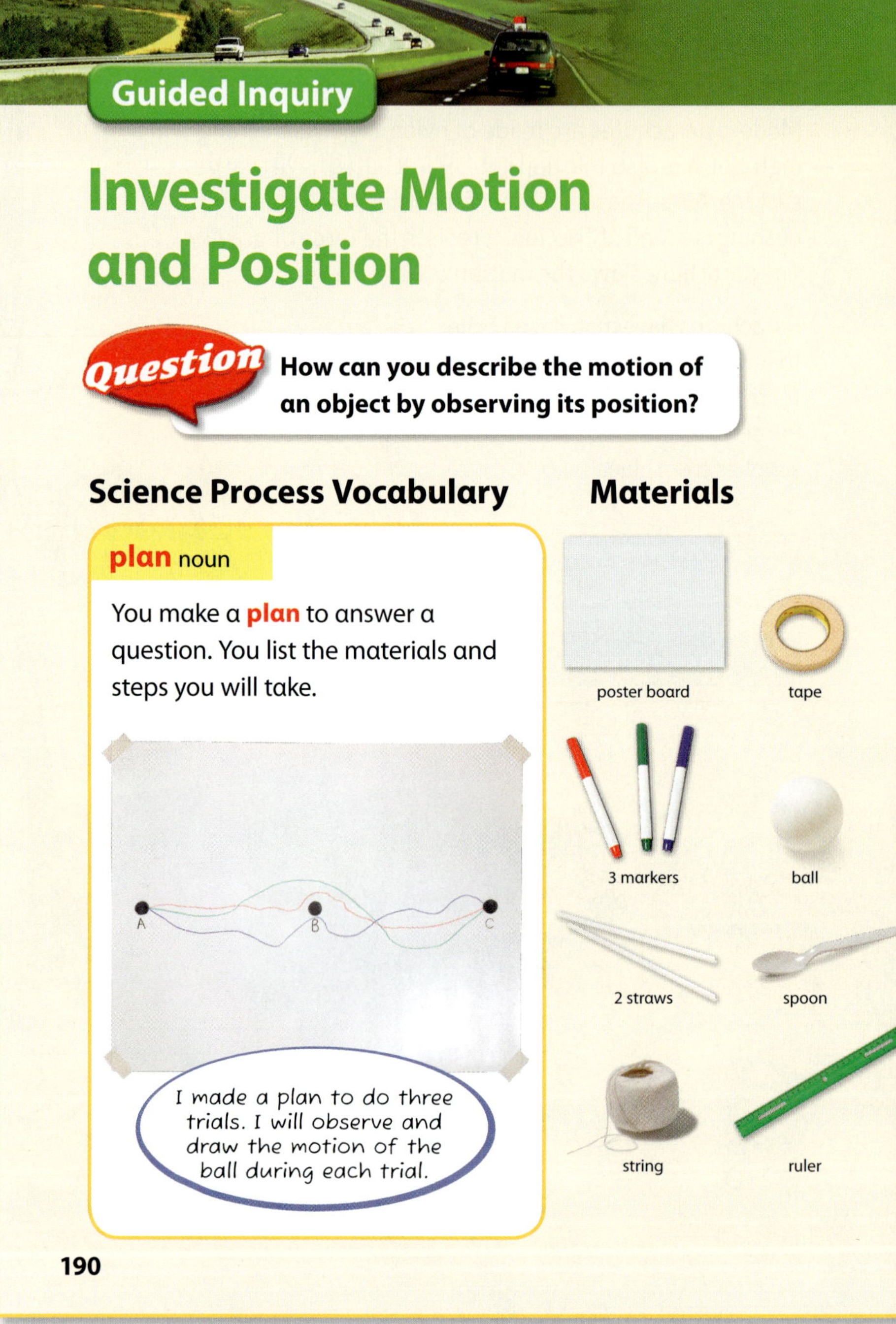

MANAGING THE INVESTIGATION

Time

 30 minutes

Groups

Small groups of 4

What to Expect

- Students will move the ball from circle A around circle B to circle C by pushing or pulling it. They will observe the variations in the path the ball takes in three trials.
- Students may move the ball by pushing it with a spoon, by taping string to it and pulling on the string, by pushing the ball with one or two straws, or by pushing it with their breath by blowing through a straw.

What to Do

1 Tape the poster board to a flat surface. Make 3 circles on the poster board as shown in the photo. Label the circles **A**, **B**, and **C**.

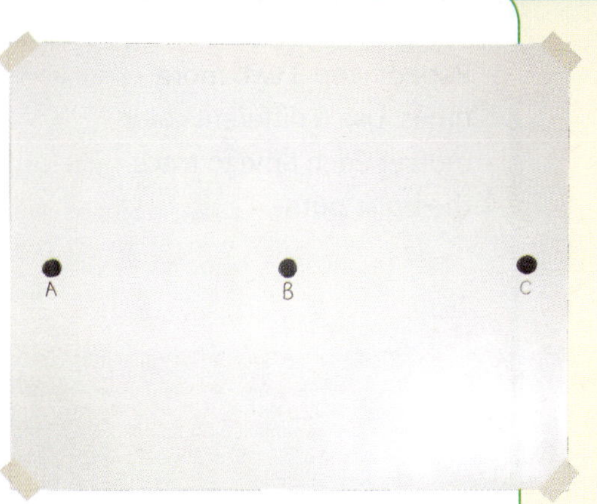

2 Make a **plan** to move the ball from circle A to circle C. You must move the ball as close to circle B as possible, but the ball may not roll over it. You can choose 2 straws, a spoon, or string to move the ball. Record your plan in your science notebook.

3 Place the ball at circle A. Test your plan. As you move the ball, have a partner use a marker to trace the ball's path on the poster board.

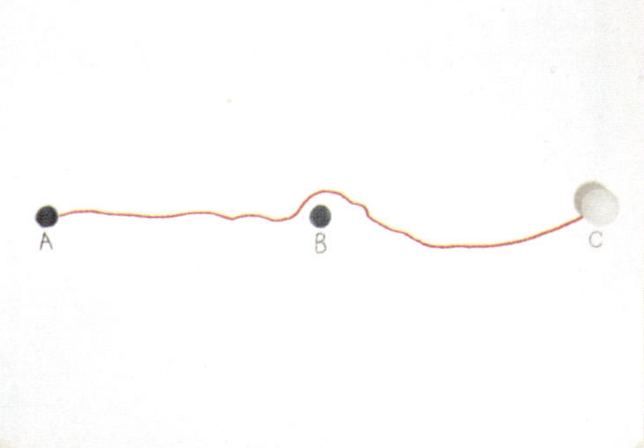

191

Teaching Tips

- Before students begin the activity, discuss words that describe position, such as *in front of*, *beside*, and *above*. List these words on the board.
- Students should draw a circle near each edge of the poster board and another circle in the middle.
- Make sure the investigation is done on a flat surface.
- Allow each student one or two practice runs at moving the ball with the tool he or she chose. This will help students control the ball's motion before tracing its position.

Introduce, continued

Connect to the Big Idea

- Review the Big Idea Question, *How does force change motion?* Explain to students that in this inquiry they will use force to move a ball from one position to another and describe its motion.
- Have students open their Science Inquiry and Writing Books to page 190. Read the Question and invite students to share ideas about how to describe a ball's motion by observing its position.

❷ Build Vocabulary

Science Process Word: plan

Use this routine to introduce the word.

1. Pronounce the Word Say **plan**. Have students repeat the word.

2. Explain Its Meaning Choral read the sentences. Ask students for another word or phrase that means the same as **plan**. (something that tells how to do something and what you need to do it)

3. Encourage Elaboration Ask: **Why do you think it is important to plan a trip?** (to know where to go and what to pack)

ELL Use Cognates
The Spanish word for **plan** is identical to the English word. Ask Spanish-speaking students what they do when they make a **plan**.

❸ Guide the Investigation

- Distribute materials. Read the steps on pages 191–192 together with students. Move from group to group and clarify steps if necessary.
- Explain to students that there is no right or wrong plan for moving the ball. Say: **Your goal is to trace and describe the ball's changing position—its motion—accurately.**
- Have students within each group discuss their plans. Encourage them to listen to each other's ideas, and discuss the importance of listening to each group member. Tell groups to write down their plans in their science notebooks. Remind them to include the materials they will use and the steps they will take.

Guide the Investigation, continued

- As students move the ball, have them note its path. Ask: **What could you do to have more control over its motion?** (Possible answers: use less force with the spoon; hold the ball more tightly with the straws) Encourage students to seek the help of others.

❹ Explain and Conclude

- Guide students to record their measurements in a chart similar to the one on page 193. Ask: **In which trial did you get the ball closest to circle B?** (Answers will vary.)

- Encourage students to share observations with the class. Have students who used similar methods to move the ball compare results. Ask: **Which method moved the ball closest to circle B? How did your results differ? Why?** Help students understand that similar investigations may not always produce similar results.

- Have students use qualitative as well as quantitative data as they describe the motion of the ball over time. They should describe the direction of movement and measure how far (in cm) the ball moved and how far it was from circle B.

- Guide students to describe the motion of the ball in terms of its position (in front of, beside, above) relative to the three circles. Point to spots on the traced path and ask: **Where was the ball in relation to circle A? Where was it in relation to circle B?**

- Help students to draw conclusions. Ask: **How can you describe *position*?** (by comparing an object to something in the background or to other objects) **What happens when the position of an object is changing?** (It is moving.) **Can an object move without changing its position?** (No.)

Answers

1. Possible answer: The ball moved away from circle A and toward circle C.

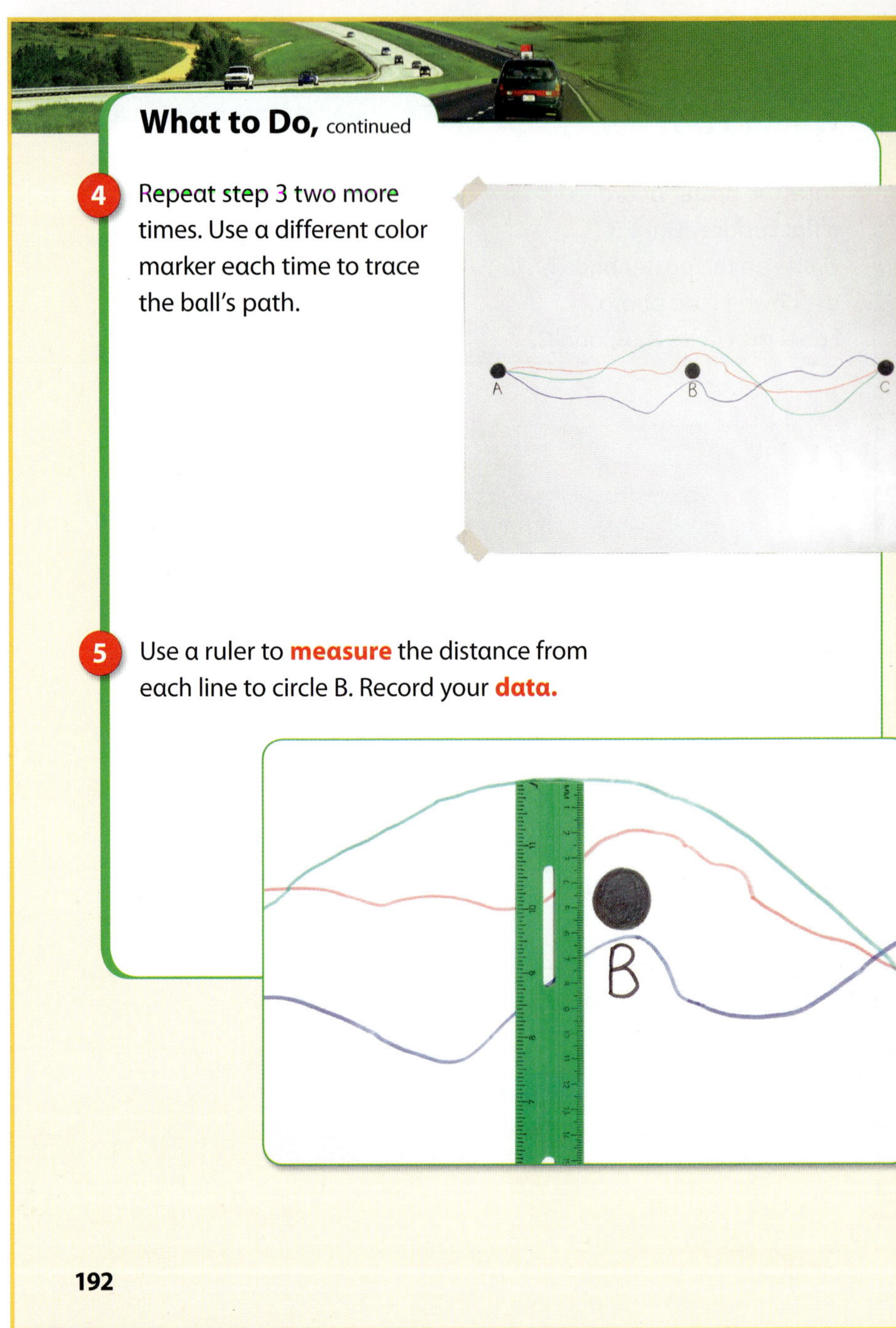

What to Do, continued

4 Repeat step 3 two more times. Use a different color marker each time to trace the ball's path.

5 Use a ruler to **measure** the distance from each line to circle B. Record your **data**.

192

Differentiated Instruction

ELL **During the Investigation**

BEGINNING	INTERMEDIATE	ADVANCED
To give students language for describing position, ask yes/no questions such as: • **Did the ball move above circle B?**	Give students language frames to describe position: • *The ball moved ______ circle B.*	Give students language frames to elaborate their ideas: • *The ball moved ______ circle B and then ______ circle C.*

Record

Write in your science notebook.
Use a table like this one.

Path of Ball

Trial	Distance from Circle B (cm)
1	
2	
3	

Explain and Conclude

1. How did the ball's position change in relation to circle A? How did it change in relation to circle C?

2. **Compare** your **data.** In which trial did the ball come closest to circle B? Describe the path of the ball in that trial. Describe the ball's position in relation to circle B.

3. **Conclude** how you can use position to describe motion.

Think of Another Question

What else would you like to find out about motion and position? How could you find an answer to this new question?

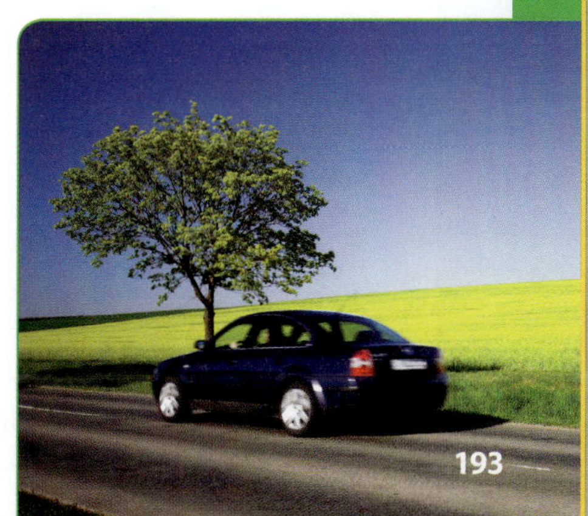

The car is moving toward the tree.

193

Inquiry Rubric at ✈ myNGconnect.com	Scale			
The student made a **plan** to **investigate** motion and position.	4	3	2	1
The student **observed** and described motion in relation to position.	4	3	2	1
The student **measured** the distance of the closest path of motion to a target circle and recorded the **data** in a table.	4	3	2	1
The student **compared** the results of three trials.	4	3	2	1
The student **shared** and compared observations and **conclusions** with others.	4	3	2	1
Overall Score	4	3	2	1

Explain and Conclude, continued

2. Possible answer: The ball came closest to circle B in trial 2. The ball started out from circle A and moved toward circle B. As the ball passed circle B, it was 2 cm above it. Then the ball moved away from circle B toward circle C.

3. Motion can be described by how an object's position changes.

❺ Find Out More

Think of Another Question

- Students should use their observations to generate questions. For example: *What would happen if we made a pencil move from circle A to circle C?*

- Record student questions for possible future investigations. Discuss what students could do to find answers to the new questions.

❻ Reflect and Assess

- To assess student work with the Inquiry Rubric below, see Assessment Handbook, page 213, or go online at ✈ **myNGconnect.com**

- Have students use the Inquiry Self-Reflection on Assessment Handbook, page 222, or at ✈ **myNGconnect.com**

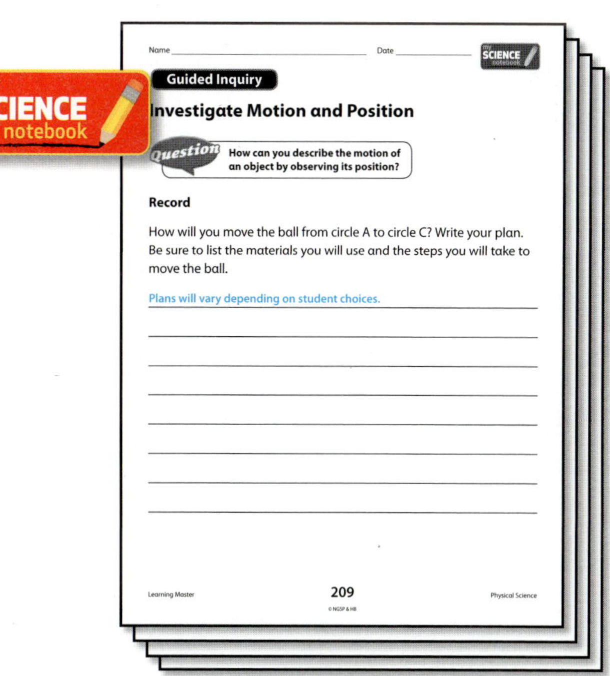

Learning Masters 209–212 or at ✈ myNGconnect.com

LESSON 9 □ Conclusion and Review

PROGRAM RESOURCES

- Chapter 3 Test, Assessment Handbook, pages 97–100, or at ⊙ **myNGconnect.com**
- NGSP ExamView CD-ROM

❶ Sum Up the Big Idea

- Display the chart from page T72–T73. Read the Big Idea Question. Then add new information to the chart.

- Ask: **What did you find out in each section? Is this what you expected to find?**

How Does Force Change Motion?

Position
Mike: Position describes where an object is.

Motion
Nate: Things that are in motion change position.

Force
Maria: Forces are pushes and pulls that change the motion of objects.

Gravity
Sasha: Gravity is a force that pulls objects toward the center of Earth without touching them.

Magnetism
Mari: Magnetism pulls some metal objects toward a magnet without touching them.

Conclusion

Objects in motion change their position. Speed and direction are ways to describe motion. A force is a push or a pull. Forces change the motion of objects. Friction is a force that works against the motion of objects. It affects objects that rub together. Earth's gravity is force that pulls objects to the center of Earth. Magnets can pull some metals without touching them.

Big Idea Forces are pushes and pulls that can change the motion of objects.

OR

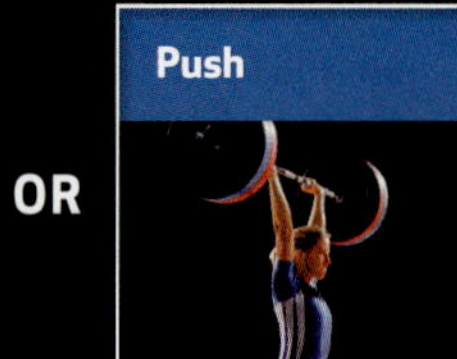

= FORCE

Vocabulary Review

Match the following terms with the correct definition.

A. force
B. friction
C. gravity
D. magnetism
E. motion
F. speed

1. The distance an object moves in a period of time
2. A force that pulls things to the center of Earth
3. A force that acts when two surfaces rub together
4. A force between magnets and the objects magnets attract
5. A push or a pull
6. Describes an object that is moving

96

Review Academic Vocabulary

Academic Vocabulary Tell students that scientists use special words when describing their work to others. For example:

motion   speed   force

friction   gravity   magnetism

Have students write the list in their science notebook and use each word in a sentence that tells about how force changes motion. They should then share their sentences with a partner.

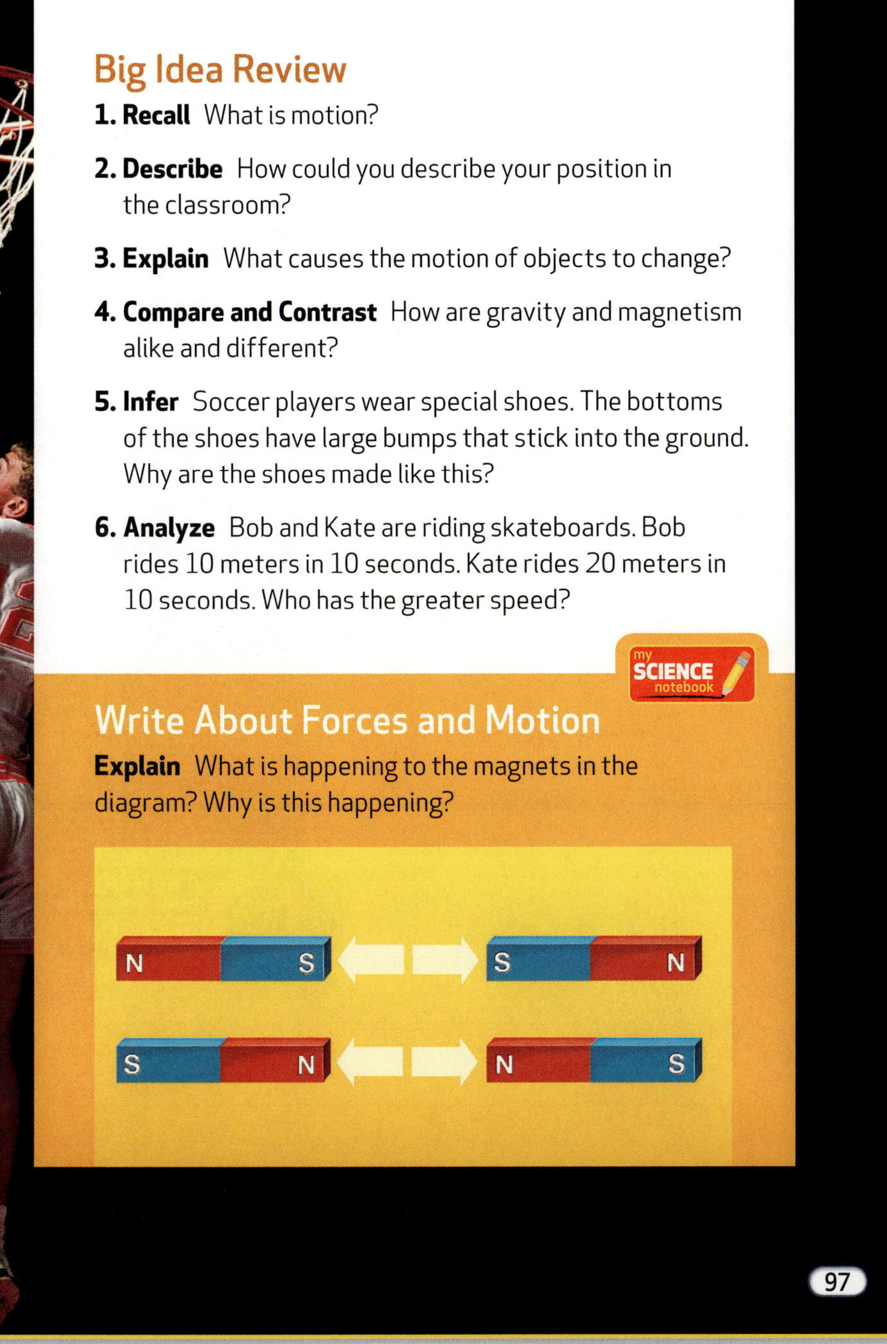

Big Idea Review

1. **Recall** What is motion?

2. **Describe** How could you describe your position in the classroom?

3. **Explain** What causes the motion of objects to change?

4. **Compare and Contrast** How are gravity and magnetism alike and different?

5. **Infer** Soccer players wear special shoes. The bottoms of the shoes have large bumps that stick into the ground. Why are the shoes made like this?

6. **Analyze** Bob and Kate are riding skateboards. Bob rides 10 meters in 10 seconds. Kate rides 20 meters in 10 seconds. Who has the greater speed?

Write About Forces and Motion

Explain What is happening to the magnets in the diagram? Why is this happening?

97

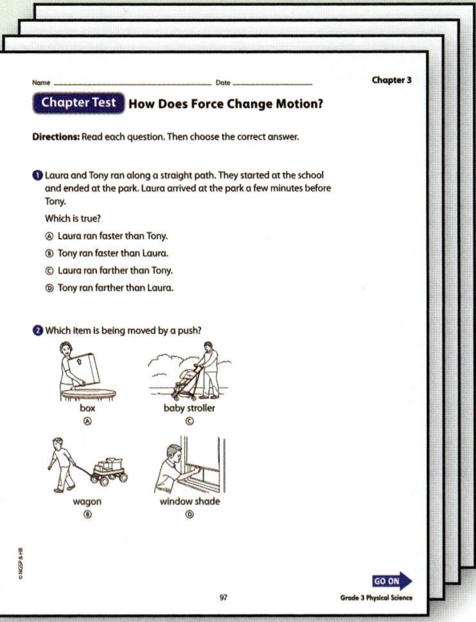

Chapter Test How Does Force Change Motion?

Directions: Read each question. Then choose the correct answer.

1. Laura and Tony ran along a straight path. They started at the school and ended at the park. Laura arrived at the park a few minutes before Tony.
   Which is true?
   Ⓐ Laura ran faster than Tony.
   Ⓑ Tony ran faster than Laura.
   Ⓒ Laura ran farther than Tony.
   Ⓓ Tony ran farther than Laura.

2. Which item is being moved by a push?
   box Ⓐ baby stroller Ⓒ
   wagon Ⓑ window shade Ⓓ

GO ON

97 Grade 3 Physical Science

❷ Discuss the Big Idea

Notetaking 

Have students write in their science notebook to show what they know about the Big Idea. Have them:

1. Describe the position and motion of objects.

2. Write about forces that change an object's motion by touching the object.

3. Write about forces that change an object's motion without touching the object.

❸ Assess the Big Idea

Vocabulary Review

1. F   2. C   3. B   4. D   5. A   6. E

Big Idea Review

1. Possible answer: Motion describes an object that is moving.

2. Answers may vary. Students will use direction words such as "in front of," "behind," "between," "on the right," and "on the left" to compare themselves to another object.

3. Forces, or pushes and pulls, cause the motion of objects to change.

4. Answers may vary. Possible answer: Gravity and magnetism are both forces that pull objects together. They both can act on objects without touching them. Gravity pulls all objects together, but magnetism only pulls magnets and some metals together. Magnetism is also a pushing force between like poles of magnets.

5. Possible answers: The shoes are made to increase the force of friction. There is more friction between rough surfaces. They will help keep the soccer players from slipping.

6. Kate has the greater speed. Bob's speed is 1 meter per second. Kate's speed is 2 meters per second.

Assess Student Progress Chapter 3 Test

Have students complete their Chapter 3 Test to assess their progress in this chapter.

Chapter 3 Test, Assessment Handbook, pages 97–100, or at 🔵 **myNGconnect.com**

or NGSP ExamView CD-ROM

Write About Forces and Motion 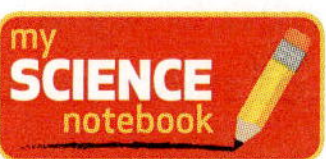

The force of magnetism is pushing the magnets away from each other. Two poles that are alike repel, or push away, each other.

LESSON 10 ▫ Physical Science Expert

Objectives

Students will be able to:

- Recognize how different scientists' work has contributed to general scientific understanding.

PROGRAM RESOURCES

- Big Ideas Book: *Physical Science*
- Big Ideas Book: *Physical Science* **eEdition** at ⊘ **myNGconnect.com**
- **Digital Library** at ⊘ **myNGconnect.com**

❶ Introduce

Tap Prior Knowledge

- Ask students if they have ever looked closely at the different parts of a computer. Ask: **What different types of materials were used to make the parts?** (Possible answers: plastic, metal, glass)

Set a Purpose and Read

- Read the heading. Tell students that they will read about how a materials scientist studies materials that are used to make different products. They make sure the right materials are used to make them.
- Have students read pages 98–99.

❷ Teach

Describe Tools Used in Materials Science

- Explain: **At one time, many years ago, computers were as large as an entire room. Now, we use computers that can fit on our desk or that we can carry with us. One way scientists have been able to make these machines smaller is by using new kinds of materials that work even when the parts are extremely small.**
- Say: **Ainissa needs to see what materials look like when they are very small.** Ask: **What kind of microscope does Ainissa use?** (an Atomic Force Microscope)
- Say: **An atom is the smallest part of any material.** Ask: **If this kind of microscope is "atomic," how small are the objects that it can "feel"?** (the size of atoms)

NATIONAL GEOGRAPHIC

CHAPTER 3 PHYSICAL SCIENCE EXPERT: MATERIALS SCIENTIST

Ainissa Ramirez is an Associate Professor of Mechanical Engineering at Yale University. She is a scientist that works with other scientists to study the properties of different materials and the ways those materials respond to different forces.

Q: What do you study?

My research group and I are looking at how materials behave when they are really really small—less then 1/100,000 the thickness of your hair. Since these objects are too small to "see," we use a machine, called an Atomic Force Microscope, which "feels" atoms with a very sharp tip. When the tip runs over the surface, forces between the tip and the atoms are fed into a computer and become images of atoms that we can examine.

Q: Why is what you study important?

These studies are important because properties of materials change when you change the size of the material. And, in order to use these materials in future appliances, like cell phones and video games, we must map how they behave.

98

NATIONAL GEOGRAPHIC — Raise Your SciQ!

Nanotechnology One area of science that has recently seen an explosion of development is nanotechnology, or the study of materials and devices on an atomic or molecular scale. Scientists have learned to change the atoms or molecules of certain materials to use them for new purposes. One field where nanotechnology has become increasingly important is in electronics. Complex circuits can be made even tinier, allowing devices like personal computers and cell phones to become more powerful. One way to make circuits is to use an Atomic Force Microscope as a "writing" tool to print circuits in a pattern with chemicals.

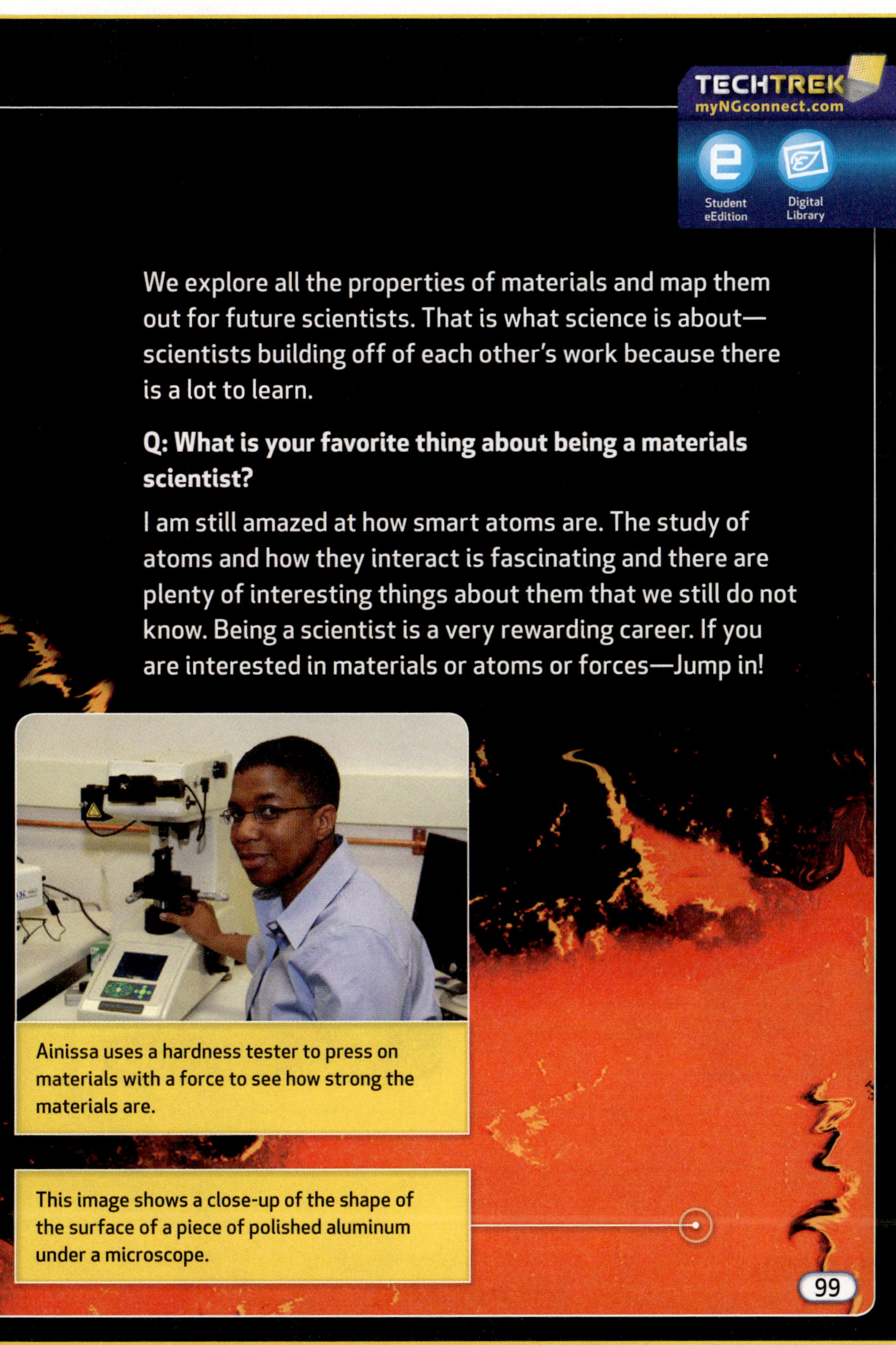

We explore all the properties of materials and map them out for future scientists. That is what science is about—scientists building off of each other's work because there is a lot to learn.

Q: What is your favorite thing about being a materials scientist?

I am still amazed at how smart atoms are. The study of atoms and how they interact is fascinating and there are plenty of interesting things about them that we still do not know. Being a scientist is a very rewarding career. If you are interested in materials or atoms or forces—Jump in!

Ainissa uses a hardness tester to press on materials with a force to see how strong the materials are.

This image shows a close-up of the shape of the surface of a piece of polished aluminum under a microscope.

99

Differentiated Instruction

ELL **Language Support for Describing the Work of a Materials Scientist**

BEGINNING	INTERMEDIATE	ADVANCED
Help students describe what a materials scientist does by asking yes/no questions. **Does a materials scientist study the properties of materials? Do scientists use the work of other scientists?**	Help students describe a materials scientist using Academic Language Frames. • *A materials scientist studies the _____ of materials.* • *Materials the size of atoms and molecules can be studied using an _____ force microscope.*	Help students elaborate on expert materials scientist with this Academic Language Stem: *A materials scientist can …*

Teach, continued

Describe the Work of a Materials Scientist

- Say: **Ainissa studies the properties of materials when they are very small. What are some properties of some materials that we learned about in this chapter?** (Some metals are magnetic. Some materials have more mass than others.)

- Say: **Look at the background picture on page 99. When we look at polished aluminum with only our eyes, it looks very smooth and shiny. How does it look different when we see it up close?** (It looks rough or bumpy.)

myNGconnect.com

Have students use the 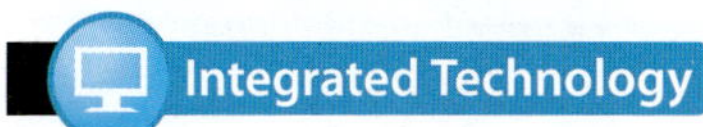

to find photos of a materials scientist at work.

Integrated Technology

Digital Booklet Students can use the photos to make a digital booklet with captions that describe what the materials scientist is doing.

❸ Assess

1. **Recall** As a materials scientist, what does Ainissa Ramirez study? (She studies how materials behave when they are very small. She uses special machines that help her study these materials by making images of atoms, which are too small to "see.")

2. **Explain** Why is Ainissa's work important? (The properties of materials change when they become very small. If a material is going to be used in a product like a cell phone or a video game, you need to know how the material is going to behave.)

3. **Draw Conclusions** What can a materials scientist learn from the work of another materials scientist? (Sample answer: One materials scientist can learn more about a certain material by building on the work of other scientists who have already studied the properties of that material.)

LESSON 10 ▫ Become an Expert

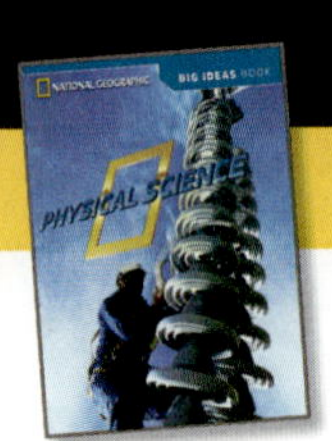

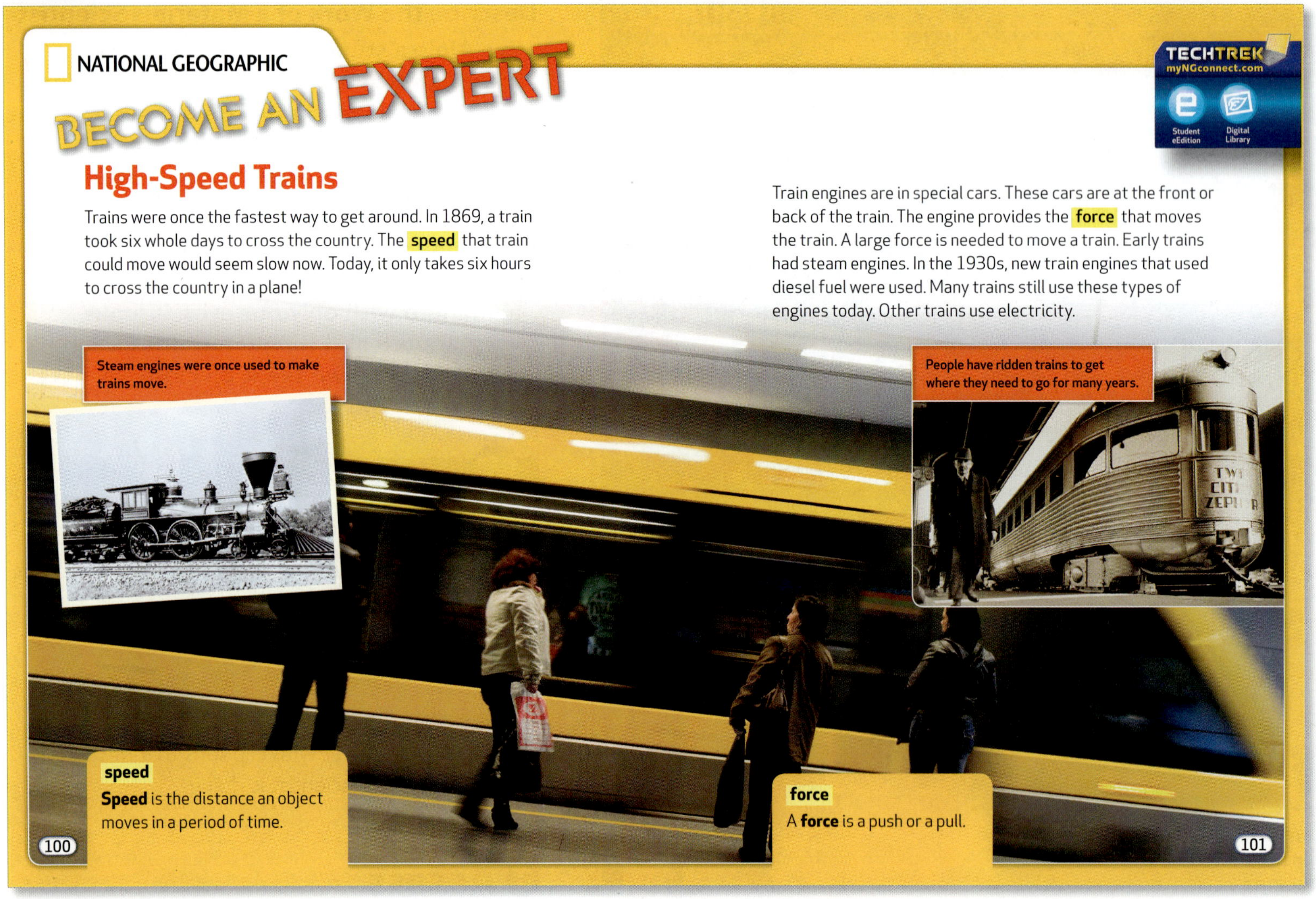

NATIONAL GEOGRAPHIC

BECOME AN EXPERT

TECHTREK myNGconnect.com Student eEdition Digital Library

High-Speed Trains

Trains were once the fastest way to get around. In 1869, a train took six whole days to cross the country. The **speed** that train could move would seem slow now. Today, it only takes six hours to cross the country in a plane!

Train engines are in special cars. These cars are at the front or back of the train. The engine provides the **force** that moves the train. A large force is needed to move a train. Early trains had steam engines. In the 1930s, new train engines that used diesel fuel were used. Many trains still use these types of engines today. Other trains use electricity.

Steam engines were once used to make trains move.

People have ridden trains to get where they need to go for many years.

speed
Speed is the distance an object moves in a period of time.

force
A **force** is a push or a pull.

100 101

PROGRAM RESOURCES

- Big Ideas Book: *Physical Science*
- Big Ideas Book: *Physical Science* eEdition at myNGconnect.com
- Digital Library at myNGconnect.com

Access Science Content

Recognize the Speed of Trains

- Have students read pages 100–101. Point out the word **speed**. Say: **We describe a train's speed as the distance it can travel in a certain period of time.**
- Have students observe the photos on pages 100–101. Explain: **Trains are a form of transportation that has changed over time.** Ask: **How has the speed of transportation changed over time?** (Possible answer: We can travel at much faster speeds now.)
- Point out the word **force**. Ask: **Where is a train's engine located when it uses a pulling force?** (at the front of the train) **Where is a train's engine located when it uses a pushing force?** (at the back of the train)

Notetaking

my SCIENCE notebook

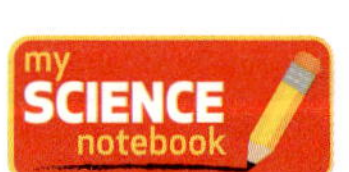

Have students make a timeline that shows the types of trains used at three different times in history—1870s, 1930s, and present day. The timeline can show what each type of train used for fuel.

BECOME AN EXPERT

NATIONAL GEOGRAPHIC

Bullet Trains Some trains move very fast. They are called bullet trains. Some of these trains have a top speed that is over 320 kilometers (200 miles) per hour! Most of these trains have electric engines. How does a train get electricity while it is moving? There are wires above the train tracks. The trains have special parts on top. These parts allow the train to get electricity from the wires. In this way, the train can get electricity while it is in **motion**.

The Acela train is the fastest train in the United States. It takes people between Boston and New York City. It also goes between Washington, D.C., and New York City. Its top speed is about 240 kilometers (150 miles) per hour. It only goes this fast where the train track is straight. It slows down when the track changes direction.

TECHTREK myNGconnect.com Digital Library

This train gets electricity from wires above the tracks.

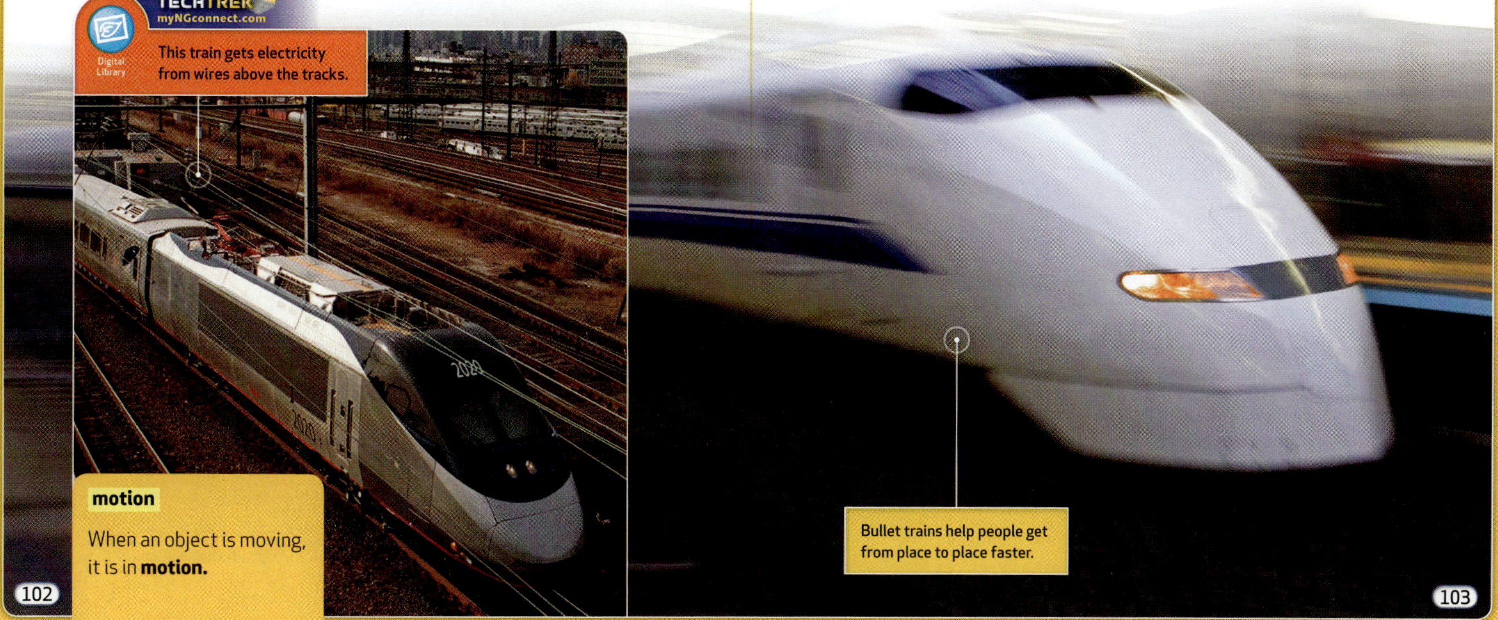

motion

When an object is moving, it is in **motion.**

Bullet trains help people get from place to place faster.

102 103

Access Science Content

Describe the Motion of Bullet Trains

- Have students study the photos and read the captions on pages 102–103. Ask: **What do you think you will learn about on these pages?** (Possible answer: bullet trains, fast trains, trains that get electricity from wires above the tracks)
- Have students read pages 102–103. Ask: **How can a bullet train get electricity while it is in motion?** (It has special parts on top that get electricity from wires above the train tracks.)
- Ask: **How can you describe the motion of the Acela train?** (Possible answers: as the fastest in U.S.; speeds up to 240 kilometers, or 150 miles, per hour; moves between Boston and New York City and between Washington, D.C. and New York City; goes fast where the track is straight, slows down when the track changes direction)

Digital Library

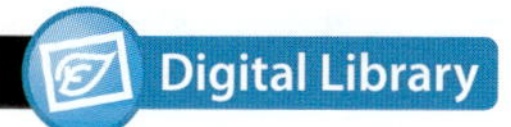

myNGconnect.com

Have students use the **Digital Library** to find photos of bullet trains and other modern trains.

Assess

1. **Contrast How do early trains differ from modern trains?** (Possible answers: Early trains used steam or diesel fuel, and modern trains use electricity. Modern trains are much faster than early trains.)
2. **Infer Are the fastest trains in the United States? Explain how you know.** (No, the fastest train in the United States has a top speed of 240 kilometers, or 150 miles, per hour. The bullet trains have a speed of over 320 kilometers, or 200 miles, per hour.)

LESSON 10 ▪ Become an Expert

BECOME AN EXPERT

Maglev Trains Some of the fastest trains in the world are in Germany. One of the trains, the Transrapid, can go up to 434 kilometers (271 miles) per hour. This train is different from electric trains. It doesn't have an engine car. It doesn't even have wheels! It glides on a thin cushion of air above the tracks. The train is a maglev train. The word "maglev" is short for "magnetic levitation." These trains are called maglev trains because the force of **magnetism** lifts them above the tracks and makes it seem like they are "floating."

A train floating in the air might seem like magic. After all, trains are big. They are very heavy. The pull of **gravity** on the train is strong. Of course, it is not magic. The magnets keep the trains up off the track. The magnets' pull is greater than the pull of gravity. These magnets keep the maglev train riding safely on the thin cushion of air above the tracks.

Maglev trains move on special train tracks.

magnetism
Magnetism is a force between magnets and the objects that magnets attract.

104

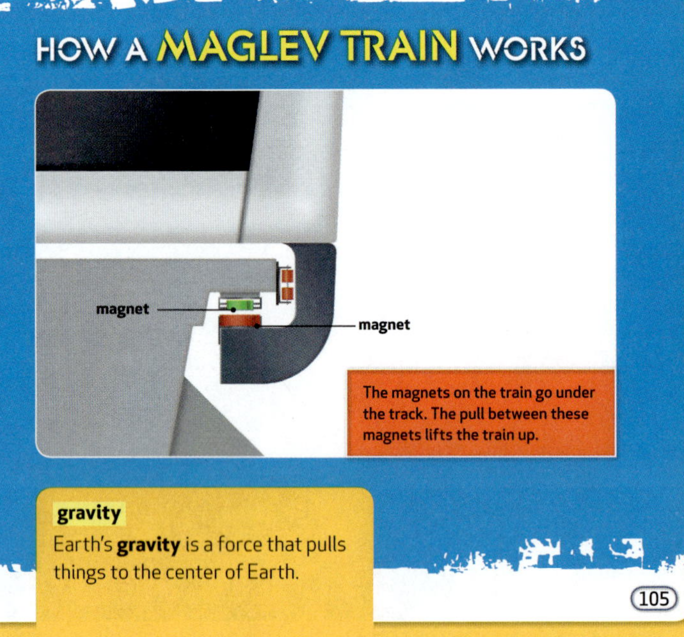

The magnets on the train go under the track. The pull between these magnets lifts the train up.

gravity
Earth's **gravity** is a force that pulls things to the center of Earth.

105

Access Science Content

Explain How Magnetism Is Used in Maglev Trains

- Have students read pages 104–105, study the photos, and read the captions. Ask: **How is magnetism used in a maglev train?** (Possible answer: It lifts the train above the track and causes the train to move.) Say: **The force of gravity pulls the train down.** Ask: **How can the train float on a cushion of air?** (The pull of the magnets is stronger than the pull of gravity.) Explain: **Special magnets that are very strong are used.**

Notetaking

Have students write a paragraph that summarizes how **magnetism** and **gravity** act on a maglev train.

Differentiated Instruction

ELL Language Support for Explaining How Magnetism Is Used in Maglev Trains

BEGINNING

Ask students either/or questions to help them explain the use of magnetism, for example:

Does magnetism lift a maglev train above or below the tracks?

INTERMEDIATE

Help students describe magnetism using Academic Language Frames, such as:

Magnetism lifts a maglev train above the ______.

ADVANCED

Help students elaborate on the use of magnetism using Academic Language Stems, such as:

A maglev train works because magnets …

Maglev trains use special tracks that have coils. The coils are turned into special magnets that use electricity. The magnets are only turned on when the train is nearby. The magnets in the track push and pull on the magnets in the train. These pushes and pulls move the train forward and stop the train as well. **Friction** is not a problem for maglev trains. Why? The trains do not rub against the tracks. They float above the tracks. Without much friction, the trains can go very fast.

The Future of Maglev Trains Maglev trains have a lot of good properties. They do not make noise. The ride from one place to another is very smooth. They are fast. But there's one problem—the cost of building tracks. Maglev train tracks are very expensive to build. People may want to spend the money to make cheaper methods of travel better. However, maglev trains will probably have a place in our high-speed future!

friction
Friction is a force that acts when two surfaces rub together.

106   107

Access Science Content

Explain the Motion of Maglev Trains

- Have students read pages 106–107, study the photos, and read the captions. Say: **Air pushing or rubbing against a surface causes friction. However, the amount of air between the tracks and the train is so small that it does not affect the motion of the train.**

- Say: **The maglev train has magnets and the track it floats above has special coils that are magnets. Ask: How do these magnets work together to cause motion?** (Possible answer: The magnets in the tracks push and pull on the magnets in the train, which moves the train forward.)

Make Inferences

Say: **The maglev tracks have coils that are only magnets when electricity is turned on. Ask: What can you infer about how electricity and magnetism are related?** (Answers will vary. Possible answer: Electricity can be changed into magnetism.)

Assess

1. **Compare** Why does friction affect the motion of other trains but not maglev trains? (Possible answer: Maglev trains float above the track, but other trains sit on the track and rub against it as they move.)

2. **Draw Conclusions** Why haven't people built more maglev trains? (Possible answer: The tracks are very expensive to build.)

Share and Compare

Turn and Talk

Ask students to turn to partners and talk about the advantages and disadvantages of high-speed trains. Prompt students by asking:

1. **Recall** **What are two kinds of high-speed trains?** (bullet trains and maglev trains)

2. **Explain** **What special parts are needed to move high-speed trains?** (Answers may vary. Sample answer: For bullet trains—cars with electric engines and tracks with wires that provide electricity; For maglev trains—special magnets and tracks with coils that become magnets when electricity flows through them)

3. **Compare** **How fast can high-speed trains move?** (Sample answer: Bullet trains have a top speed of 320 kilometers, or 200 miles, per hour. A maglev train in Germany can go up to 434 kilometers, or 271 miles, per hour.)

Read

Ask students to select two pages in this section and practice reading the pages. Then ask them to read the pages aloud to a partner and talk about why the pages are interesting.

Have students write a conclusion that tells the main ideas of what they have learned about high-speed trains. Ask them to include a statement of what they think is the Big Idea of this section. Then ask them to share what they wrote with a classmate and compare their conclusions.

Have students draw a picture of a high-speed train. Have students include labels of the parts of the train. Then have students share their drawings and combine them to show the different ways that forces make high-speed trains speed up or slow down.

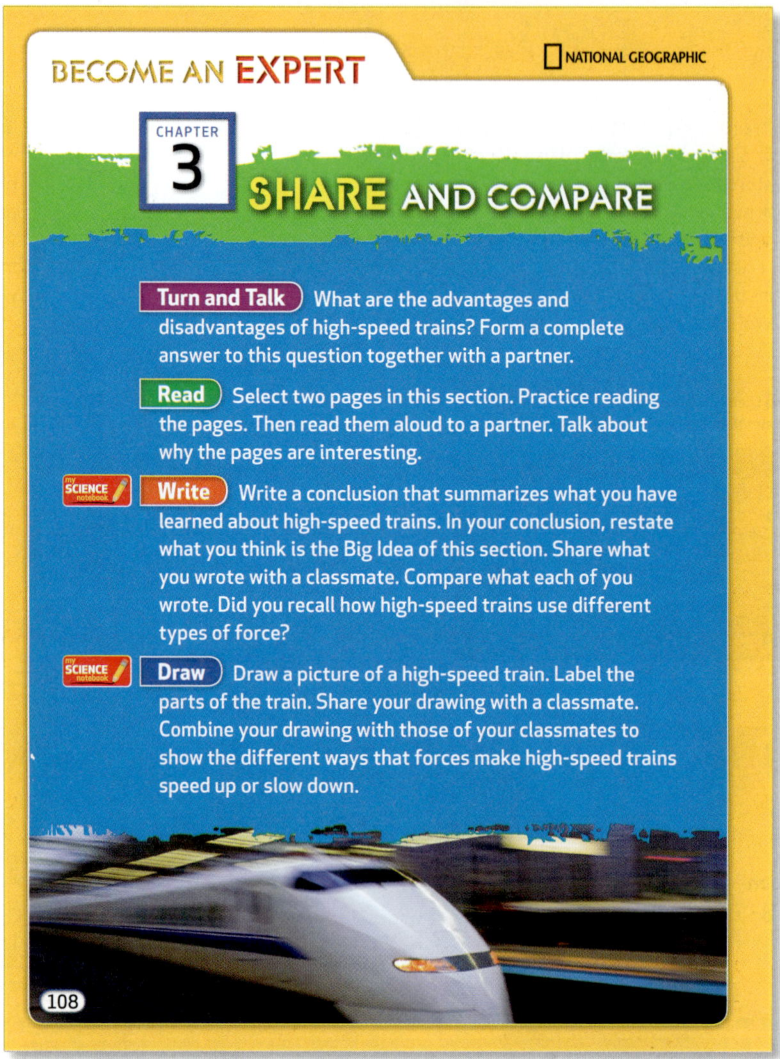

❯ Sum Up

Tell students that to sum up a text helps them pull together the ideas of the text. Remind students that they summed up the Become an Expert lesson in the Write section of Share and Compare. Have students take turns reading the conclusions they wrote to a partner. Invite them to compare and contrast their conclusions.

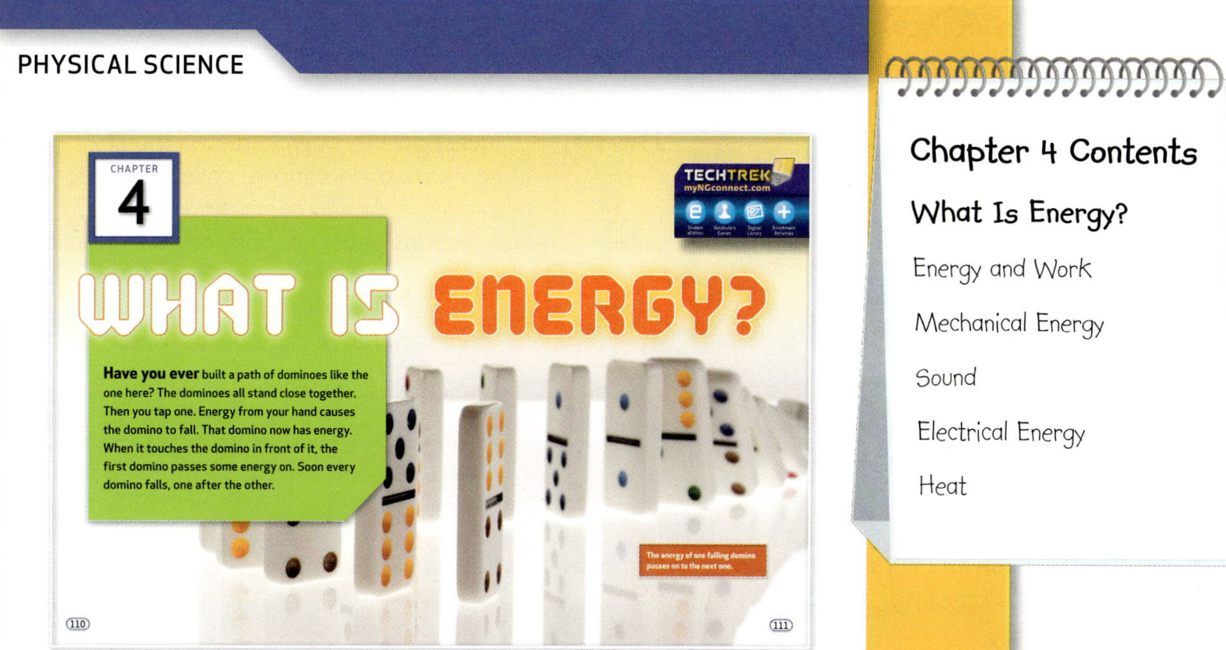

After reading Chapter 4, you will be able to:

- Identify and describe different forms of energy, including mechanical, sound, electrical, and heat. **MECHANICAL ENERGY, SOUND, ELECTRICAL ENERGY, HEAT**

- Recognize that energy has the ability to cause motion or create change. **ENERGY AND WORK, MECHANICAL ENERGY, ELECTRICAL ENERGY, HEAT**

- Identify and describe that vibrations are the source of sound energy. **SOUND**

- Distinguish materials that heat flows easily through and those it does not. **HEAT**

- Demonstrate that rubbing objects together produces heat. **HEAT**

- **Science in a Snap!** Demonstrate that rubbing objects together produces heat. **HEAT**

109

〉 Preview and Predict

Tell students that they can preview a text by looking over it to get an idea of what the text is about and how it is organized. Tell students that predicting is forming ideas about what they will read. Students should confirm their ideas with others.

Have students preview and make predictions about the chapter. Remind them to use section heads, vocabulary words, pictures, and their background knowledge to make predictions.

CHAPTER 4 □ What Is Energy?

LESSON	PACING	OBJECTIVES
Directed Inquiry **1** *Investigate Energy of Motion* pages T109e–T109h **Science Inquiry and Writing Book** pages 194–197	**20** minutes	Investigate through Directed Inquiry (answer a question; make and compare observations; collect and record data and observations; generate explanations and conclusions based on evidence; share findings; ask questions based on observations to increase understanding). Describe mechanical energy. Use quantitative and qualitative data to produce reasonable explanations. Evaluate reasonableness of an explanation. Identify similarities and differences between explanations received from others or personal observations or understandings.
2 **Big Idea Question and Vocabulary** pages T110-T111–T112-T113 **Energy and Work** pages T114–T115	**30** minutes	Recognize that energy has the ability to cause motion or create change.
3 **Mechanical Energy** pages T116–T119	**25** minutes	Identify mechanical energy.
4 **Sound** pages T120–T125	**30** minutes	Identify sound energy. Describe how vibrations are the source of sound energy.
5 **Electrical Energy** pages T126–T129	**25** minutes	Identify electrical energy.
6 **Heat** pages T130–T133	**25** minutes	Identify heat energy. Distinguish materials that heat flows easily through. Demonstrate that rubbing objects together produces heat.
7 NATIONAL GEOGRAPHIC **Wind Power: Energy in the Air** pages T134–T135	**20** minutes	Identify wind as a source of mechanical energy.

VOCABULARY	RESOURCES	ASSESSMENT
account compare	Science Inquiry and Writing Book: *Physical Science* Science Inquiry Kit: *Physical Science* Directed Inquiry: Learning Masters 213–216	Inquiry Rubric: Assessment Handbook, page 214 Inquiry Self-Reflection: Assessment Handbook, page 223 Reflect and Assess, page T109h
energy	Vocabulary: Learning Master 217 Big Ideas Book: *Physical Science*	Assess, page T115
mechanical energy	Extend Learning: Learning Master 218	Assess, page T119
sound	Science Inquiry and Writing Book: *Physical Science*	Assess, page T125
electricity		Assess, page T129
heat		Assess, page T133
	Share and Compare: Learning Master 219	Assess, page T135

TECHNOLOGY RESOURCES

STUDENT RESOURCES

myNGconnect.com

- Student eEdition
- Big Ideas Book
- Science Inquiry and Writing Book
- Explore on Your Own Books
- Read with Me
- Vocabulary Games
- Digital Library
- Enrichment Activities

National Geographic Kids

National Geographic Explorer!

TEACHER RESOURCES

myNGconnect.com

- Teacher eEdition
- Teacher's Edition
- Science Inquiry and Writing Book
- Explore on Your Own Books
- Online Lesson Planner
- National Geographic Unit Launch Videos
- Assessment Handbook
- Presentation Tool
- Digital Library

NGSP ExamView CD-ROM

IF TIME IS SHORT...

FAST FORWARD.

CHAPTER 4 □ What Is Energy?

LESSON	PACING	OBJECTIVES
Guided Inquiry **8** *Investigate Vibrations and Sound* pages T135a–T135d **Science Inquiry and Writing Book** pages 198–201	**30** minutes	Investigate through Guided Inquiry (answer a question; make and compare observations; collect and record data and observations; generate explanations and conclusions based on evidence; share findings; ask questions based on observations to increase understanding; adjust explanations based on findings and new ideas). Identify sound as a form of energy and relate fast or slow vibrations to pitch. Conduct an investigation using appropriate tools and instruments. Share findings with others and actively seek their interpretations and ideas.
9 **Conclusion and Review** pages T136–T137	**15** minutes	
10 NATIONAL GEOGRAPHIC **PHYSICAL SCIENCE EXPERT** *Product Designer* pages T138–T139 NATIONAL GEOGRAPHIC **BECOME AN EXPERT** *Use Some Energy: Have Some Fun!* pages T140-T141–T148	**35** minutes	Identify scientists' contributions to science and technology by using and exploring energy. Identify and describe different forms of energy including mechanical, sound, electrical, and heat.

FAST FORWARD ▶▶▶
ACCELERATED PACING GUIDE

DAY 1  **20** minutes

Directed Inquiry

Investigate Energy of Motion, page T109e

DAY 2 **35** minutes

NATIONAL GEOGRAPHIC **PHYSICAL SCIENCE EXPERT** *Product Designer,* page T138

NATIONAL GEOGRAPHIC **BECOME AN EXPERT** *Use Some Energy: Have Some Fun!,* page T140–T141

VOCABULARY	RESOURCES	ASSESSMENT
observe **infer**	Science Inquiry and Writing Book: *Physical Science* Science Inquiry Kit: *Physical Science* Guided Inquiry: Learning Masters 220–222	Inquiry Rubric: Assessment Handbook, page 214 Inquiry Self-Reflection: Assessment Handbook, page 224 Reflect and Assess, page T135d
		Assess the Big Idea, page T137 Chapter 4 Test: Assessment Handbook, pages 102–104 NGSP ExamView CD-ROM
		Assess, pages T139, T142–T143, T146–T147

TECHNOLOGY RESOURCES

STUDENT RESOURCES

myNGconnect.com

Student eEdition

Big Ideas Book

Science Inquiry and Writing Book

Explore on Your Own Books

Read with Me

Vocabulary Games

Digital Library

Enrichment Activities

National Geographic Kids

National Geographic Explorer!

TEACHER RESOURCES

myNGconnect.com

Teacher eEdition

Teacher's Edition

Science Inquiry and Writing Book

Explore on Your Own Books

Online Lesson Planner

National Geographic Unit Launch Videos

Assessment Handbook

Presentation Tool

Digital Library

NGSP ExamView CD-ROM

DAY 3   30 minutes

Guided Inquiry

Investigate Vibrations and Sound, page T135a

Objectives

Students will be able to:

- Investigate through Directed Inquiry (answer a question; make and compare observations; collect and record data and observations; generate explanations and conclusions based on evidence; share findings; ask questions based on observations to increase understanding).
- Describe energy of motion.
- Use quantitative and qualitative data to produce reasonable explanations.
- Evaluate reasonableness of an explanation.
- Identify similarities and differences between explanations received from others or personal observations or understandings.

Science Process Vocabulary

count, compare

PROGRAM RESOURCES

- Science Inquiry and Writing Book: *Physical Science*
- Science Inquiry and Writing Book **eEdition** at **myNGconnect.com**
- **Inquiry eHelp** at **myNGconnect.com**
- Science Inquiry Kit: *Physical Science*
- Learning Masters Book, pages 213–216, or at **myNGconnect.com**
- Inquiry Rubric: Assessment Handbook, page 214, or at **myNGconnect.com**
- Inquiry Self-Reflection: Assessment Handbook, page 223, or at **myNGconnect.com**

MATERIALS

Kit materials are listed in italics.

safety goggles; *string; paper clip;* masking tape; *5 metal washers;* ruler; *stopwatch*

For teacher use: scissors

❶ Introduce

Tap Prior Knowledge

- Discuss the pendulums in grandfather clocks. Ask: **What does the pendulum do?** (It moves back and forth.) Ask: **What else moves like a pendulum?** (a swing)

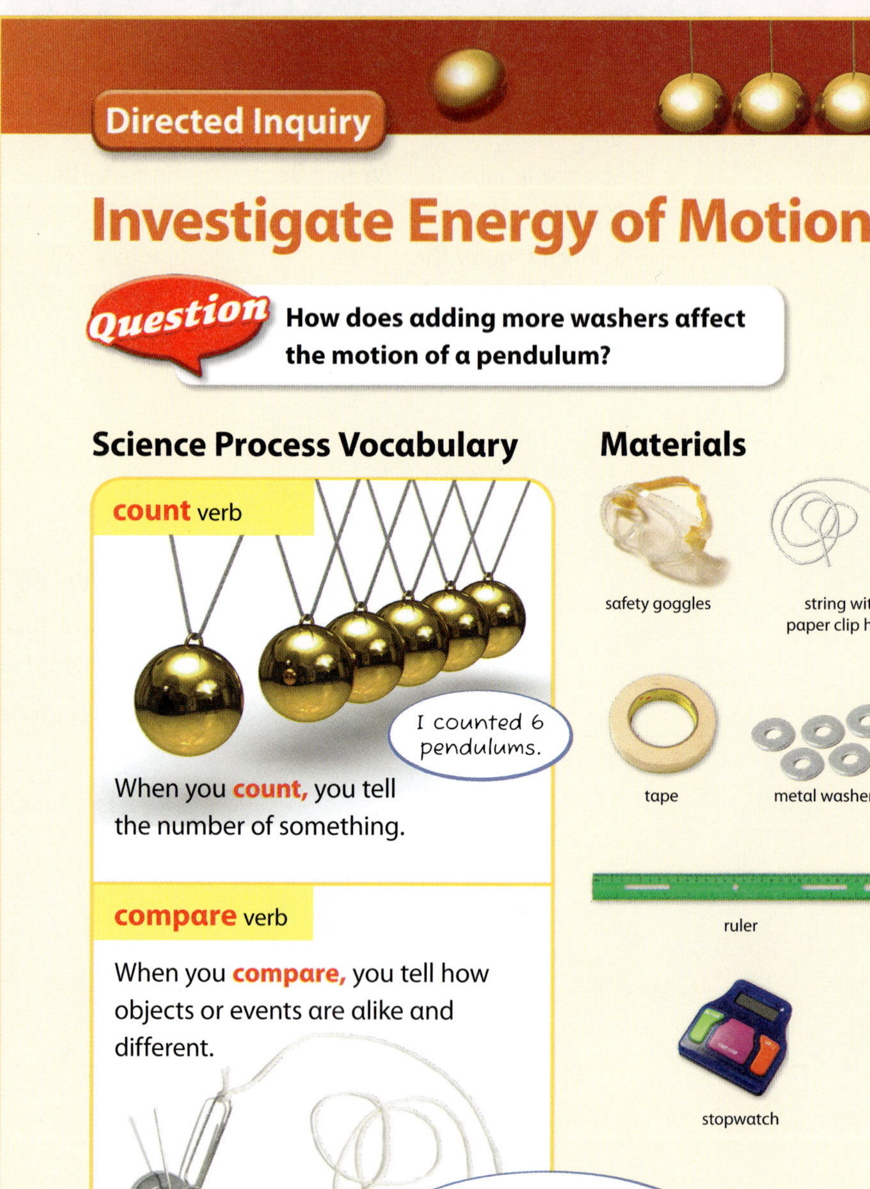

MANAGING THE INVESTIGATION

Time

20 minutes

Groups

Small groups of 4

Advance Preparation

- Cut the string into 70-cm lengths.
- For each group, bend a paper clip into a hook and tie one end of the string to it. Make sure the string is tied tightly.
- Clear the area around the tables used for this activity.

What to Do

1. Put on your safety goggles. Tape the free end of the string to the edge of a table. Put a metal washer on the paper clip hook. The metal washer should be close to the floor but not touching it. You have made a pendulum.

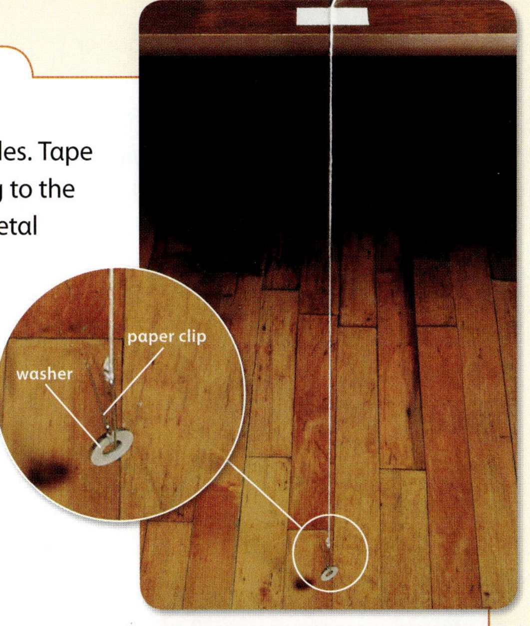

2. Pull the pendulum back 10 cm, and then let it go. Use the stopwatch to **count** the number of swings the pendulum makes in 1 minute. One swing is when the pendulum moves back and forth one time. Record your **data** in your science notebook.

195

Teaching Tips

- Make sure the paper clip is bent to no more than a 45° angle.
- Make sure students tape the string securely to the top edge of the table.

What to Expect

- Students will observe energy of motion of a moving pendulum. Students should observe that the number of swings does not change when the number of washers changes. Students will describe changes in the motion of the pendulum as it swings.

Introduce, continued

Connect to the Big Idea

- Review the Big Idea Question, *What is energy?* Explain to students that this inquiry will demonstrate energy of motion.

- Have students open their Science Inquiry and Writing Books to page 194. Read the Question and invite students to share ideas about forms of mechanical energy they have experienced.

❷ Build Vocabulary

Science Process Words: count, compare

Use this routine to introduce the words.

1. **Pronounce the Word** Say **count**. Have students repeat it.

2. **Explain Its Meaning** Choral read the sentence. Ask students for another word or phrase that means the same as **count**. (tell the number of something)

3. **Encourage Elaboration** Ask: **How many washers do you count?** (Answers will vary.)

ELL **Vocabulary Support**

Write: *When I count something, I tell how many there are.* Ask: **How can I tell how many times a pendulum swings?** (I can **count** how many times the pendulum moves back and forth.)

Repeat for the word **compare**. To encourage elaboration, ask: **What will you compare in this activity?** (I will **compare** how many times a pendulum swings in 1 minute.)

❸ Guide the Investigation

- Distribute materials. Have students match each pictured item to the real item.

- Read the inquiry steps on pages 195–196 together with students. Move from group to group and clarify steps if necessary.

- Before step 2, demonstrate one swing of the pendulum to the students. One swing is when the pendulum moves back and forth one time. Remind students to count only complete swings. Have students observe how the motion of the pendulum changes as it swings.

❹ Explain and Conclude

- Students should find relationships in the data. Ask: **What do you observe about the number of swings?** (The number remains about the same.)

- Ask students to explain their results with others. Tell them to use qualitative and quantitative data such as how the washers affected the motion of the pendulum and the number of swings the pendulum made. Ask students to evaluate whether the answers of other students are reasonable based on their evidence.

- Explain to students that an increase in the number of washers on the pendulum does not affect the number of times it swings in one minute. Ask: **If I put 10 washers on the pendulum, would that affect the number of swings?** (No, the number of swings would probably remain about the same.)

- Explain to students that an unsupported prediction does not mean that the investigation failed. Ask: **What other changes could you make to change the speed of the pendulum?** (make the string longer or shorter)

- Explain that the moving pendulum demonstrates energy of motion. Ask: **How would you describe the motion of the pendulum at different parts of its swing?** (The pendulum moves faster approaching the bottom of the swing. It slows down, then changes direction at the top of the swing.)

- Have students compare their results. Ask: **Did all groups have the same results? What could have caused any differences?** If results vary, remind students not to change their data. Instead, they should come up with an explanation for their results.

Answers

1. Possible answer: My predictions were not supported because I predicted that adding washers would result in more swings in one minute.

2. As the number of washers increased, the number of swings did not change.

3. Possible answer: The pendulum moves fastest as at the very bottom of its swing. It moves slowest at the very top of its swing.

What to Do, continued

3 Add 2 more washers to the paper clip hook. How do you think adding more washers will affect the number of pendulum swings in 1 minute? Record your **prediction.** Repeat step 2. Be sure to pull the pendulum back 10 cm before letting it go.

4 Add 2 more washers to the paper clip hook. How do you think adding even more washers will affect the number of pendulum swings in 1 minute? Record your prediction. Repeat step 2.

5 Use your data to make a bar graph to **compare** the pendulum swings. Look for a **pattern** in the graph.

196

Think Like a Scientist **How Scientists Work**

Did the Investigation Fail? Students may have predicted that a change in the number of washers on the pendulum would change the speed of the pendulum or the number of swings. Many students will find that their predictions were not supported.

Scientists can often learn more from predictions that are not supported than by predictions that are supported. When hypotheses or predictions are not supported, further investigation is needed.

Record

Write in your science notebook. Use your data to make a bar graph like the one shown here.

Pendulum Swings

Number of Washers	Prediction	Number of Swings
1		
3		
5		

Explain and Conclude

1. Did your results support your **predictions?** Explain.

2. **Compare** the number of swings the pendulum made with 1, 3, and 5 washers. How did increasing the number of washers affect the motion of the pendulum?

3. Describe the motion of the pendulum. Where in its swing does the pendulum move fastest? Where does it move slowest?

Think of Another Question

What else would you like to find out about how adding more washers can affect the swing of a pendulum? How could you find an answer to this new question?

197

---

❺ Find Out More

Think of Another Question

- Students should use their observations to generate new questions that arise from their investigations. Have them use these question stems to form testable questions: *What happens when…? What would happen if…? How can I/we…?* For example, students may ask: *What would happen if the string was longer?*

- Record student questions for possible future investigations. Discuss what students could do to find answers to the new questions.

❻ Reflect and Assess

- To assess student work with the Inquiry Rubric shown below, see Assessment Handbook, page 214, or go online at ⏺ **myNGconnect.com**

- Have students use the Inquiry Self-Reflection on Assessment Handbook, page 223, or at ⏺ **myNGconnect.com**

❼ Extend

- Repeat the activity using two strings of different lengths. Use the same number of washers at the end of each string. Students will find that the shorter pendulum will swing more times than the longer pendulum.

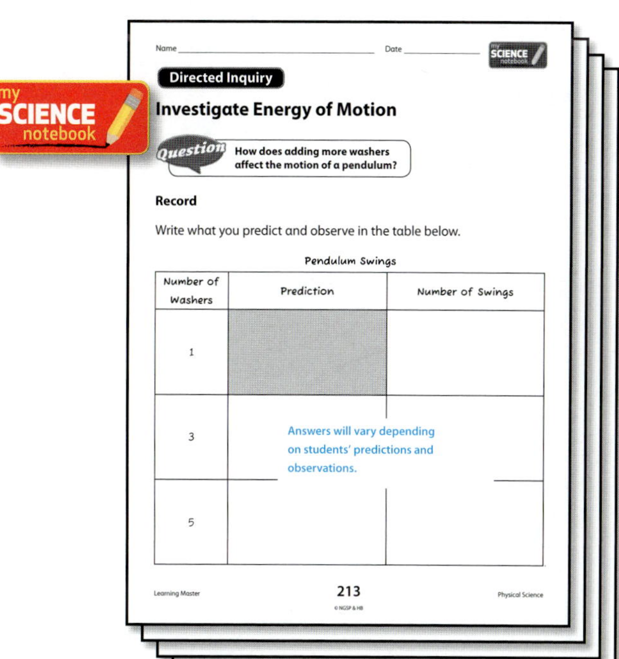

Learning Masters 213–216 or at ⏺ myNGconnect.com

Inquiry Rubric at ⏺ myNGconnect.com	Scale			
The student collected and recorded **data** by **counting** the number of swings of a pendulum.	4	3	2	1
The student **compared** the **predictions** with the results.	4	3	2	1
The student compared **observations** and results with those of other groups.	4	3	2	1
The student described the motion of the pendulum.	4	3	2	1
The student **shared conclusions** with other students.	4	3	2	1
Overall Score	4	3	2	1

Objectives

Students will be able to:

- Recognize that energy has the ability to cause motion or create change.

Science Academic Vocabulary

energy, mechanical energy, sound, electricity, heat

PROGRAM RESOURCES

- Big Ideas Book: *Physical Science*
- Big Ideas Book: *Physical Science* **eEdition** at **myNGconnect.com**
- **Vocabulary Games** at **myNGconnect.com**
- **Digital Library** at **myNGconnect.com**
- **Enrichment Activities** at **myNGconnect.com**
- **Read with Me** at **myNGconnect.com**
- Learning Masters Book, page 217, or at **myNGconnect.com**

❶ Introduce

Tap Prior Knowledge

- Ask students to describe how they move when they play soccer, ride a bike, or do some other activity that they enjoy. Discuss what gives people the power to move.

❷ Focus on the Big Idea

Big Idea Question

- Read the Big Idea Question aloud and have students echo it.
- Preview pages 114–115, 116–119, 120–125, 126–129, 130–133, linking the headings with the Big Idea Question.

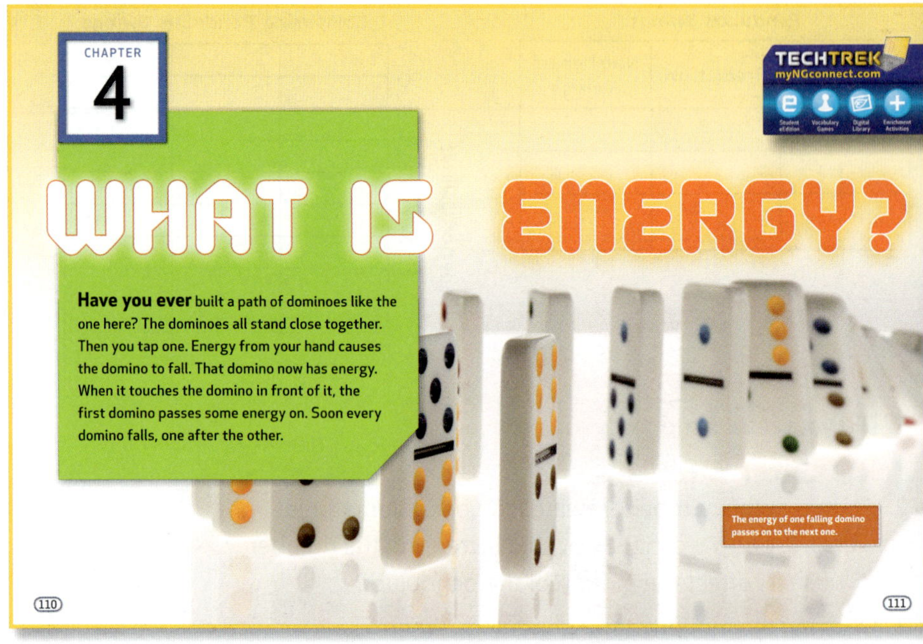

Differentiated Instruction

ELL Vocabulary Support

BEGINNING	INTERMEDIATE	ADVANCED

Have students chant rhymes such as the following to familiarize them with the word *energy*.
 	
Energy helps us run
  And push and pull
    And have some fun.
 	
Energy also gives us light
  And keeps us warm
    Throughout the night.
	Have students draw a picture of themselves using energy (e.g., riding a bike, kicking a soccer ball). Have partners share their drawing and tell about it using this Academic Language Frame:
	
	I use energy to ______.
		Have partners list ways they use energy (e.g., running, jumping, walking). Have them use these Academic Language Stems to write about how they use energy.
		
		• *I use energy when I . . .*
		• *Energy helps me . . .*
		• *Every day I use energy to . . .*

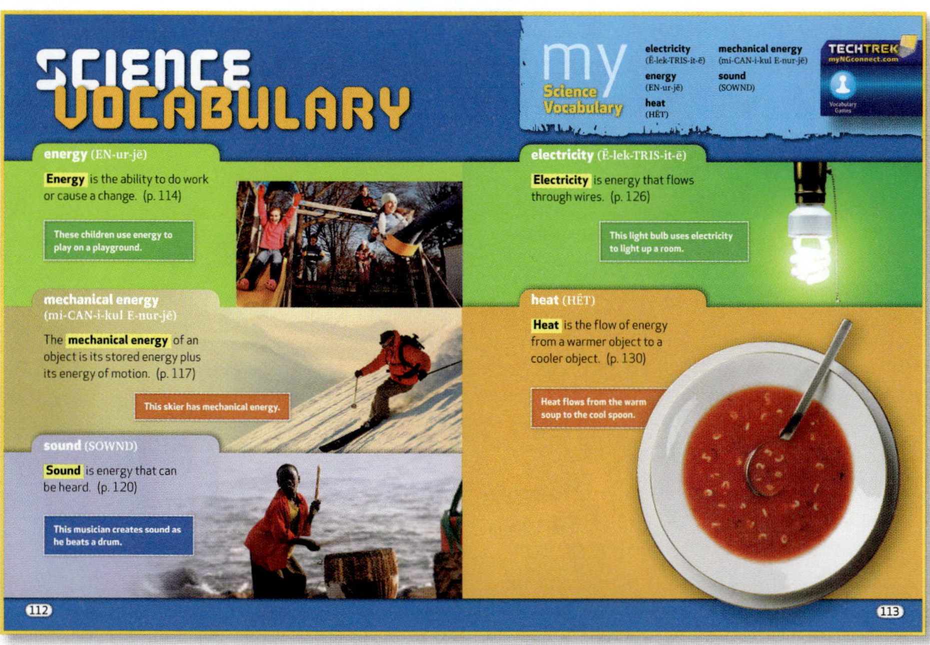

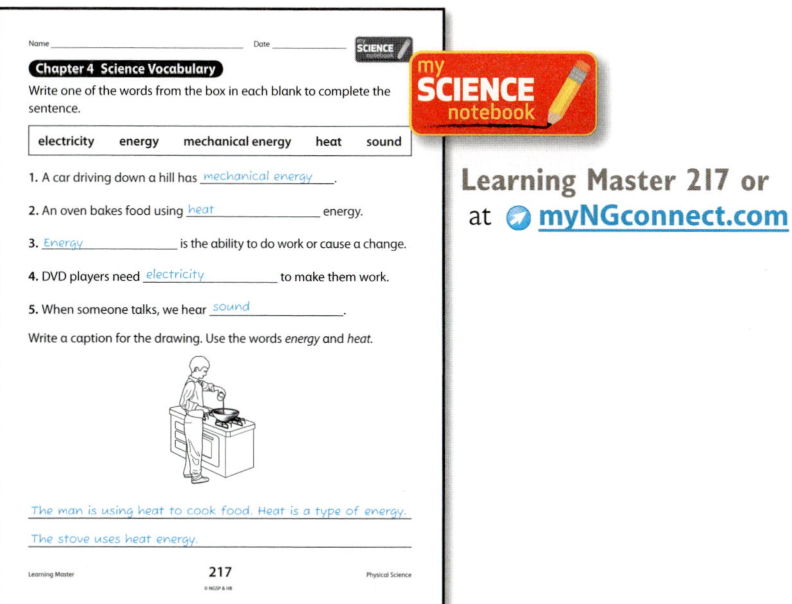

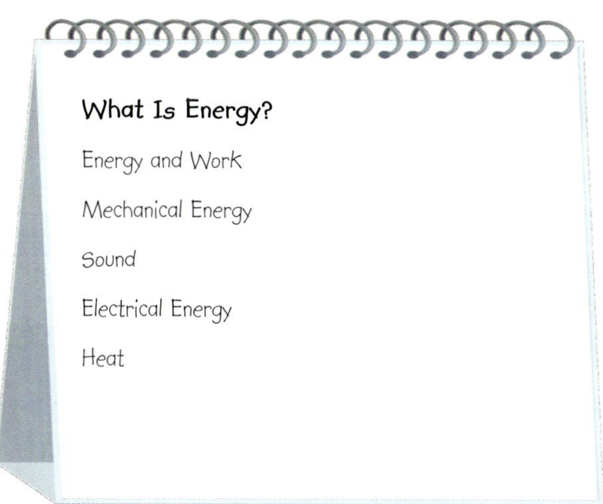

Focus on the Big Idea, continued

- With student input, post a chart that displays the chapter headings. Have students orally share what they expect to find in each section. Then read page 110 aloud.

❸ Teach Vocabulary

Have students look at pages 112–113, and use this routine to teach each word. For example:

1. **Pronounce the Word** Say **energy** and have students repeat it.

2. **Explain Its Meaning** Read the word, definition, and sample sentence, and use the photo to explain the word's meaning. Ask: **How are these children using their energy?** (by sliding; by swinging; by balancing or climbing)

3. **Encourage Elaboration** Have students look at what else the photo shows. Ask: **How is each child using energy to cause motion?** (by pushing off the top of the slide; by moving their legs back and forth to swing; by moving their arms and legs to climb)

Repeat for the words **mechanical energy, sound, electricity,** and **heat** using the following Elaboration Prompts:

- **What makes up mechanical energy?** (stored energy plus energy of motion)

- **How can you recognize sound energy?** (You can hear it, or feel its vibrations.)

- **How is electricity related to wires?** (It flows through wires.)

- **How does heat energy flow?** (from a warmer object to a cooler object)

Learning Master 217 or at 🌐 myNGconnect.com

Objectives

Students will be able to:

- Recognize that energy has the ability to cause motion or create change.

Science Academic Vocabulary

energy

PROGRAM RESOURCES

- **Digital Library** at ⊘ **myNGconnect.com**

❶ Introduce

Tap Prior Knowledge

- Ask students to share how they put objects in motion every day.

- Have students use the photos on pages 114–115 to describe what causes objects to move. (A moving ball causes the pins to move. The children's movments cause the swings to move.) Record their ideas.

Set a Purpose and Read

- Read the heading. Then tell students they will read about **energy** and what it has to do with work.

- Have students read pages 114–115.

❷ Teach

Academic Vocabulary: *energy*

- Point out the word **energy**. Say: **Energy is the ability to do work or cause a change.**

- Ask: **When do you use energy to cause a change or do work?** (Possible answers: when I walk, bike, turn on a light, talk) **What things do you use that require energy?** (any machine, tool, or appliance)

Energy and Work

Look at the children on the swings. They are doing work. When scientists talk about work, they use this word differently than you might. In science, work means using force to make something move. These children use force to pump back their legs back and forth. They need **energy** to do this. Energy is the ability to do work or cause a change.

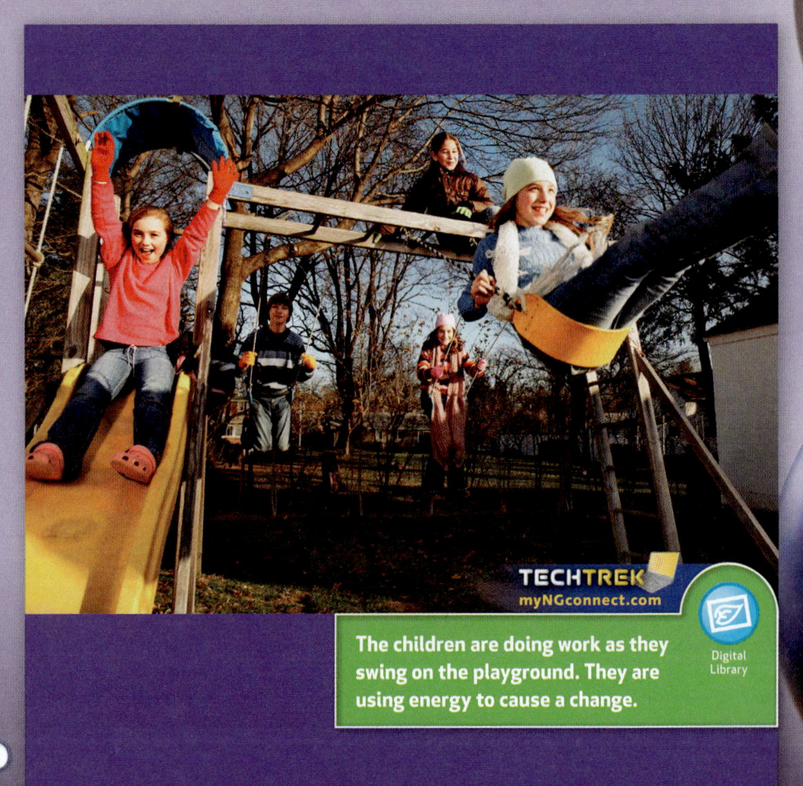

The children are doing work as they swing on the playground. They are using energy to cause a change.

114

Science Misconceptions

Energy Energy is often difficult for students to grasp. Students may think that energy is a thing rather than the ability to do work or cause a change. They may think that energy is found only in living things, but everything has energy. Students may associate energy with movement (kinetic energy), but nonmoving objects have stored (potential) energy. They may think that energy is a fuel. Fuels, such as coal and oil, are sources of energy, but they are not energy itself.

In bowling, you give energy to a ball by rolling it down the lane. The ball hits the pins and knocks them down. Energy has gone all the way from you, to the ball, and down the lane to the pins.

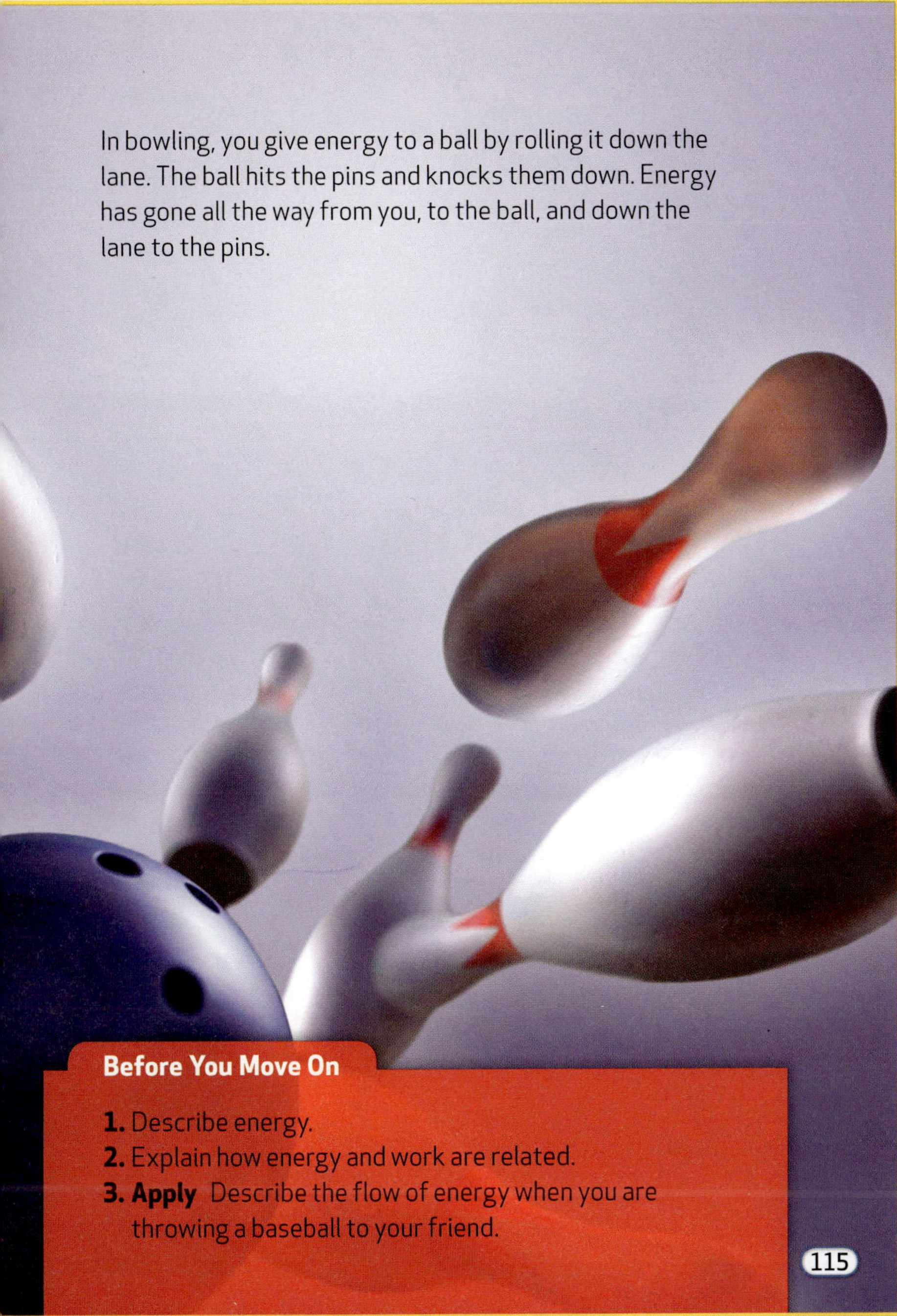

Before You Move On

1. Describe energy.
2. Explain how energy and work are related.
3. **Apply** Describe the flow of energy when you are throwing a baseball to your friend.

115

Describe Energy

- Have students look at the children on the swings, and ask: **How do you know the children have energy?** (They are doing work.) **What work are they doing?** (pumping their legs, moving the swings)

- Display a playground ball and a few tall, narrow children's blocks placed upright. Have students predict: **What will happen if I use my energy to push this ball into the blocks?** (The blocks will fall down.) Then push the ball and ask: **How did the ball use energy?** (It rolled toward the blocks.) **What change was caused by the energy of the moving ball?** (The blocks fell down.)

- Have students draw a picture of your demonstration. Have them use their drawings to describe to a partner how energy can cause objects to move and change position.

 Digital Library

myNGconnect.com

Have students use the  Digital Library to find photos of children on a playground.

Integrated Technology

Whiteboard Presentation Students can make a whiteboard presentation of the photos. They should add labels to identify the motions that different playground equipment can make when energy is applied.

❸ Assess

» Before You Move On

1. **Describe** Describe energy. (Energy is the ability to do work or cause a change.)

2. **Explain** Explain how energy and work are related. (If an object has energy, it has the ability to do work.)

3. **Apply** Describe the flow of energy when you are throwing a baseball to your friend. (Sample answer: The energy from my arm flows to the ball as I let go of it. The energy in the moving ball flows to my friend's hands when she catches it.)

Differentiated Instruction

ELL **Language Support for Describing Energy**

BEGINNING	INTERMEDIATE	ADVANCED
Have students draw a picture of a bowling ball knocking down bowling pins. Have them answer yes or no to these questions about energy: **Does the ball have energy? Is the ball making the pins fall down? Is the ball doing work?**	Have students draw a picture of a bowling ball knocking down bowling pins. Provide Academic Language Frames to help them describe energy: • *The _____ has energy.* • *The energy of the ball forces the pins to _____.* • *The ball is doing _____.*	Have students draw a picture of a person with a bat hitting a baseball. Provide Academic Language Stems to help them describe energy: • *The person uses energy to …* • *The bat …* • *The baseball …*

Objectives
Science
Students will be able to:
• Identify mechanical energy.

Science Academic Vocabulary
mechanical energy

PROGRAM RESOURCES
• Learning Masters Book, page 218, or at
   **myNGconnect.com**

❶ Introduce

Tap Prior Knowledge
• Ask students to describe what happens when they
  sled down a snow-covered hill or ride down a ramp
  on a skateboard. Discuss how energy makes it
  possible for them to move.

Preview
• Read the heading. Have students preview pages
  116–119. Ask them to describe what they see in
  the photos, and have volunteers read the photo
  captions aloud.

Academic Language: *mechanical energy*
• Point to the term **mechanical energy** on page
  117 and read the definition. Display this equation
  as students read:
  mechanical energy = stored energy + motion

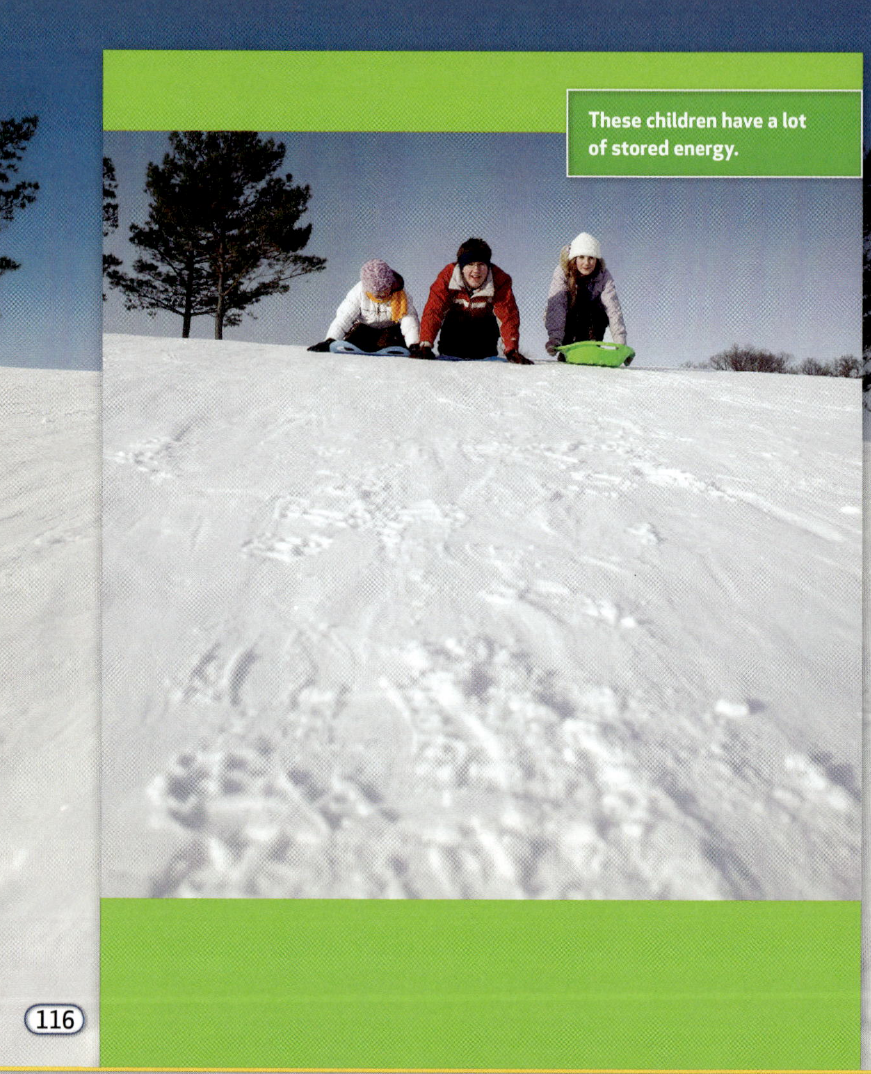

Mechanical Energy

Look at the picture below. The children ready to sled down
the hill have a lot of stored energy. Stored energy is energy
that is ready to be used.

116

Extend Learning

MANAGING THE INVESTIGATION

Time
 20 minutes

Groups
 Small groups of 4

PROGRAM RESOURCES
• Learning Master 218

MATERIALS
• 1 ball
• 1 meterstick

Investigate Mechanical Energy

Preview	What To Do
Question **How high will a ball bounce?** Students will investigate that the height at which a ball is held above the floor will affect the height of its bounce. They will observe that the greater the height of the ball (stored energy), the higher it will bounce (energy of motion).	1. Have one student hold a ball 0.5 meters above the floor. Have the group predict how high the ball will bounce. 2. Have a student drop the ball as the group observes how high it bounces. 3. Repeat, holding the ball at 1 meter and 1.5 meters. Have students record and discuss their conclusions.

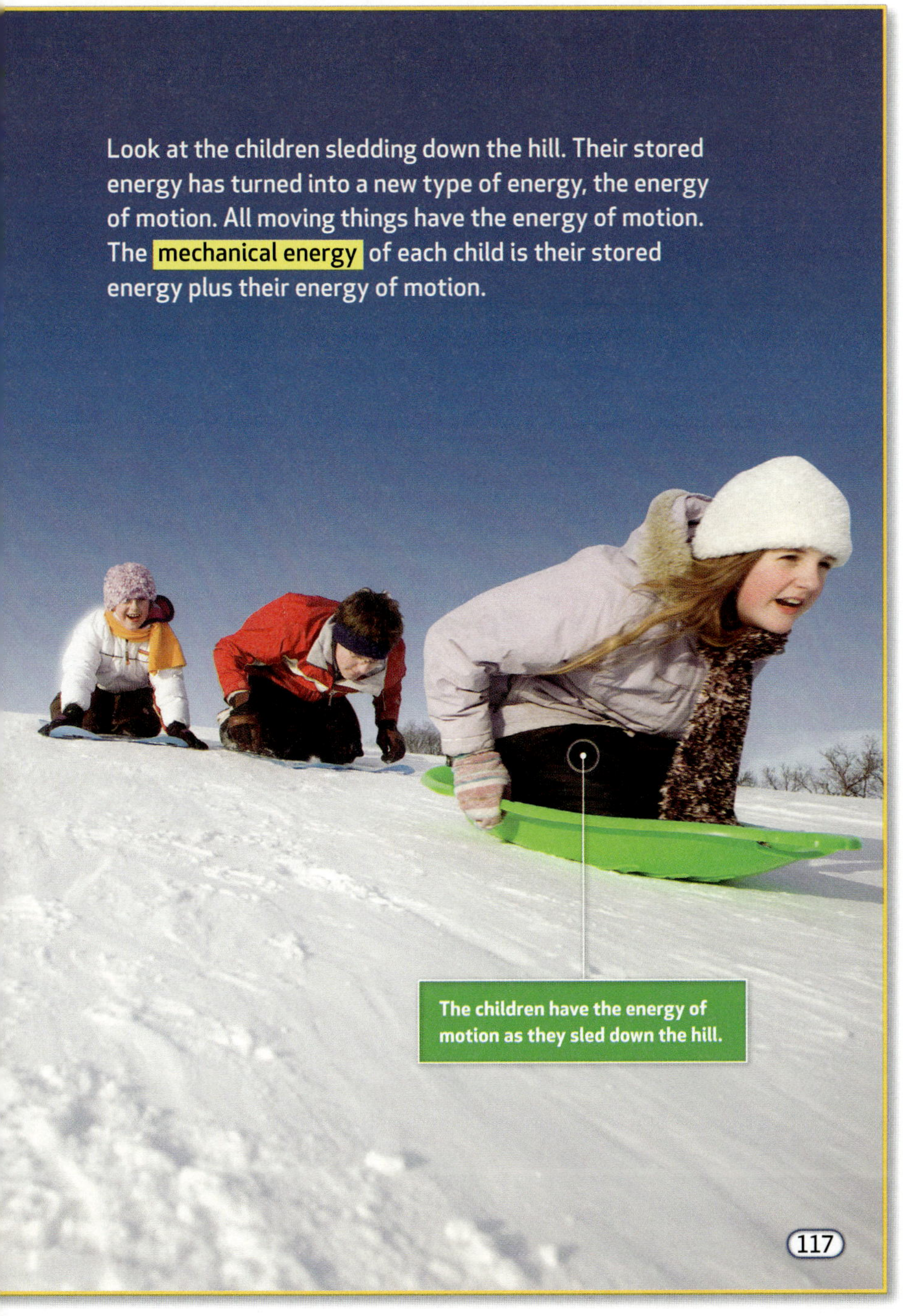

Look at the children sledding down the hill. Their stored energy has turned into a new type of energy, the energy of motion. All moving things have the energy of motion. The mechanical energy of each child is their stored energy plus their energy of motion.

117

❷ Teach

Identify Mechanical Energy

- Have students read pages 116–117.

- Display a playground ball, and say: **All objects, including this ball, have energy.** Ask: **What kind of energy does this ball have when it's resting in my hands?** (stored energy) **What other objects in this room have stored energy?** (Students may name any objects in the room that are not moving, such as desks, chairs, and books.)

- Drop the ball to the floor, and ask: **What kind of energy does the ball have when it's falling?** (energy of motion and stored energy) **As the ball falls, what happens to the amount of stored energy?** (It gets less the closer the ball is to the floor.)

- Hold the ball, and have students say: **Stored energy.** Drop the ball, and say as it falls: **Energy of motion and stored energy.** Repeat the process with other durable objects.

- Say: **This combination of stored energy and energy of motion is called mechanical energy.**

- Ask: **When do the objects have stored energy?** (when you're holding them up, and when they are still above the ground or floor) **When do the balls have energy of motion?** (when they're falling or in motion) **What is mechanical energy?** (stored energy plus energy of motion)

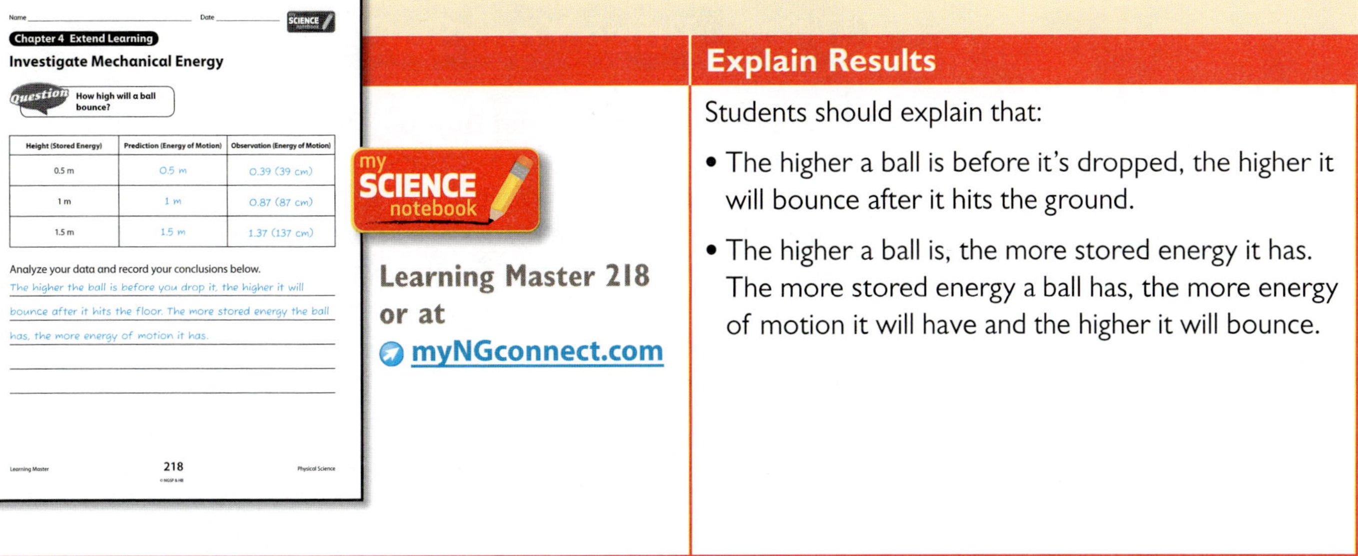

**Learning Master 218
or at
 myNGconnect.com**

Explain Results

Students should explain that:

- The higher a ball is before it's dropped, the higher it will bounce after it hits the ground.

- The higher a ball is, the more stored energy it has. The more stored energy a ball has, the more energy of motion it will have and the higher it will bounce.

Objectives
Science
Students will be able to:
• Identify mechanical energy.

PROGRAM RESOURCES
• **Digital Library** at **myNGconnect.com**

Teach, continued

Identify Mechanical Energy
• Have students read pages 118–119.

• Ask: **What is stored energy?** (the kind of energy an object has when it's not moving; energy that is ready to be used) **What is energy of motion?** (the kind of energy an object has when it is moving)

• Ask: **What is mechanical energy?** (stored energy plus energy of motion)

Text Features: Compare and Contrast Photos
• Have students compare and contrast the energy in the photos on pages 118–119. Say: **Both photos show skiers with mechanical energy.** Then ask: **Which photo shows a skier with stored energy but no energy of motion?** (page 118) **How do you know?** (The skier is standing still at the top of the hill.)

• Ask: **What kinds of energy make up the mechanical energy of the skier on page 119?** (stored energy and energy of motion) **How do you know?** (The skier is moving down the mountain.)

Let's look at another example of mechanical energy. A skier starts at the top of a mountain. She has stored energy. She is not moving. She does not have any energy of motion.

(118)

NATIONAL GEOGRAPHIC Raise Your SciQ!

Energy-Generating Speed Bumps The city of London plans to test a new way of generating electricity, using the energy of motion of moving cars. These new "green" speed bumps will be embedded into a road, flush with the road's surface. When cars drive over the road, the bumps will move up and down, turning a wheel under the road. The wheel then causes a motor to turn and generate electricity. The city will use the electricity to operate traffic signals and streetlights.

Look at this skier. He is in motion. His stored energy changes to energy of motion as he moves down the mountain. The skier's stored energy plus his energy of motion is his mechanical energy.

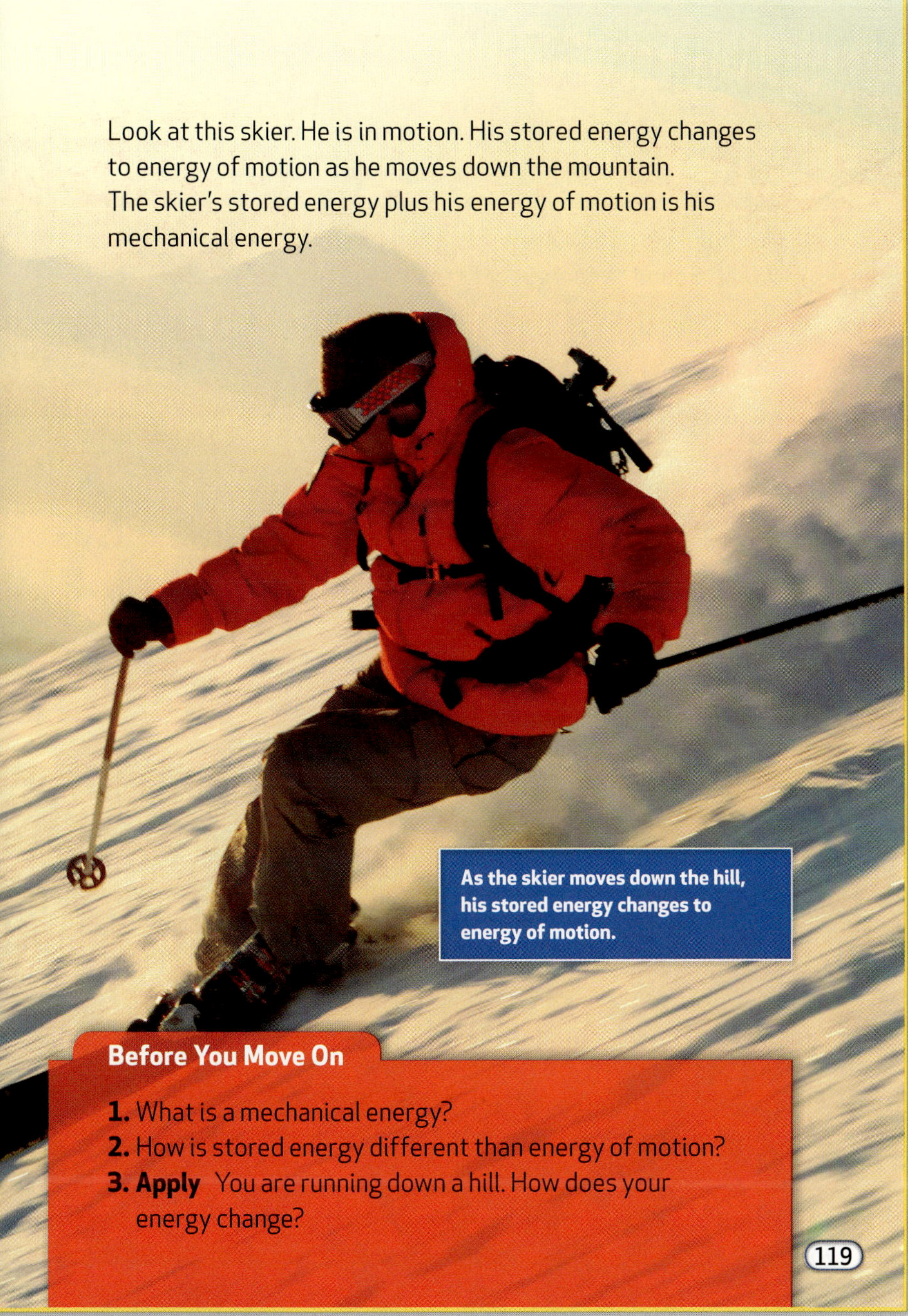

Before You Move On

1. What is a mechanical energy?
2. How is stored energy different than energy of motion?
3. **Apply** You are running down a hill. How does your energy change?

(119)

Differentiated Instruction

ELL **Language Support for Describing Changes in Mechanical Energy**

BEGINNING	INTERMEDIATE	ADVANCED
Have students look at the photos on pages 118–119 as they answer these questions, the first two for page 118 and the last two for page 119: **Is this skier moving? Does she have stored energy? Is this skier moving? Does he have energy of motion?**	Have partners complete Academic Language Frames to describe energy differences between the two skiers: • *This skier is not moving. She has _____ energy.* • *This skier is moving down the hill. He has energy of _____.*	Have students use Academic Language Stems to describe changes in mechanical energy: • *The skier on page 118 has…* • *She could ski down the hill. Then she would have …* • *The skier on page 119 has …*

myNGconnect.com

Have students use the **Digital Library** to find other images of downhill skiing for comparison.

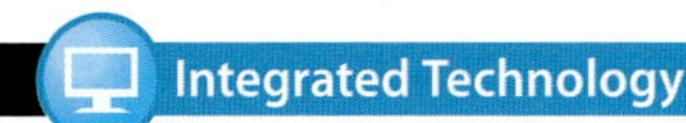

Digital Slide Show Students can use the photos to give a slide show that contrasts the different movements and speeds of various skiers.

Make Inferences

Ask students to infer how the skier who is not moving could change her stored energy into energy of motion. Then ask them to infer if the stored energy of the moving skier increases, decreases, or stays the same as the skier moves downhill.

Describe Changes in Mechanical Energy

• Challenge students to describe other examples of changes in mechanical energy in their environment, such as a bird sitting on a tree branch (stored energy) and then flying away (energy of motion).

• Have each student fold a sheet of paper in half and then label one half *Stored Energy* and the other half *Energy of Motion*. Have them draw an object with stored energy only and that same object with energy of motion.

• Have volunteers share their pictures with the class, using the terms *stored energy* and *energy of motion*.

❸ Assess

》 Before You Move On

1. **Recall** **What is mechanical energy?** (Mechanical energy is stored energy plus energy of motion.)

2. **Contrast** **How is stored energy different than energy of motion?** (Stored energy is energy that objects have when they are not moving. Energy of motion is energy that objects have when they are moving.)

3. **Apply** **You are running down a hill. How does your energy change?** (As you run down the hill, your stored energy is changed to energy of motion.)

LESSON 4 □ Sound

Objectives

Students will be able to:

- Identify sound energy.
- Describe how vibrations are the source of sound energy.

Science Academic Vocabulary

sound

❶ Introduce

Tap Prior Knowledge

- Have students close their eyes and listen for **sounds**. Ask: **What do you hear?** (Possible answers: footsteps in the hall, a bell ringing, cars moving outside, a bird chirping)

Set a Purpose

- Say: **In this lesson, you'll learn about sound—what causes sound, how loud and soft sounds are made, and what makes a sound high or low.**

❷ Teach

Academic Vocabulary: *sound*

- Point out the word **sound**. Say: **Sound is energy that can be heard.**

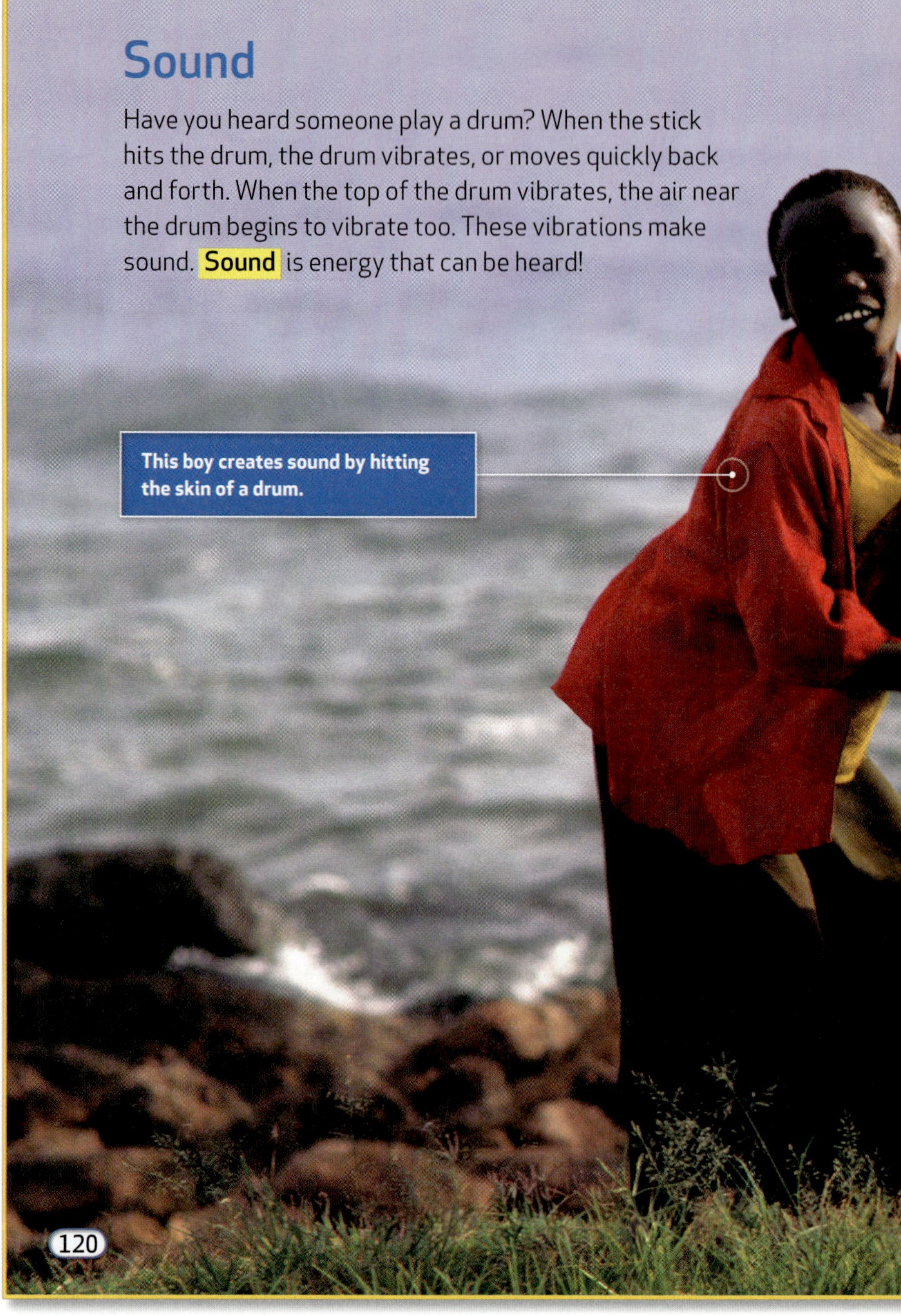

Sound

Have you heard someone play a drum? When the stick hits the drum, the drum vibrates, or moves quickly back and forth. When the top of the drum vibrates, the air near the drum begins to vibrate too. These vibrations make sound. **Sound** is energy that can be heard!

Extend Learning

Making Music

Find Out	**Think and Do**	**Describe and Compare**
Have small groups of students use the ■ Digital Library to find photos of musical instruments. Have each group choose an instrument and research how that instrument is played and which part of it vibrates to create sound. Have students use books from your school's library in their research.	 In their science notebook, have students draw and label the instrument they researched. Have them write a few sentences that describe how the musician plays the instrument and what part of the instrument vibrates to create sound.	Have groups take turns sharing their findings with the class, including: • How the musician plays the instrument • What part of the instrument vibrates to create sound

You hear sound when sound travels to your ear. Look at your ear in a mirror. It is made to collect sound vibrations. The vibrations move to your ear, and you hear the sound.

121

Identify and Describe Sound Energy

- Have students read pages 120–121.

- Ask: **What is sound energy?** (energy that can be heard) **How does a stick hitting a drum produce sound energy?** (The drum vibrates, creating vibrations that travel to your ear.)

- Have volunteers describe their experiences playing a drum or hearing the sound made by a drum in music class. Ask: **How does the sound made by a drum help the music?** (It keeps the beat.)

Describe How Vibrations Make Sound

- Write *vibrate,* pronounce it, and say: **Vibrate means to move back and forth very quickly.**

- Demonstrate vibrating by blowing air through your mouth so your lips vibrate. Have students try it. Say: **Your lips are vibrating.** Ask: **Do you hear a sound?** (yes) Say: **The vibration of your lips causes the sound that you hear. When you hear a sound, something is vibrating to cause that sound. You hear the sound when the vibrations move to your ear.**

- Ask: **What other objects vibrate to cause sound?** (Possible answers: vocal cords in your throat, cell phone, wind chimes)

Notetaking

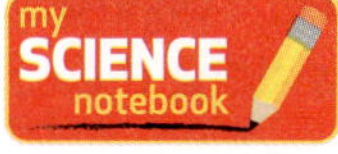

Have students write the word *sound* at the center of a page and circle it. Have them develop a Sound Concept Cluster to connect their ideas about sound as they read. Students may begin with *vibrations* and *hearing.* Invite them to elaborate on their ideas with brief descriptions.

Differentiated Instruction

ELL **Language Support for Describing How Vibrations Make Sound**

BEGINNING	INTERMEDIATE	ADVANCED
Ask either/or questions to help students describe how vibrations make sound. For example: **Are vibrations fast movements or slow movements? When something vibrates, does it light up or make a sound?**	Provide these Academic Language Frames to help students describe vibration and sound: • *When objects _____, they make sound.* • *Sound is a form of _____.* • *You can _____ sound with your ears.*	Provide these Academic Language Stems to help students describe vibration and sound: • *Sound is . . .* • *Vibrations are . . .*

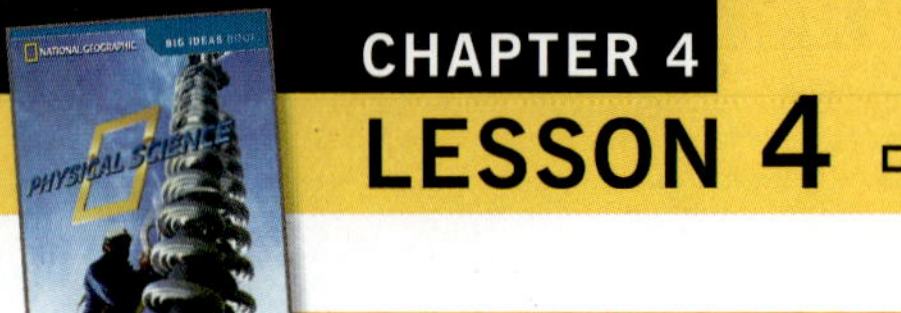

Objectives
Students will be able to:
- Identify sound energy.

PROGRAM RESOURCES
- Science Inquiry and Writing Book: *Physical Science*
- Science Inquiry and Writing Book **eEdition** at **myNGconnect.com**
- **Digital Library** at **myNGconnect.com**

Teach, continued

Describe Sound Energy: Volume

- Have students read pages 122–123.

- Remind students that *volume* can mean "the amount of space that matter takes up." Ask: **What is another meaning of *volume*?** (how loud or soft a sound is) Say: **Volume is a characteristic of sound energy.**

- Draw a continuum, and write *soft* on one end and *loud* on the other end. Ask: **What makes a soft sound?** (Possible answers: a whisper, a paper clip falling on the floor, a book being opened) **What makes a loud sound?** (Possible answers: a baby crying, a door slamming, a big truck driving by) Record students' ideas along the continuum.

- Explain to students that stronger the vibrations an object makes, the louder its sound is. Ask: **Which of these sounds are made by strong vibrations?** (the louder sounds) Write *strong vibrations* on the loud end of the continuum. **Which are made by weak vibrations?** (the softer sounds) Write *weak vibrations* on the soft end of the continuum.

Volume A jet engine is very loud when you are near it. It has a loud volume. Volume is how loud or soft a sound is. Strong vibrations make loud sounds. People working on the ground at an airport wear special headphones to protect their ears.

Think Like a Scientist Math in Science

Sound and Distance Choose a source of constant and unvarying volume, such as a ticking clock or mechanical chime or bell. Have students start off close to the source of the sound and gradually move farther away. Ask individual students to measure the distance from the sound source to the point at which they stop hearing the sound. Students then can create a graph that compares these distances from different students. Be aware of student sensitivities when comparing hearing among students.

When you speak, the vocal chords inside your throat vibrate. When you shout, your vocal chords vibrate more. This makes your voice sound louder.

When you speak quietly, your vocal chords vibrate less. This makes your voice sound softer.

Differentiated Instruction

ELL **Language Support for Describing Sound Energy**

BEGINNING	INTERMEDIATE	ADVANCED
Have students chant rhymes such as the following to identify loud and soft sounds: *I hear soft sounds* *Like gentle rain.* *I hear loud sounds* *Like a passing train.*	Have students draw a picture of something that makes either a loud sound or a soft sound. Have partners tell about their picture using this Academic Language Frame: *This _____ makes a _____ sound.*	Have students use Academic Language Stems to identify loud and soft sounds: • *The sound of a jet engine is . . .* • *The sound of a rabbit hopping is . . .* • *The sound of writing is . . .* • *The sound of clapping hands is . . .*

Teach, continued

Describe Vibrations as a Source of Sound Energy

• Have students look at the photos on page 123. Ask: **Which boy is making the louder sound?** (the boy on the left) **How can you tell?** (Possible answer: He looks like he is straining more and trying to be heard from far away.) **How would the throat of the boy on the left feel compared to the throat of the boy on the right?** (The vibrations would be bigger and more noticeable.)

Digital Library

myNGconnect.com

Have students use the **Digital Library** to find other images of loud and soft sounds being made.

Integrated Technology

Whiteboard Presentation Students can use the images to make a whiteboard presentation about different kinds of sound.

Science in a **Snap!**

Make a Sound

Science Inquiry and Writing Book: *Physical Science*, page 164

Read and follow the instructions together with students. Have students work individually. Students should record their observations and results in their science notebook.

What to Expect Students should feel vibrations in their throat as they make humming sounds. They should notice different vibrations for low versus high sounds. They also should notice greater vibrations for louder sounds.

Notetaking

Have students add *volume* to the Sound Concept Cluster they started on page T121. Invite students to elaborate on their ideas with brief descriptions.

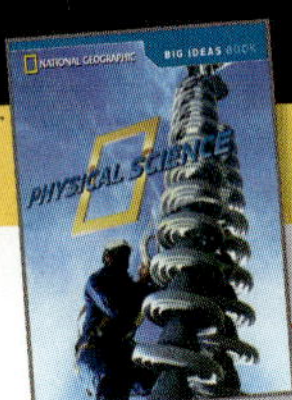

Objectives

Students will be able to:

- Identify sound energy.
- Describe how vibrations are the source of sound energy.

PROGRAM RESOURCES

- **Enrichment Activities** at 🌐 **myNGconnect.com**

Teach, continued

Describe Sound Energy: Pitch

- Have students read pages 124–125. Ask: **What is pitch?** (how high or low a sound is) Remind students that vibrations make sound. Explain: **The pitch of a sound depends upon the speed of the vibrations that cause the sound. Quick vibrations make sounds with a higher pitch, and slow vibrations produce low-pitched sounds.**

- Point out the photo of the kitten. Ask: **Does a kitten's meow have a high pitch or a low pitch?** (high) **Why?** (Its vocal cords are thin and short and vibrate quickly.)

- Ask: **Does a tiger's roar have a high pitch or a low pitch?** (low) **Why?** (Its vocal cords are thick and long and vibrate slowly.)

- Ask: **What causes the sound that both kittens and tigers make?** (vibrations of their vocal cords)

Pitch Look at the animals on this page. What kind of sounds do you think they make?

The kitten has thin, short vocal chords that vibrate quickly. It makes a high sound. The tiger has thick, long vocal chords that vibrate more slowly. It makes a low sound. Pitch is how high or low a sound is.

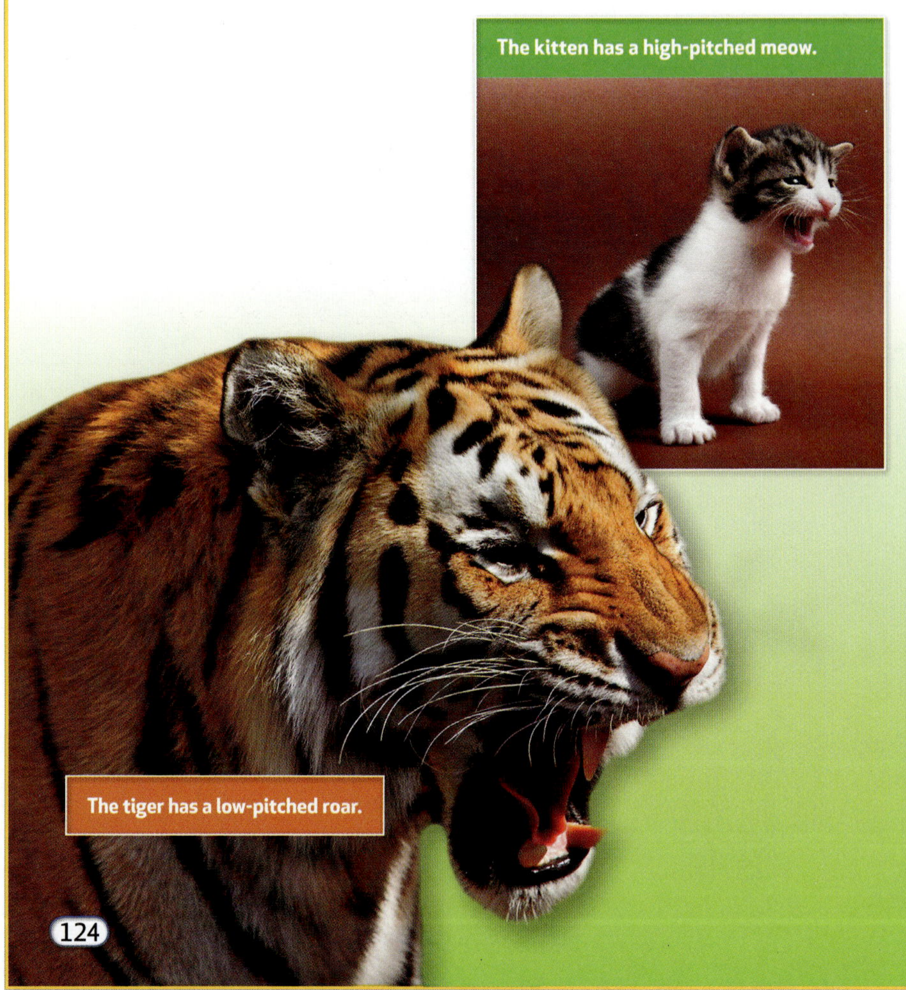

Differentiated Instruction

Extra Support

Provide partners with safety goggles and two rubber bands, one thin and one thick. Show them how to wrap one end of each band around a secure object, such as a chair leg, and pull the other end until each band is tightly stretched. Then have them pluck each band and compare the pitches. Have them discuss their observations and draw a conclusion about the relationship between the thickness of the rubber band and the pitch of the sound after plucking it.

Challenge

Provide partners with safety goggles and several rubber bands of varying thicknesses. Repeat the method above to compare the pitches. Have students place the bands in order from high to low pitch. Ask them to share their results with the class.

Look at the guitar. When the strings move, they vibrate. They make sound. A thick string vibrates slowly. The thick strings make low, deep sounds. This is a low pitch. A thin string can vibrate much faster. The thin strings make high sounds. This is a high pitch.

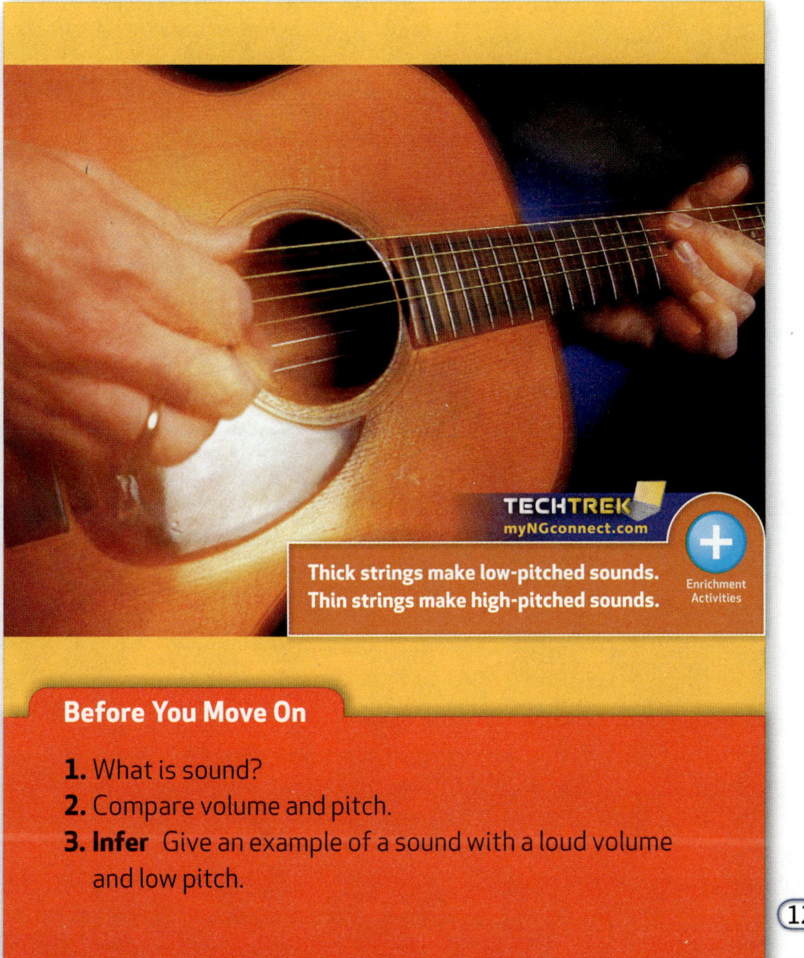

Before You Move On

1. What is sound?
2. Compare volume and pitch.
3. **Infer** Give an example of a sound with a loud volume and low pitch.

(125)

Differentiated Instruction

ELL **Language Support for Describing Sound Energy**

BEGINNING	INTERMEDIATE	ADVANCED
Have students write *High Pitch* at the top of a sheet of paper and draw a picture of something that makes a high-pitched sound. Repeat for *Low Pitch* at the bottom of the sheet. Students can share with a partner using this Academic Language Frame: *A _____ makes a _____ -pitched sound.*	Have partners use Academic Language Frames to describe pitch as portrayed in the photo on pages 124–125. • *The tiger makes a _____ -pitched sound.* • *The kitten makes a _____ -pitched sound.* • *A _____ guitar string makes a _____ -pitched sound.*	Have students use Academic Language Stems to write about pitch in their environment. • *The pitch of a sound depends on . . .* • *Things that make low-pitched sounds include . . .* • *Things that make high-pitched sounds include . .*

Teach, continued

Describe Vibrations as the Source of Sound Energy

- Have students look at the photo of the guitar on page 125. Say: **Think about what you just learned about the pitch of a kitten's meow and a tiger's roar.** Then ask: **Which strings on the guitar do you think will make a low-pitched sound?** (the thick strings) **Why?** (Thick strings vibrate slower than thin strings. The slower the vibration, the lower the pitch.)

- Ask: **What causes sound energy?** (vibrations)

Notetaking 

Have students add *pitch* to the Sound Concept Cluster they started on page T121. Invite students to elaborate on their ideas with brief descriptions.

myNGconnect.com

Have students use the **Enrichment Activities** to see how the thickness of strings affects the pitch of musical sounds.

Make Sound Files Students can use the material to make sound files to demonstrate differences in pitch.

❸ Assess

» Before You Move On

1. **Define** **What is sound?** (Sound is energy you can hear.)

2. **Compare** **Compare volume and pitch.** (Volume is how loud or soft a sound is. Pitch is how high or low a sound is.)

3. **Infer** **Give an example of a sound with a loud volume and low pitch.** (Possible answers: a lion's roar, a thick guitar string that is plucked strongly)

LESSON 5 □ Electrical Energy

Objectives

Students will be able to:

• Identify electrical energy.

Science Academic Vocabulary

electricity

❶ Introduce

Tap Prior Knowledge

• Ask: **What are some things you use in this classroom that need energy to work?** (lights, heat, air conditioning, computers) **What kind of energy do you think these things need to work?** (electricity)

Preview and Read

• Read the heading. Then preview the pictures on pages 126–129. Tell students that they will read about ways people use **electricity**.

• Have students read pages 126–129.

❷ Teach

Academic Vocabulary: *electricity*

• Pronounce **electricity** and write it on the board. Have a volunteer read the definition.

• Underline the word *electric* in **electricity**. Write *electric*. Then write the term *electrical energy*. Say: **Electricity** also can be called electrical energy.

Identify Electrical Energy

• Use an electric pencil sharpener to help students identify how **electricity** works. Attempt to use the sharpener unplugged. Ask: **What does the pencil sharpener need to work?** (electricity)

• Have students identify the outlets where objects can get **electricity**. Plug in the sharpener and show that it works. Ask: **How does the electricity flow to the sharpener?** (through the cord) Say: **The cord contains wires. Electricity** is energy that flows through wires.

Electrical Energy

A DVD player needs energy to play movies. It uses **electricity**. Electricity is energy that flows through wires. You can watch the movies and hear the sound because of electricity.

(126)

Raise Your SciQ!

Lightning As its name suggests, lightning is a burst of electrical energy that produces a bright flash of light. The average bolt of lightning contains enough electricity to keep a 100-watt light bulb glowing for three months.

Throughout the United States, lightning strikes the ground more than 25 million times each year. Worldwide, lightning strikes Earth's surface approximately 100 times every second!

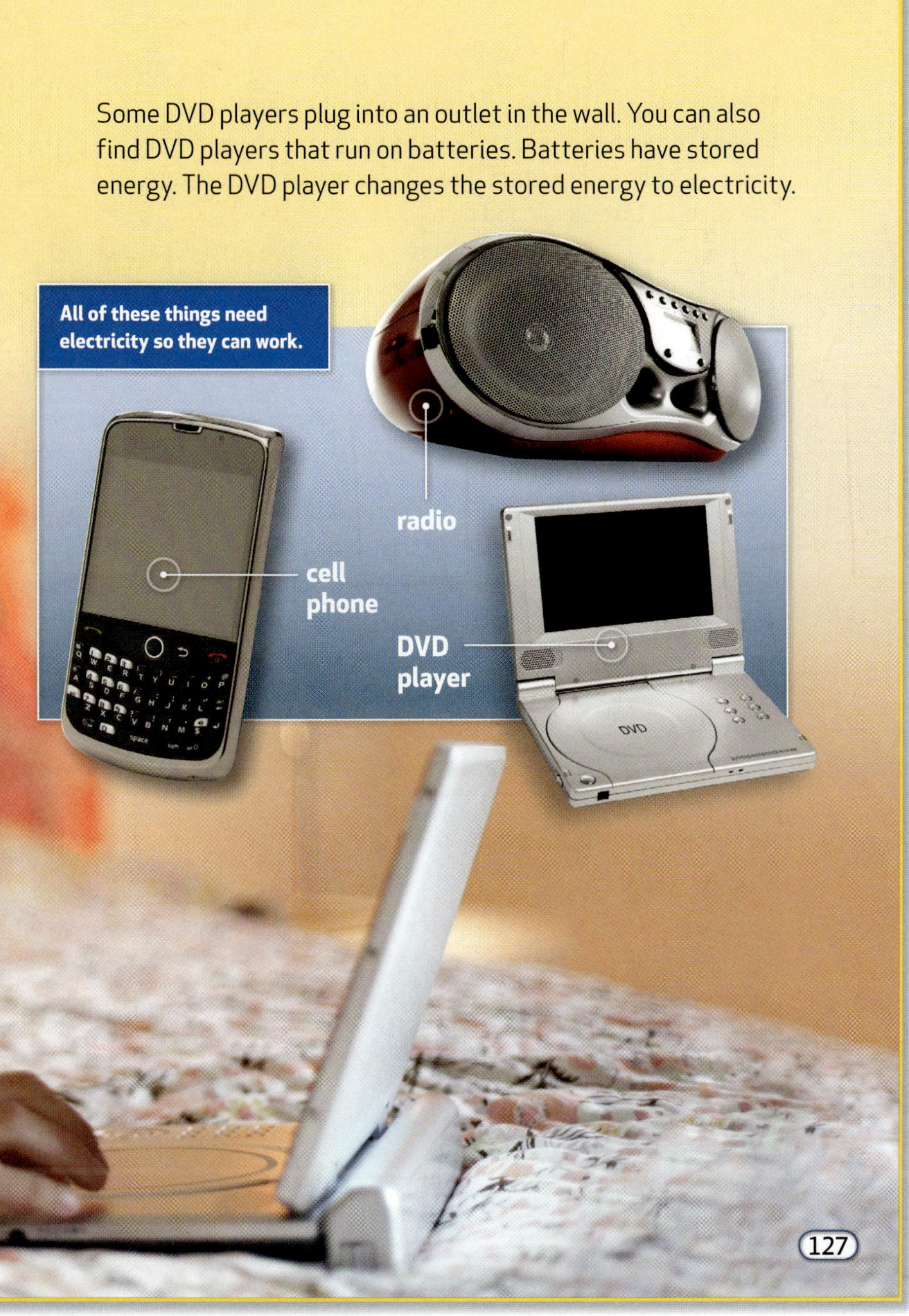
Some DVD players plug into an outlet in the wall. You can also find DVD players that run on batteries. Batteries have stored energy. The DVD player changes the stored energy to electricity.

127

Teach, *continued*

Identify Ways People Use Electricity

- Have students name the objects that use electrical energy in the photos on pages 126–127. Ask: **How does the cell phone get electricity when it is not plugged in?** (from a battery) **What kind of energy do batteries have?** (stored energy) Say: **The stored energy in batteries is changed into electrical energy.**

- Ask: **What other ways do people use electricity?** (Possible answers: video games, blenders, toasters, televisions)

- Say: **You use electricity at home, at school, at work, and many other places.** Ask: **Where have you been that there was no electricity?** (camping, hiking, canoeing)

- Have students try to remember a time when the electricity went out. Ask: **What happened around you when the electricity went out?** (Possible answers: I needed a flashlight to see in the dark. I couldn't use the computer or watch TV. I got cold when the heater turned off. The food in the refrigerator spoiled.)

Making and Recording Observations

Have students take their science notebook home with them and record everything they do that uses electricity. Students can compare their results with a partner's.

ELL **Language Support for Identifying Electrical Energy**

BEGINNING	INTERMEDIATE	ADVANCED
Help students identify objects that use electrical energy in the photos on pages 126–127 by answering these yes/no questions: **Does the radio use electricity? Do pillows use electricity? Does the DVD player use electricity?**	Have partners use these Academic Language Frames to identify uses of electricity: • *The radio uses ______.* • *The DVD player uses ______.* • *The cell phone uses ______.*	Provide Academic Language Stems to help students identify uses of electricity: • *Uses of electricity shown on this page are . . .* • *Some uses of electricity in my home are . . .* • *The radio uses ______. We use radios to . . . .*

Objectives

Students will be able to:
• Identify electrical energy.

Teach, continued

Identify Electrical Energy

• Have students look at the photo of the vacuum cleaner on page 128. Ask: **What kind of energy powers this vacuum?** (electrical energy, or electricity) **What does electrical energy flow through?** (wires) **Where are the vacuum's wires?** (inside the vacuum)

• Read aloud this sentence on page 128: **"The electricity becomes energy of motion."** Ask: **What kind of energy is energy of motion?** (mechanical energy)

• Remind students that energy of motion is the kind of energy an object has when it is moving. Say: **In this vacuum cleaner, electrical energy changes into energy of motion.**

This vacuum gets its power from electricity. Electricity is flowing through the wires inside the vacuum. The electricity becomes energy of motion. The wheels on the vacuum turn. The vacuum takes in air and dirt.

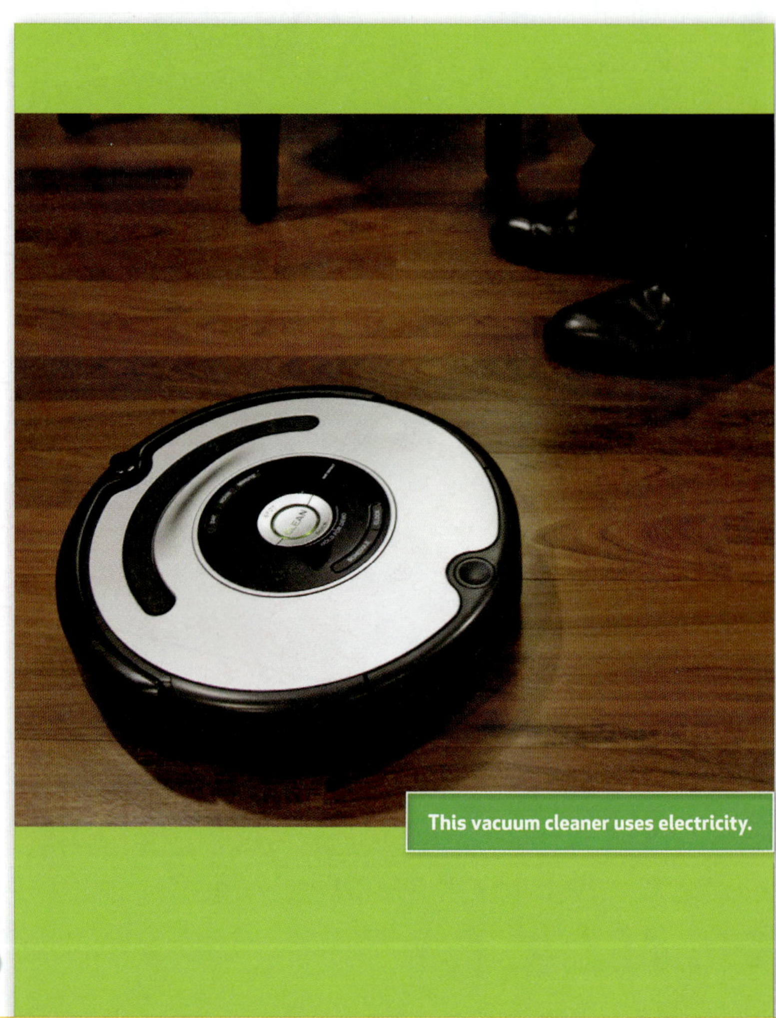

(128)

Social Studies in Science

Invention of the Light Bulb The invention of the incandescent light bulb dramatically changed people's lives by making many nighttime activities possible. Before the light bulb was invented, people used candles, torches, oil lamps, and gas lamps to light the night. Although Thomas Edison is famous in the United States for inventing the light bulb in 1879, an Englishman named Sir Joseph Swan received a British patent for his light bulb in 1878.

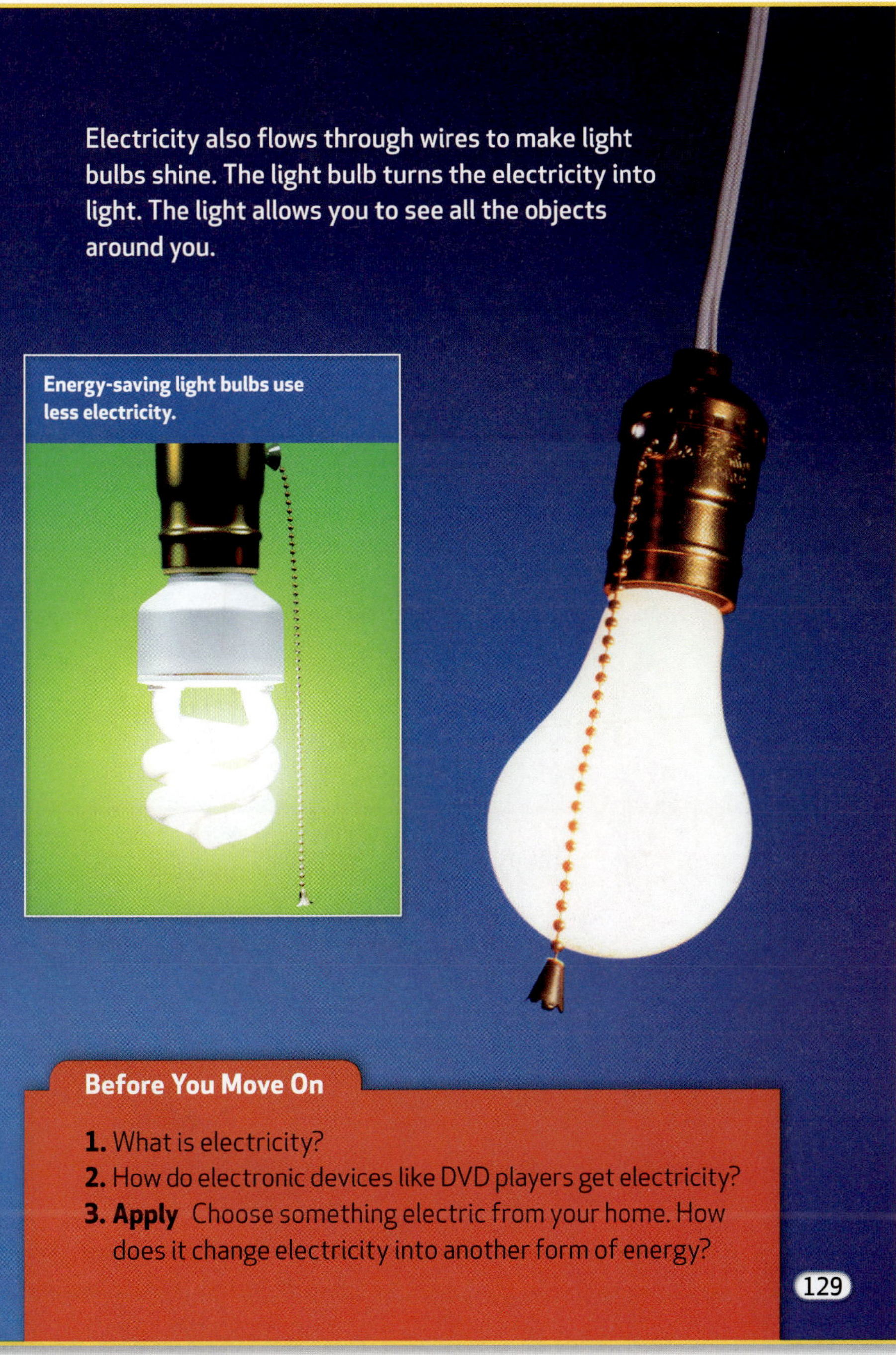

Electricity also flows through wires to make light bulbs shine. The light bulb turns the electricity into light. The light allows you to see all the objects around you.

Before You Move On

1. What is electricity?
2. How do electronic devices like DVD players get electricity?
3. **Apply** Choose something electric from your home. How does it change electricity into another form of energy?

129

Teach, continued

Identify Light Energy

- Have students look at the photos of the light bulbs on page 129. Ask: **What kind of energy powers these light bulbs?** (electrical energy, or electricity)

- Read aloud this sentence on page 129: **"The light bulb turns the electricity into light."** Ask: **What happens to electrical energy in a light bulb?** (It changes into light energy.)

- Say: **In these light bulbs, electrical energy changes into light energy.** Point out that the spiral-shaped bulb uses less electricity and will last longer than a traditional light bulb. Using these bulbs is one way that people can help protect the environment by using less energy.

- Ask: **What other objects use electricity and make light?** (Possible answers: televisions, computers, flashlights, headlights)

❸ Assess

» Before You Move On

1. **Recall** What is electricity? (Electricity is energy that flows through wires.)

2. **Explain** How do electronic devices like DVD players get electricity? (DVD players get electricity either from an electrical outlet in the wall or from batteries that store energy.)

3. **Apply** Choose something electric from your home. How does it change electricity into another form of energy? (Possible answer: A television changes electricity into light and sound energy when it is turned on.)

Differentiated Instruction

ELL **Language Support for Identifying Electrical Energy**

BEGINNING	INTERMEDIATE	ADVANCED
Help students use language about electricity. Write these words and have students choose which completes each Academic Language Frame: *energy of motion, light* • *A vacuum changes electricity to ______.* • *A light bulb changes electricity to ______*	Have partners use Academic Language Frames to describe uses of electricity: • *A vacuum changes electricity into ______.* • *A light bulb changes electricity into ______.*	Have students use Academic Language Stems to write about uses of electricity: • *A vacuum gets its power from …* • *A vacuum changes …*

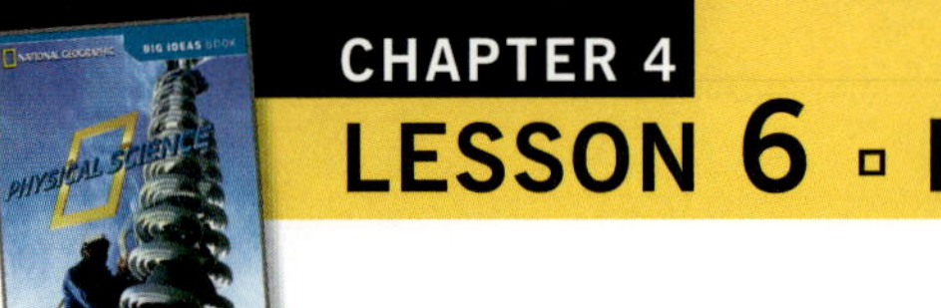

Objectives

Students will be able to:

- Identify heat energy.
- Distinguish materials that heat flows easily through.

Science Academic Vocabulary

heat

❶ Introduce

Tap Prior Knowledge

- Ask students to discuss how they keep themselves warm during cold days. (Possible responses: by wearing heavier clothing, staying indoors, moving around)

- Have students think about how their homes and the school are kept warm inside during periods of cold weather.

Set a Purpose

- Say: **In this lesson, you'll learn what heat is. You'll also learn about the kinds of materials through which heat flows easily.**

❷ Teach

Academic Vocabulary: *heat*

- Point out the word **heat**. Say: **Heat is the flow of energy from a warmer object to a cooler object.**

Distinguish Materials Through Which Heat Flows Easily

- Have students read pages 130–131.

- Direct students to the photo on page 130. Ask: **If you put a cool metal spoon into a bowl of hot soup, what will happen to the spoon?** (The spoon will get hot.) **Why?** (Heat moves from the warmer soup to the cooler spoon.)

- Have students think about the different kinds of cooking spoons they may have seen in their kitchen at home. Ask: **What different materials are cooking spoons made of?** (Possible answers: metal, wood, plastic) **Through which of these materials does heat flow easily?** (metal)

Heat

This soup was just cooked. It is very hot. The metal spoon was in a drawer. It is much cooler than the soup. When the spoon is used to stir the soup, it quickly becomes hot. Heat moves from the soup to the spoon. Heat is the flow of energy from a warmer object to a cooler object.

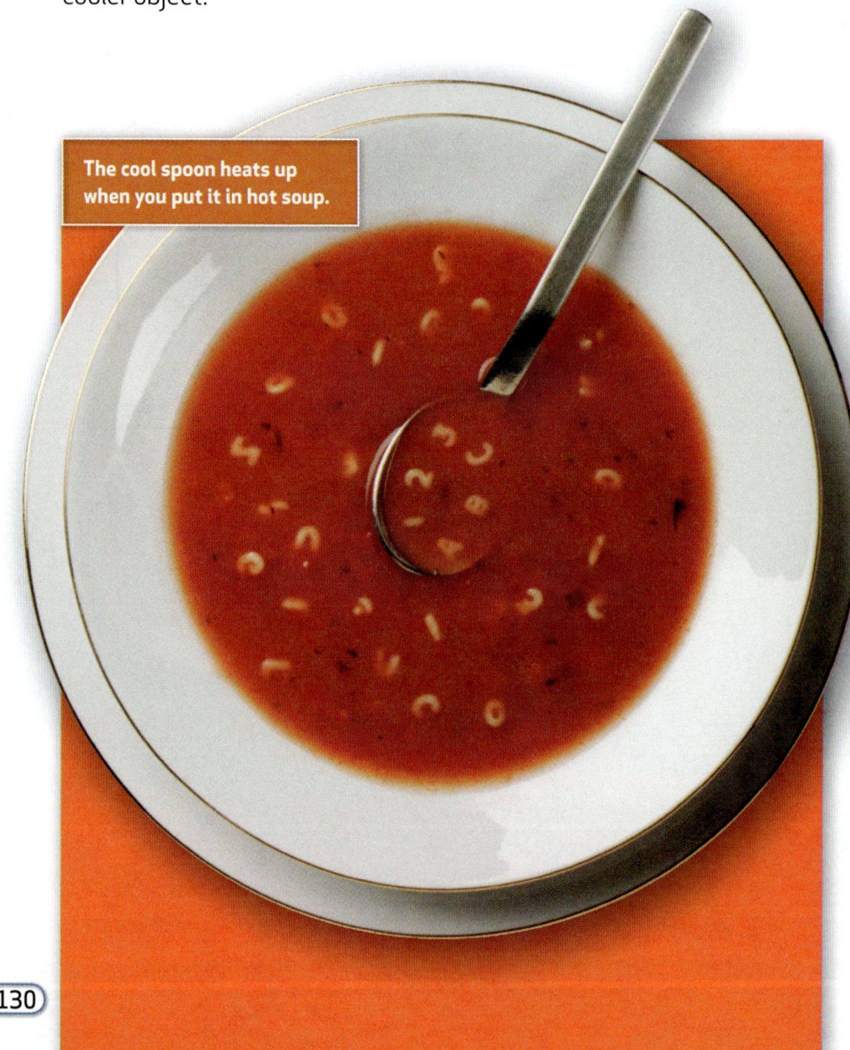

130

Raise Your SciQ!

Heat and Cold Students may think that cold, or coldness, is the opposite of heat. However, heat is a form of energy, whereas cold is not. Cold is simply the absence of heat energy. In the example of the soup and spoon, think about putting the hot spoon in a bowl of cold water. The spoon quickly gets cold, not because coldness went from the water to the spoon, but because heat went from the spoon to the water. The water gets a bit warmer and the spoon cools off.

Heat moves easily though some materials such as metal. The cookies bake on the hot metal cookie sheet. The woman wears cloth oven mitts to take the cookie sheet out of the oven. Heat does not move easily through cloth or plastic.

Differentiated Instruction

ELL **Language Support for Distinguishing Materials that Heat Flows Easily Through**

BEGINNING	INTERMEDIATE	ADVANCED
Help students identify materials through which heat flows easily by answering yes/no questions: **Does heat flow easily through a metal cookie sheet? Does heat flow easily through a cloth mitt? Does heat flow easily through a metal spoon?**	Write these words: *metal, cloth.* Have partners choose the correct word to describe how easily heat flows through these materials. • *Heat flows easily through a _____ cookie sheet.* • *Heat does not flow easily through a _____ mitt.*	Provide Academic Language Stems to help students describe how easily heat flows through materials: • *Heat flows easily through …* • *Heat does not flow easily through …*

Teach, continued

- Ask: **Suppose you're stirring hot soup in a pot. How can you protect your hand from getting burned?** (Possible answers: Use a cloth mitt while you're holding a metal spoon. Use a wooden or plastic spoon that doesn't get hot.)

- Ask: **How are the cookie sheet and the spoon on these pages alike?** (Both are made of metal, and heat flows easily through both the cookie sheet and the spoon.)

- Ask: **Why do we want heat to flow easily from the metal cookie sheet to the cookie dough?** (so the heat bakes the dough into cookies) **Why are the oven mitts made of cloth?** (Heat does not easily flow through cloth, so the cloth mitts protect a person's hands from getting burned.)

Identify Heat Energy

- Ask: **If you let hot soup sit out for too long, what can happen?** (The soup gets cold.)

- Say: **Heat does not disappear.** Ask: **Where does the heat go when hot soup cools?** (into the spoon, bowl, and air)

- Explain: **Since heat moves from warmer things to cooler things, and air is much cooler than hot soup, heat flows into the air until the soup has the same amount of heat as the air.**

- Repeat this process for the picture of the oven and cookies on page 131.

Notetaking 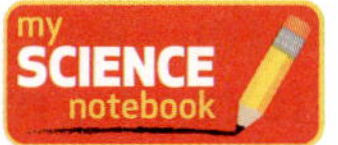

Have students make a T-chart in their science notebook that categorizes materials by how well or poorly heat flows through them. Students can name their chart *Heat Flows Through Materials* and label the left column *Easily* and the right column *Not Easily.*

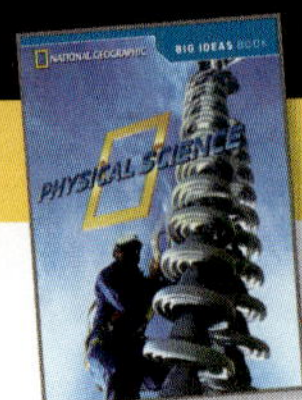

LESSON **6** ▫ Heat

Objectives

Students will be able to:
- Identify heat energy.
- Demonstrate that rubbing objects together produces heat.

Teach, continued

Identify Heat Energy

- Have students read the text and caption on page 132.

- Ask: **What is happening in this photo?** (Two rocks are being rubbed or banged together, producing heat and sparks.) **What kinds of energy can you see are produced?** (heat energy and light energy)

- Ask: **How can you infer from the photo that heat is produced?** (The sparks show that heat is produced.)

- Discuss other ways to produce heat energy. Ask: **How do you feel when you sit near a campfire on a cool evening?** (warm) **What kind of energy does a campfire produce?** (heat energy) Say: **Burning objects, such as the sticks in a campfire, produce heat energy.**

- Say: **Some chemicals mixed together can produce heat as well.** If possible, demonstrate adding a seltzer tablet to water. Say: **You can see the bubbles this tablet produced. It also gives off some heat. You can't feel the heat from the glass, but it can be measured with instruments.**

❯ Monitor and Fix Up

Ask students if they understand that heat is produced when the two rocks are rubbed together. Discuss possible fix-up strategies. For example, students may do the Science in a Snap activity on page 133, titled *Warm Hands*.

Science in a Snap!

Warm Hands 

Materials your hands, timer

Have students read and follow the instructions on page 133 of the Big Ideas Book. Students should record their observations and results in their science notebook.

What to Expect Before doing the activity, students may describe their hands as being cool or at room temperature. After the activity, the temperature of students' hands should have increased because the rubbing motion generates heat to warm their hands.

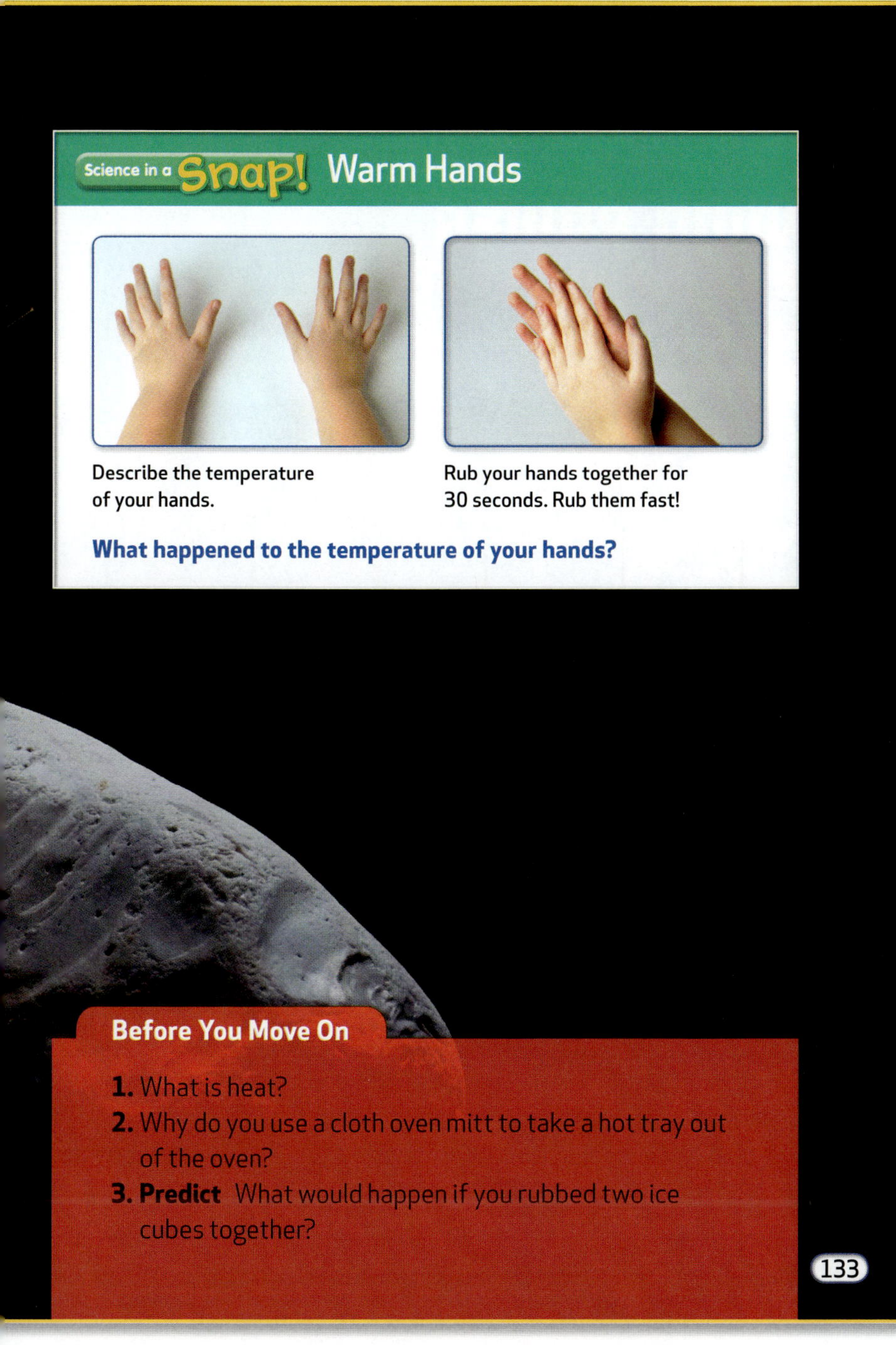

Before You Move On

1. What is heat?

2. Why do you use a cloth oven mitt to take a hot tray out of the oven?

3. Predict What would happen if you rubbed two ice cubes together?

133

Teach, continued

- Say: **Other forms of energy are involved when the rocks are struck together. They include mechanical energy, sound, and light.** Ask: **What kind of energy allowed the rocks to move toward each other and rub against each other?** (mechanical energy) **What kind of energy do the glowing sparks give off?** (heat and light energy) **What other kind of energy do you think this makes?** (sound energy)

- Have students do the Science In a Snap activity on page 133.

- After students have completed the Science In a Snap activity, ask: **Why did your hands get warmer when you rubbed them together?** (Some of the mechanical energy was changed to heat energy.) **What other kind of energy was produced when you rubbed your hands together?** (sound)

❸ Assess

» Before You Move On

1. **Define** **What is heat?** (Heat is the flow of energy from a warmer object to a cooler object.)

2. **Explain** **Why do you use a cloth oven mitt to take a hot tray out of the oven?** (Cloth is a material through which heat does not flow easily, so cloth protects your hands from being burned by the heat of the tray.)

3. **Predict** **What would happen if you rubbed two ice cubes together?** (The ice cubes would warm up and start to melt.)

Quick Questions Ask students the following questions:

- **What happened to the temperature of your hands when you rubbed them together?** (My hands got warmer.)

- **What do you think would happen if you rubbed your hands more slowly?** (They would not get as warm.) **More quickly?** (They would get even warmer than before.)

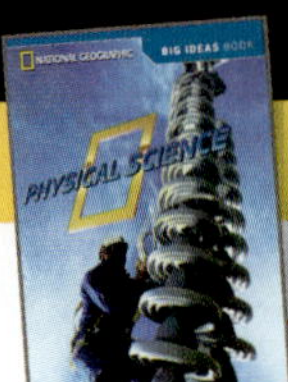

LESSON 7 ▫ Wind Power: Energy in the Air

Objectives
Students will be able to:
- Identify wind as a source of mechanical energy.

PROGRAM RESOURCES
- Learning Masters Book, page 219, or at 🌐 **myNGconnect.com**

❶ Introduce

Tap Prior Knowledge
- Have students think about the power of the wind. Have a volunteer relate an experience they have had with powerful wind.
- Ask: **How do people use wind power?** (Possible answers: moving sail boats, wind surfing, flying kites, powering windmills)

Preview and Read
- Read the heading and preview the pictures on pages 134–135.
- Have students read pages 134–135.

❷ Teach

Text Feature: Photos and Captions
- Point out the Egyptian drawing on page 134, and read aloud the caption. Ask: **How are these ancient Egyptian people using wind power?** (to help move their ship)
- Point out the windmills in the background photo and read aloud its caption. Ask: **What was the purpose of these windmills?** (to grind grain) **According to the caption, do people still use these windmills?** (no) **How can you see that people don't use these windmills anymore by looking at the photo?** (There aren't any blades to capture the wind.)

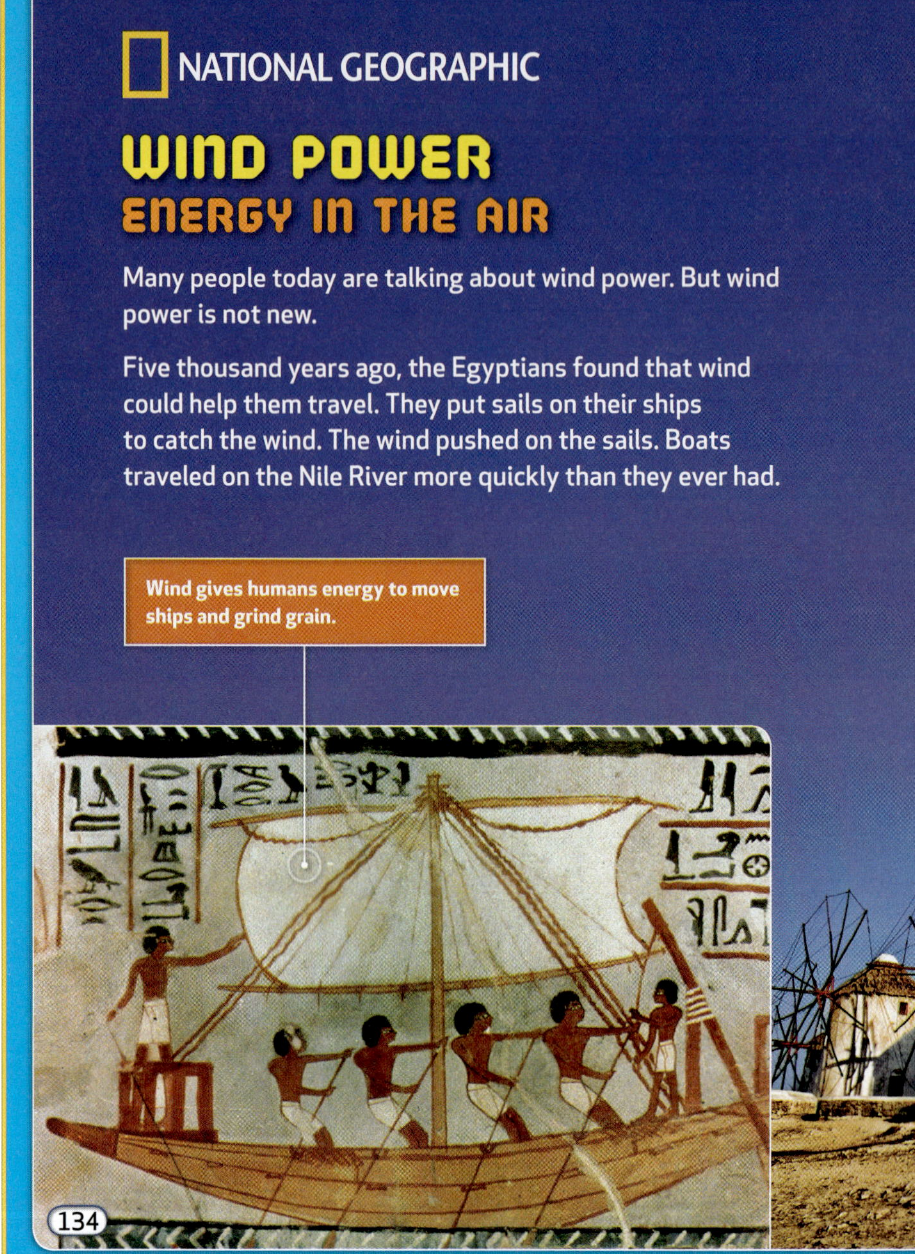

Social Studies in Science

Wind Farms Ask students if they have ever seen a wind farm. Explain that a wind farm is a group of wind turbines located in the same area to produce electricity. Point out the wind farm in the inset photo on page 135. Explain that a wind turbine is a tower with two or three blades at the top. When the wind turns the blades, a rod, or shaft, spins. The shaft is connected to a generator that makes electricity. Wind is a renewable resource. Concerns about climate change and the nonrenewable nature of carbon-based fuels has resulted in the recent construction of many wind farms around the world.

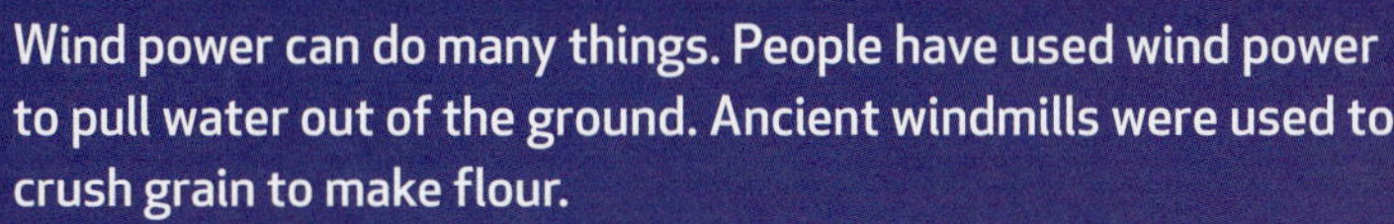

Wind power can do many things. People have used wind power to pull water out of the ground. Ancient windmills were used to crush grain to make flour.

Today windmills change the motion of wind into other kinds of energy such as electricity. The electricity made by windmills is used to power homes, office buildings, and schools. People will continue to use wind power because it is a clean, renewable source of energy.

135

Teach, continued

- Explain: **When the arms of these windmills had blades, the blades gave the wind a surface to push against. The wind pushing on the blades turned the arms. The arms of each windmill were connected to a large, stone grinding wheel inside. When the arms turned, the grinding wheel moved. As the grinding wheel moved, it crushed grain into flour.**

Identify Wind as a Source of Electrical Energy

- Have students compare and contrast the photo of the old windmills with the inset photo of the modern wind farm.

- Ask: **How are these two groups of windmills alike?** (They both use the energy of wind to do work.) **How are they different?** (The wind farm changes wind energy into electrical energy. The old windmills changed wind energy into mechanical energy to crush grain into flour.)

- Say: **Many wind farms are built to produce electricity.** Ask: **What is the source of energy on a wind farm?** (wind, or moving air) Say: **Wind is a renewable source of energy. People won't run out of wind.**

❸ Assess

1. **Identify** How do people use wind as a source of energy? (Wind is used to move sailboats and produce electricity.)

2. **Contrast** How are the windmills on wind farms used differently from ancient windmills? (Ancient windmills used wind to do things like grind grain. Today's wind farms use windmills to make electricity.)

3. **Infer** Why is wind power important today? (Wind is a renewable source of energy. It is clean and it won't run out like some other sources of energy will.)

Share and Compare

What Is Energy?

Give students the Learning Master. Have them list on the chart the different types of energy explored in Chapter 4. Then have them describe each type of energy and list examples of how people use each type. Check their work. Then have them write a summary statement about how people use energy. Have partners discuss ways that energy improves their lives.

Learning Master 219 at 🌐 myNGconnect.com

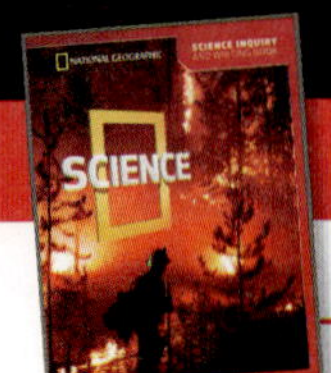

Objectives

Students will be able to:

- Investigate through Guided Inquiry (answer a question; make and compare observations; collect and record data and observations; generate explanations and conclusions based on evidence; share findings; ask questions based on observations to increase understanding; adjust explanations based on findings and new ideas).
- Identify sound as a form of energy and relate fast or slow vibrations to pitch.
- Conduct an investigation using appropriate tools and instruments.
- Share findings with others and actively seek their interpretations and ideas.

Science Process Vocabulary

observe, infer

PROGRAM RESOURCES

- Science Inquiry and Writing Book: *Physical Science*
- Science Inquiry and Writing Book **eEdition** at **myNGconnect.com**
- **Inquiry eHelp** at **myNGconnect.com**
- Science Inquiry Kit: *Physical Science*
- Learning Masters Book, pages 220–222, or at **myNGconnect.com**
- Inquiry Rubric: Assessment Handbook, page 214, or at **myNGconnect.com**
- Inquiry Self-Reflection: Assessment Handbook, page 224, or at **myNGconnect.com**

MATERIALS

Kit materials are listed in italics.

2 tuning forks; ruler; *clear plastic cup (9 oz); plastic wrap; salt (½ tsp); rubber band; eraser*

❶ Introduce

Tap Prior Knowledge

- **How do you think sound is created when you play the piano, or the violin, or a guitar?** (The strings are plucked or struck, and they make sounds with different pitches.) Discuss how musical instruments make sounds by producing different vibrations.

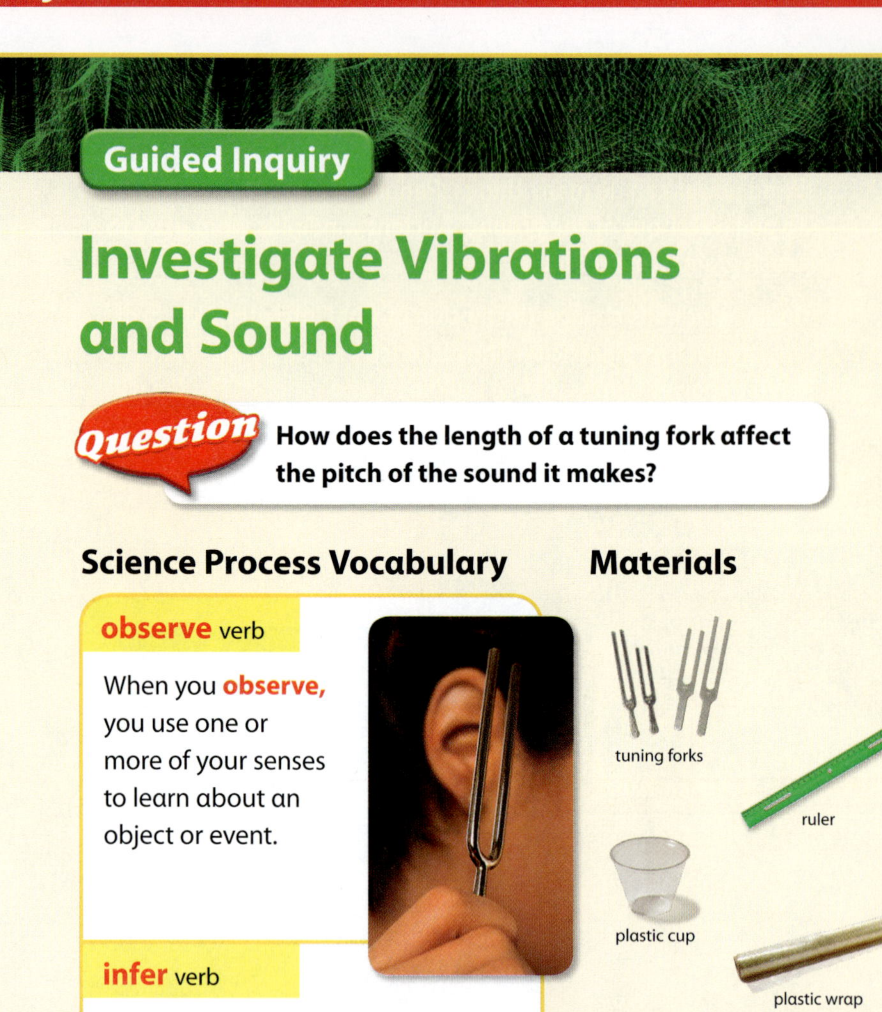

MANAGING THE INVESTIGATION

Time

 30 minutes

Groups

Small groups of 4

Advance Preparation

- Measure ½ tsp salt for each group.

What to Expect

- Students should observe that the tuning forks vibrate and cause the salt on the plastic wrap to move. They should understand that the vibrations of the tuning forks also cause the sounds they hear. Students should observe that the shorter tuning fork makes a higher-pitched sound than the longer tuning fork.

What to Do

1. Choose a set of tuning forks to test. **Measure** the length of the tines on each tuning fork. Record your measurements in your science notebook.

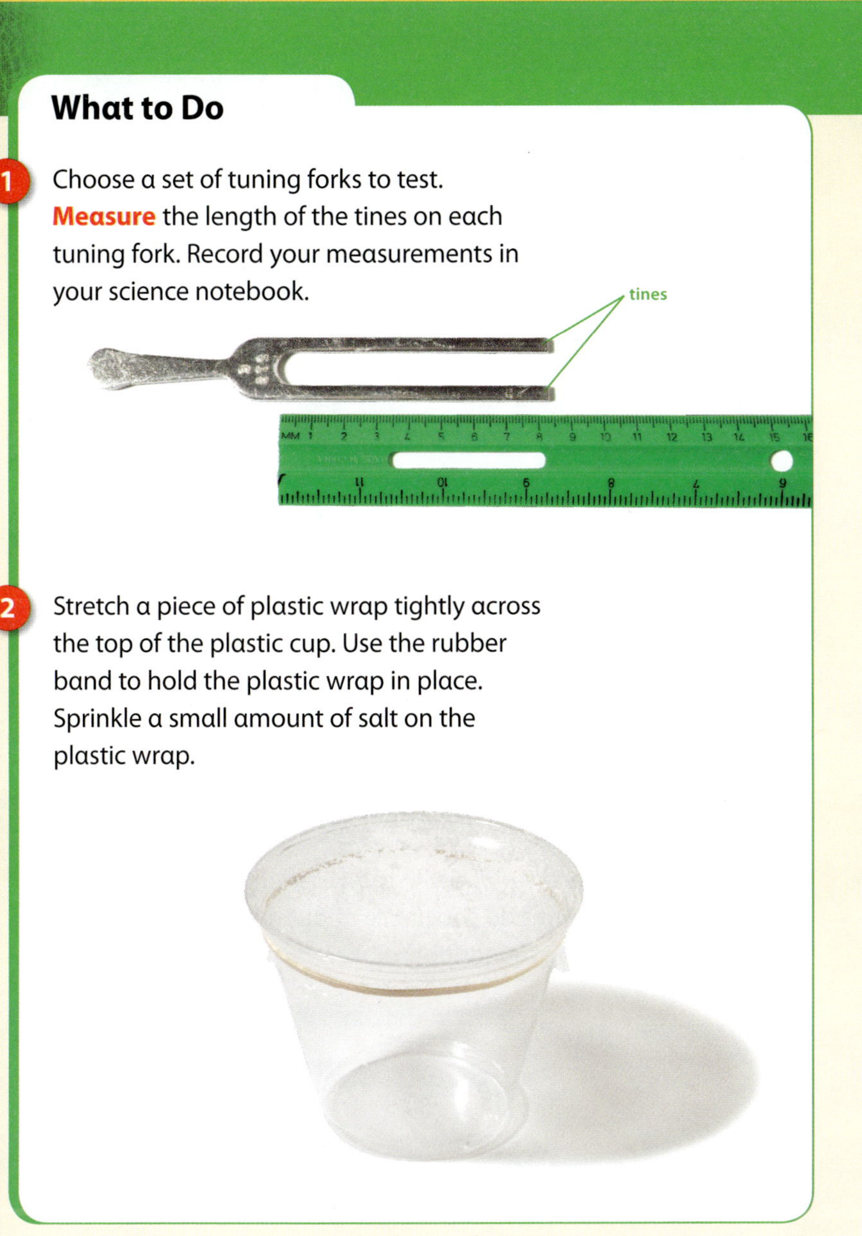

2. Stretch a piece of plastic wrap tightly across the top of the plastic cup. Use the rubber band to hold the plastic wrap in place. Sprinkle a small amount of salt on the plastic wrap.

199

Teaching Tips

- Make sure that each group chooses two tuning forks with different lengths. Groups can share the tuning forks if necessary.

- Model striking the tuning forks with the same amount of force each time. Stress that students must be careful not to use too much force or the tuning forks or plastic wrap might break.

- If necessary, help students stretch the plastic wrap across the cup.

- Tell students to measure the lengths of the tuning forks to the nearest centimeter.

- Students may need to repeat trials for comparison to determine which tuning fork produced a higher pitch.

Introduce, continued

Connect to the Big Idea

- Review the Big Idea Question, *What is energy?* Explain to students that they will conduct an investigation about sound energy and how it is related to vibrations.

- Have students open their Science Inquiry and Writing Books to page 198. Read the Question and ask how the length of a tuning fork might affect the pitch of the sound it makes when it vibrates.

❷ Build Vocabulary

Science Process Words: observe, infer

Use this routine to introduce the words.

1. **Pronounce the Word** Say **observe**. Have students repeat it in syllables.

2. **Explain Its Meaning** Choral read the sentence. Ask students for another word or phrase that means **observe**. (to look, smell, taste, listen, or touch to learn about something)

3. **Encourage Elaboration** Ask: **How could you observe a string of an instrument such as a guitar?** (touch the string, pluck the string, look for vibrations, listen for sound)

   ELL Vocabulary Support

   Write *I can* **observe** _____ *with my* _____. Point to your ears and ask: **What can you observe with your ears?** Repeat with the other senses.

Repeat for the word **infer**. To encourage elaboration, ask: **If you hear a large crash, what could you infer?** (Possible answer: Something fell.)

❸ Guide the Investigation

- Distribute materials. Read the steps on pages 199–200. Clarify steps if necessary.

- Preview the investigation by showing students the tuning forks. Point out the tines and the handle. Demonstrate how the tuning fork is used and have students listen to the sound. Explain that tuning forks are used to tune musical instruments, such as pianos.

- Ask: **Why do you think you need to stretch the plastic wrap tightly across the top of the cup?** (If it is not tight, it will sag, and the salt will collect together.)

Guide the Investigation, continued

- At step 3, ask: **Can you see the tuning fork vibrate?** It might look blurry, or students may not be able to see it vibrate. Have students observe the salt as they repeat the step. Ask: **How does the salt on the plastic wrap help you know that the tuning fork is vibrating?** (The salt bounces on the plastic wrap.)

- Help students make the connection that sounds are caused by vibrations. Ask: **Why did you hear a sound when you struck the tuning fork?** (because the tuning fork was vibrating)

- After step 4, tell students to touch the tuning fork to feel its vibrations. Ask: **What happens to the sound when you touch the tuning fork?** (The sound stops.)

> ⚠ **SafetyFirst**
>
> **Students should not strike the tuning forks anywhere except on the eraser.**

❹ Explain and Conclude

- Make sure students record their data accurately and precisely. Encourage students to share observations and seek the interpretations and ideas of others. Ask: **How are your observations similar and different?**

- Guide students as they find patterns in their data. Ask: **How is the length of a tuning fork related to the sound it makes?** (A shorter tuning fork makes a higher sound.)

- Ask students to make connections between the vibrations and the pitch. Ask: **Which tuning fork produced a higher pitch?** (shorter tuning fork) **Does a shorter tuning fork vibrate faster or slower than a longer tuning fork?** (faster)

- As students present their findings, encourage them to summarize their inferences in complete sentences and use their analyzed data as evidence.

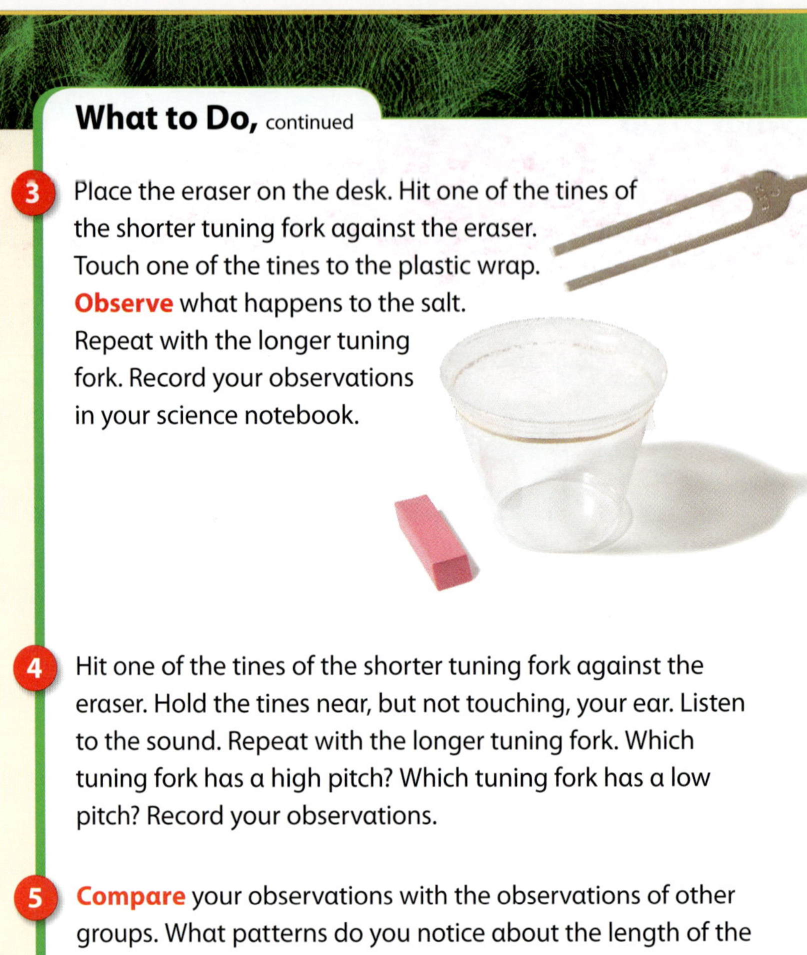

What to Do, continued

3 Place the eraser on the desk. Hit one of the tines of the shorter tuning fork against the eraser. Touch one of the tines to the plastic wrap. **Observe** what happens to the salt. Repeat with the longer tuning fork. Record your observations in your science notebook.

4 Hit one of the tines of the shorter tuning fork against the eraser. Hold the tines near, but not touching, your ear. Listen to the sound. Repeat with the longer tuning fork. Which tuning fork has a high pitch? Which tuning fork has a low pitch? Record your observations.

5 **Compare** your observations with the observations of other groups. What patterns do you notice about the length of the tuning forks and the sounds they make?

200

> **NATIONAL GEOGRAPHIC** — **Raise Your SciQ!**
>
> **Musical Tuning** Musical instruments are tuned to specific frequencies to match the notes in the scale. For example, the note A has a frequency of 440 hertz, B has a frequency of 494 hertz, and so on. Sharps and flats are used to indicate the frequencies between the notes. When playing in an ensemble, musicians must tune their instruments to make sure they are playing at a frequency that matches the rest of the group.

Record

Write in your science notebook.
Use a table like this one.

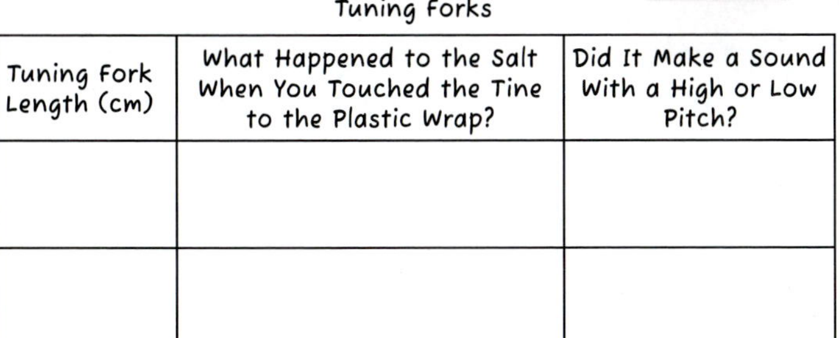

	Tuning Forks	
Tuning Fork Length (cm)	What Happened to the Salt When You Touched the Tine to the Plastic Wrap?	Did It Make a Sound With a High or Low Pitch?

Explain and Conclude

1. What happened to the salt when you touched the tuning forks to the plastic wrap? Explain why you think this happened.
2. **Compare** the sounds made by the tuning forks. How did the length of the tuning fork affect the pitch of the sound it made?
3. When an object vibrates faster, it makes a sound with a higher pitch than an object that is vibrating more slowly. Use your **observations** to **infer** which of your tuning forks vibrated faster.

Think of Another Question

What else would you like to find out about vibrations and sound? How could you find an answer to this new question?

This musician is using a tuning fork to tune the piano.

201

Inquiry Rubric at 🔗 myNGconnect.com	Scale			
The student **measured** tuning forks and **observed** the effects of their vibrations.	4	3	2	1
The student recorded **data** and observations in a table.	4	3	2	1
The student **compared** the sounds made by the tuning forks.	4	3	2	1
The student **inferred** which tuning fork vibrated more quickly after observing its pitch.	4	3	2	1
The student **shared** and compared observations and **conclusions** with others.	4	3	2	1
Overall Score	4	3	2	1

Explain and Conclude, continued
Answers

1. The salt moved when I touched the tuning fork to the plastic wrap. I think this happened because the tuning fork made the plastic vibrate and caused the salt to move.
2. The longer tuning fork made a sound with a lower pitch than the shorter tuning fork.
3. Possible answer: I infer that the shorter tuning fork vibrated more quickly than the longer tuning fork. The shorter tuning fork made a sound with a higher pitch.

❺ Find Out More

Think of Another Question

- Students should use observations made in this investigation to generate other questions that could be studied using readily available materials. Record student questions and discuss how to find answers.

❻ Reflect and Assess

- To assess student work with the Inquiry Rubric below, see Assessment Handbook, page 214, or go online at 🔗 myNGconnect.com
- Have students use the Inquiry Self-Reflection on Assessment Handbook, page 224, or at 🔗 myNGconnect.com

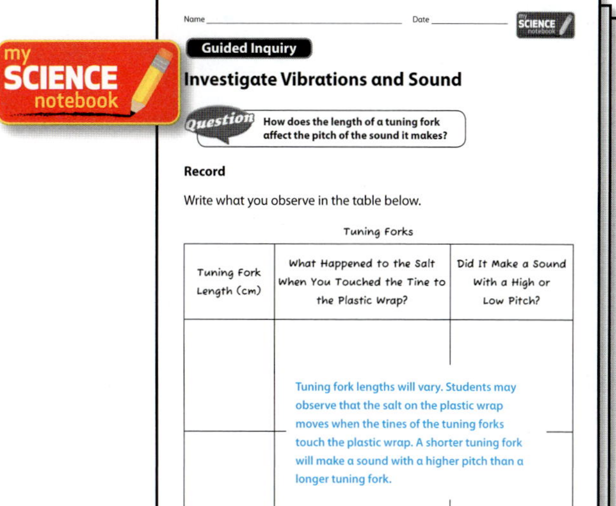

**Learning Masters 220–222
or at 🔗 myNGconnect.com**

PROGRAM RESOURCES

- Chapter 4 Test, Assessment Handbook, pages 102–104, or at 🌐 **myNGconnect.com**
- NGSP ExamView CD-ROM

❶ Sum Up the Big Idea

- Display the chart from page T112–T113. Read the Big Idea Question. Then add new information to the chart.

- Ask: **What did you find out in each section? Is this what you expected to find?**

> **What is Energy?**
>
> **Energy and Work**
> **Sofia:** Energy is the ability to do work or cause a change.
>
> **Mechanical Energy**
> **Dylan:** Mechanical energy is stored energy plus energy of motion.
>
> **Sound**
> **Ella:** Sound is energy that comes from vibrations. It is energy that you can hear.
>
> **Electrical Energy**
> **Norberto:** Electrical energy is energy that flows through wires.
>
> **Heat**
> **Alexa:** Heat is the flow of energy from a warmer object to a cooler object.

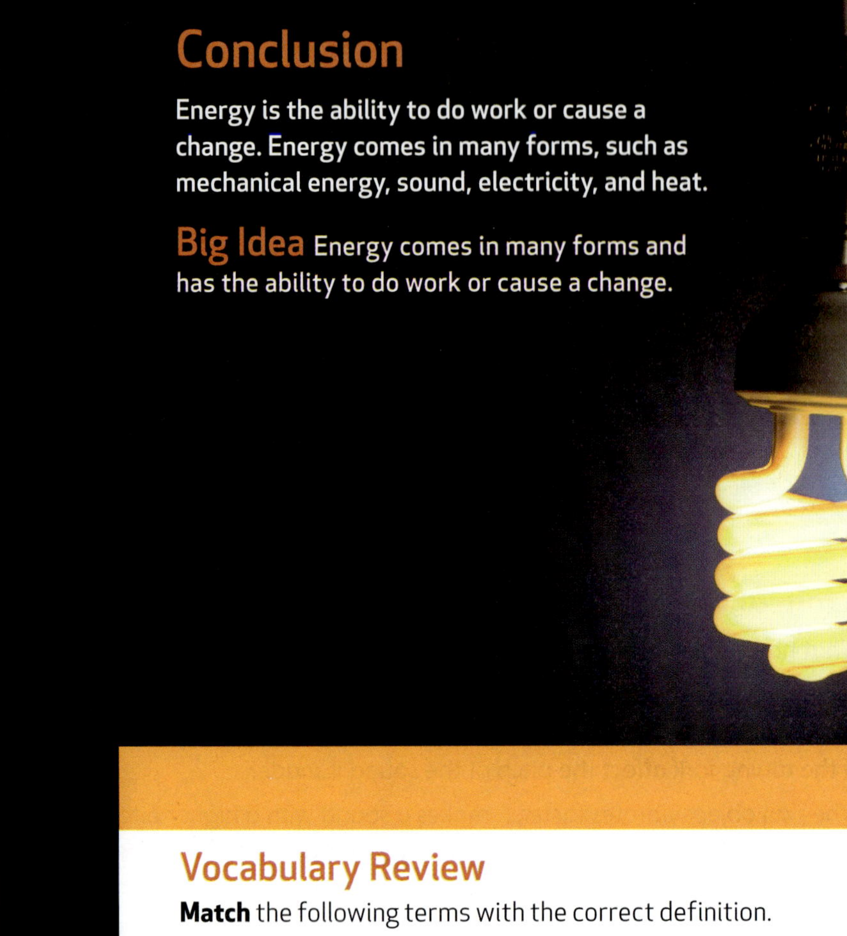

Conclusion

Energy is the ability to do work or cause a change. Energy comes in many forms, such as mechanical energy, sound, electricity, and heat.

Big Idea Energy comes in many forms and has the ability to do work or cause a change.

Vocabulary Review

Match the following terms with the correct definition.

A. energy
B. sound
C. heat
D. electricity
E. mechanical energy

1. The flow of energy from a warmer object to a cooler object.
2. The ability to do work or cause a change
3. An object's stored energy plus its energy of motion
4. Energy that can be heard
5. Energy that flows through wires

136

Review Academic Vocabulary

Academic Vocabulary Tell students that scientists use special terms and words when describing their work to others. For example:

energy   sound   mechanical energy

electricity   heat

Have students write the list in their science notebook and use each word in a sentence that tells how energy does work and causes change. Then have them share their sentences with a partner.

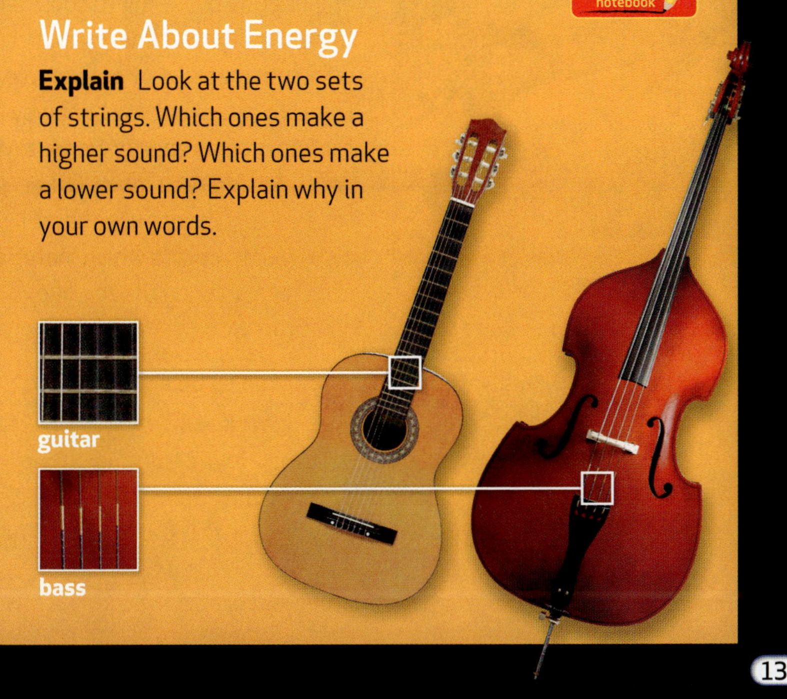

Big Idea Review

1. **Recall** What is energy?

2. **Describe** Use your own words to describe stored energy and energy of motion.

3. **Explain** Explain why sound is a form of energy.

4. **Compare** How do a fan and a heater use electricity?

5. **Predict** What happens when you rub two objects together?

6. **Apply** Name one way that electricity makes your life easier and explain why.

Write About Energy

Explain Look at the two sets of strings. Which ones make a higher sound? Which ones make a lower sound? Explain why in your own words.

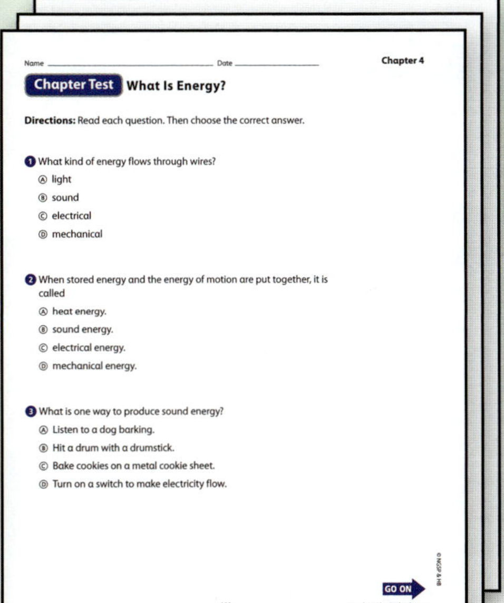

Assess Student Progress Chapter 4 Test

Have students complete their Chapter 4 Test to assess their progress in this chapter.

Chapter 4 Test, Assessment Handbook, pages 102–104, or at ⊘ **myNGconnect.com**

or NGSP ExamView CD-ROM

❷ Discuss the Big Idea

Notetaking

Have students write in their science notebook to show what they know about the Big Idea. Have them:

1. Write about what energy is.

2. Explain how the different forms of energy can be produced.

3. Explain how the different forms of energy affect them every day.

❸ Assess the Big Idea

Vocabulary Review

1. C   2. A   3. E   4. B   5. D

Big Idea Review

1. Energy is the ability to do work or cause a change.

2. Possible answer: An object has stored energy when it is not moving. An object has energy of motion when it is moving.

3. Sound is caused by vibrations that cause air to move. When an object vibrates and the air moves, work has been done. So sound is a form of energy.

4. A fan turns electricity into mechanical energy. A heater turns electricity into heat energy.

5. When you rub two objects together, heat is produced.

6. Possible answer: I use an electric lamp. It is safer than a candle. The lamp makes it easier to see when I am doing my homework.

Write About Energy

The thick strings on the bass make a lower sound. The thin strings on the guitar make a higher sound. The thick strings vibrate slowly. The thin strings vibrate faster.

LESSON 10 ▫ Physical Science Expert

Objectives

Students will be able to:

- Identify scientists' contributions to science and technology by using and exploring energy.

PROGRAM RESOURCES

- Big Ideas Book: *Physical Science*
- Big Ideas Book: *Physical Science* **eEdition** at ⊘ **myNGconnect.com**
- **Digital Library** at ⊘ **myNGconnect.com**

❶ Introduce

Tap Prior Knowledge

- Ask: **What kinds of jobs can you think of that involve energy?** (Possible answers: building houses, driving a truck, using a computer; almost all jobs involve energy of some sort)

Preview and Read

- Read the heading. Have students preview pages 138–139. Say: **This is an interview with a product designer.** Ask: **What does the letter Q represent?** (The interviewer's questions.)
- Have students read pages 138–139.

❷ Teach

Analyze Word Parts

- Write *design* and *designer* on the board. Say: **A *design* is a type of outline or drawing of something before it is built, which shows how it is supposed to look or work.**
- Underline the suffix *-er*. Say: **-er at the end of a word means "a person or thing that does what that word is about." So, a *designer* is someone who makes designs.**

Identify How Designers Can Use Science and Energy

- Ask: **What does Judy Lee do?** (designs products
- Ask: **How does she use physical science in her work?** (Possible answer: to figure out which materials work best, how something will best use energy, and which forms of energy it will use)

Product designers make all sorts of things that use energy including toys! Judy Lee uses her creativity and knowledge of energy to design new toys for kids of all ages.

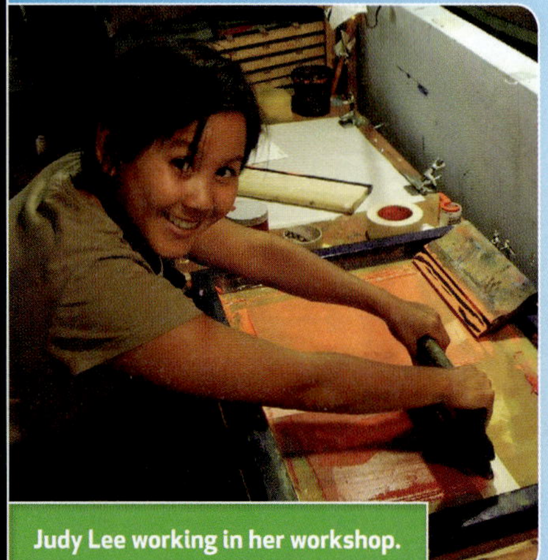

Judy Lee working in her workshop.

Q: What is your job?

I'm a product designer. I have designed toys, pet products, and even food. Product Design involves engineering by having to consider things like type of energy and different materials. For example, I have to figure out the best way for a toy to use energy. Some toys need electricity to move. Other toys you move with your hand. I also have to decide which material is best for making the toy, like plastic or wood.

Q: When did you first know you wanted to be a product designer?

When I was a kid, I loved to build things. I loved figuring out how things worked. Then, I would start thinking about how I could make things better. I still think the same way today. I guess that's why I'm a Product Designer.

(138)

NATIONAL GEOGRAPHIC Raise Your SciQ!

Designing Video Games Video games are virtual toys that have become a multibillion dollar industry. Students who really like video games, play them as much as possible, have good problem-solving skills, and are capable of working at a very fast pace may want to explore this relatively new industry for a possible career. Some colleges now offer degrees in game design and programming.

Q: What is a typical day like for you?

Each day is different. When I start a project, I talk to lots of different people. I'll take their ideas and create something, such as a toy. I'll take the toy back to them and hear what they think. Then, I'll use their feedback to make the toy better. I spend a lot of time talking with people and making things.

Q: What is your favorite thing about being a designer?

I love solving problems. I love building things that use energy in different ways. It's fun to make new things.

Judy Lee builds a new toy.

139

- Ask: **What are the different forms of energy used by some of the products she works on?** (mechanical energy, electrical energy)

Technology in Physical Science

- Have students study the photo and caption at the bottom of page 139. Ask: **What example of technology do you see?** (a tool helping to make toys) **What form of energy might this tool use?** (Possible answer: electricity)

Find Out More

- Ask: **Is this a career you would find interesting? Why or why not?** (Accept all answers, positive or negative.)

- Encourage interested students to research what other things product designers create. Also have them find out about the kinds of technology and tools product designers use.

 **Digital Library**

myNGconnect.com

Have students use the **Digital Library** to research other energy-related careers.

 **Integrated Technology**

Computer Presentation Students can use the information to make a computer presentation about how different jobs use energy.

❸ Assess

1. **Recall** What does a product designer do? (creates toys, pet products, and other things)

2. **Explain** How does this product designer do her work? (She talks to many people to get ideas. Then she creates something and takes it back to the people to see what they think. Then she makes changes to make the creation better.)

3. **Draw Conclusions** What would you have to enjoy doing to be a product designer? (Possible answers: building things, learning how things work, talking to people, solving problems, improving things until they work well)

Differentiated Instruction

ELL Language Support for Identifying How Scientists Use Energy

BEGINNING	INTERMEDIATE	ADVANCED
Help students identify how product designers use energy to make toys move. Have them answer yes/no questions about energy and toys: **Do some toys use electrical energy to move? Do some toys need mechanical energy to move?**	Provide Academic Language Frames to help students identify how toy designers use energy to make toys move. • *Some toys use ______ energy to move.* • *Other toys use ______ energy to move.*	Provide Academic Language Frames to help students elaborate on how toy designers use energy to make toys move. • *A toy that you move with your hand uses ______ energy to move.* • *A toy that moves without a push uses ______ energy.*

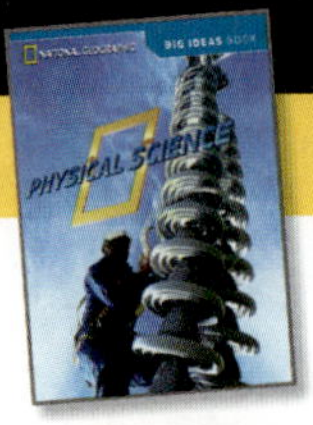

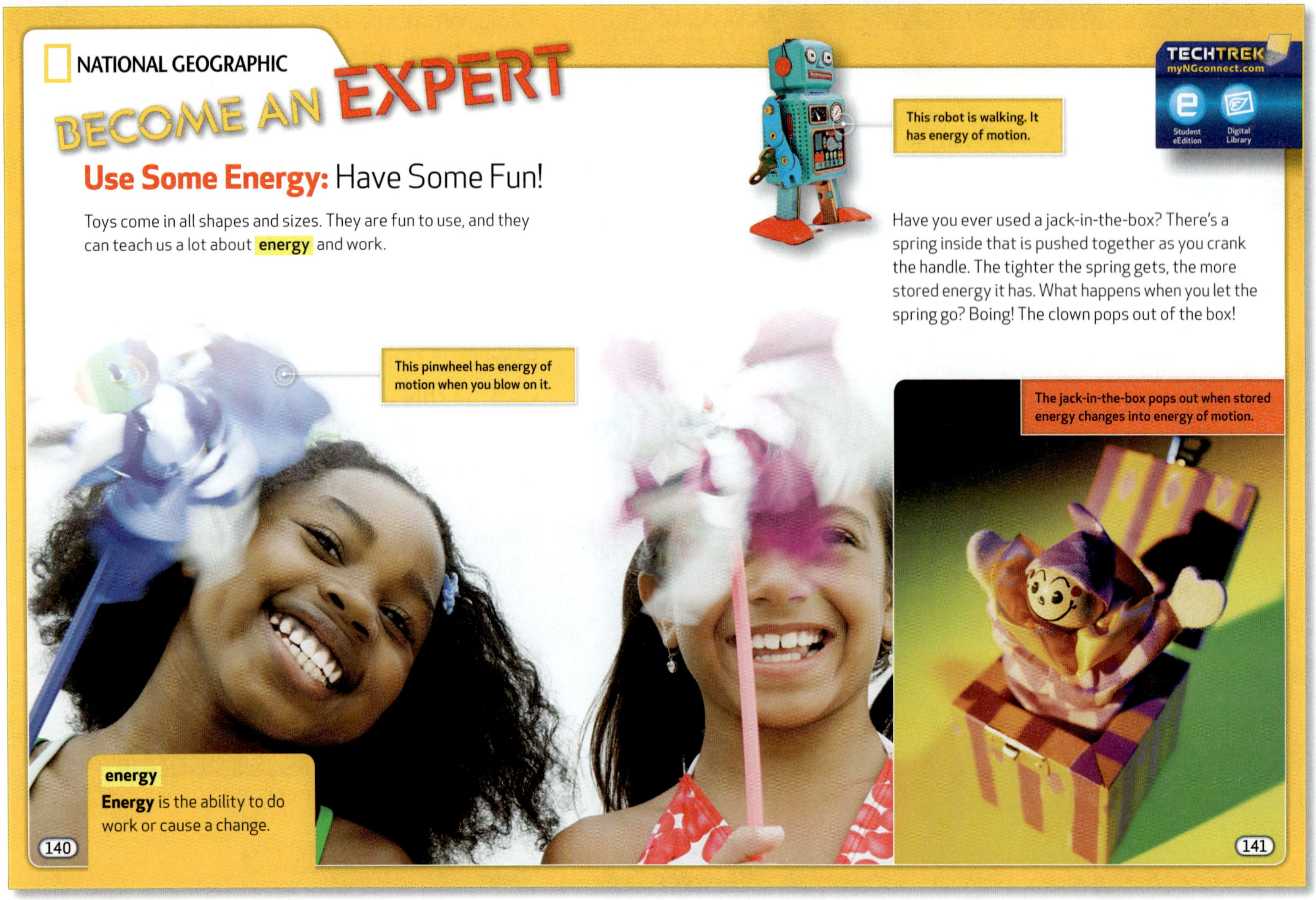

NATIONAL GEOGRAPHIC
BECOME AN EXPERT

Use Some Energy: Have Some Fun!

Toys come in all shapes and sizes. They are fun to use, and they can teach us a lot about **energy** and work.

This robot is walking. It has energy of motion.

Have you ever used a jack-in-the-box? There's a spring inside that is pushed together as you crank the handle. The tighter the spring gets, the more stored energy it has. What happens when you let the spring go? Boing! The clown pops out of the box!

This pinwheel has energy of motion when you blow on it.

The jack-in-the-box pops out when stored energy changes into energy of motion.

energy
Energy is the ability to do work or cause a change.

140

141

PROGRAM RESOURCES

- Big Ideas Book: *Physical Science*
- Big Ideas Book: *Physical Science* **eEdition** at **myNGconnect.com**
- **Digital Library** at **myNGconnect.com**

Access Science Content

Describe Energy Used by Toys

- Have students read pages 140–141 and study the photos and captions. Ask: **What kind of energy does the spinning pinwheel have?** (energy of motion) **What force causes the pinwheel to do work?** (the push of moving air; wind) **When does the robot have stored energy?** (before it starts to move) **When does the jack-in-the-box have stored energy?** (before it pops out of the box)

❯ Make Inferences

Ask: **Why are people surprised when the jack-in-the-box pops out of the box?** (Help students understand that the change from stored energy to energy of motion happens suddenly.) **Why does a robot stop walking after awhile?** (All of its stored energy has become energy of motion.)

Notetaking

Have students make a 3-column chart in their science notebook. Have them list each toy pictured on pages 140–141, the kind of **energy** the toy uses to move, and what force changes the stored energy into the energy of motion. Then have them complete this sentence: *These toys use _____ **energy** to move.* (mechanical)

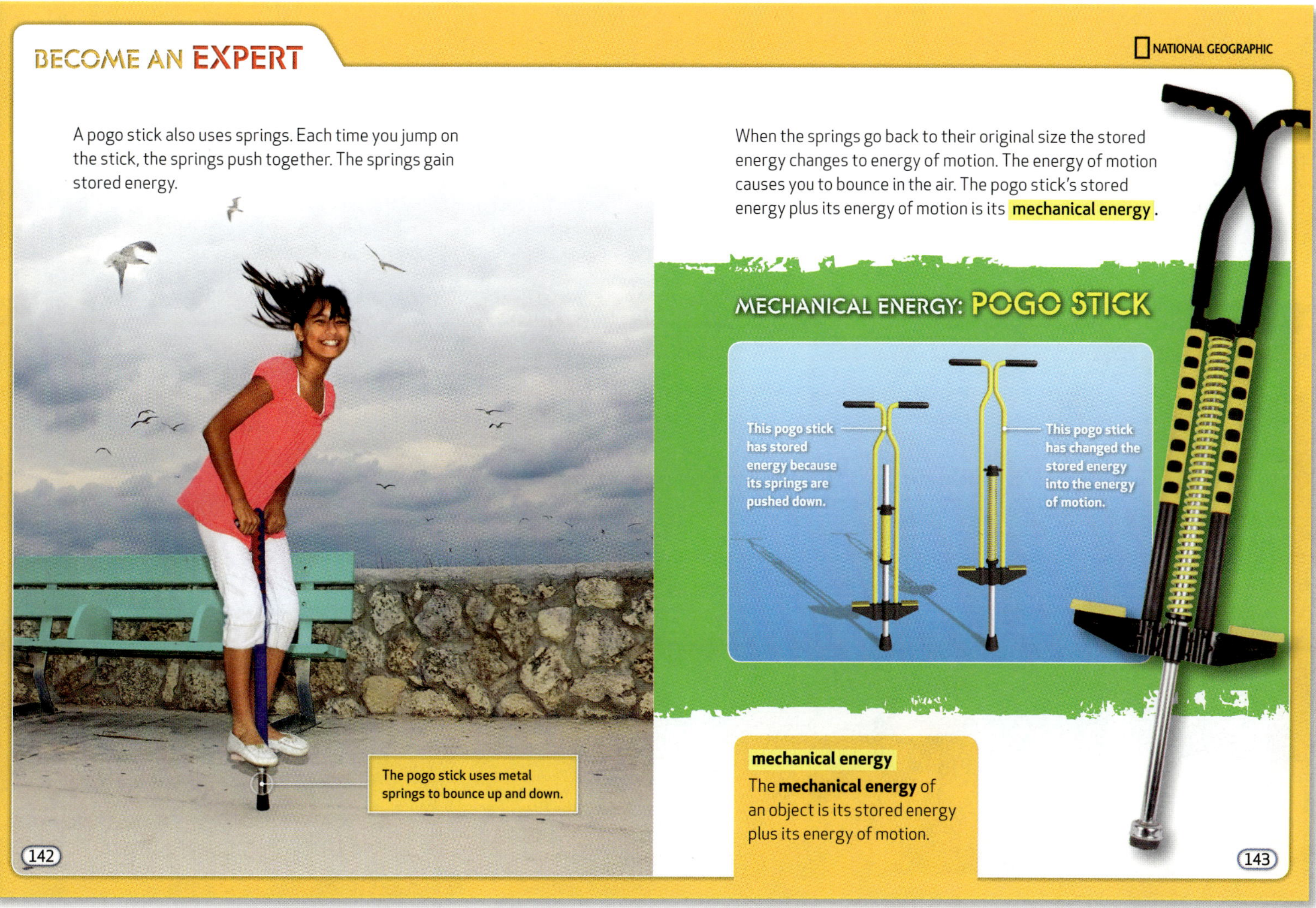

Access Science Content

Identify Mechanical Energy

- Have students read pages 142–143 and study the pogo stick on page 143. Ask: **What kind of energy is the pogo stick using?** (mechanical energy) **When does the pogo stick have the most stored energy?** (When the spring is pushed in, or down.)

ELL Language Support for Describing Mechanical Energy

BEGINNING

Help students identify mechanical energy by answering yes/no questions: **Does the pogo stick have mechanical energy when the spring is down? Does it have mechanical energy when the spring is up?**

INTERMEDIATE

Provide Academic Language Frames to help students describe mechanical energy.

- *The pogo stick has _____ energy when the spring is down.*
- *The pogo stick has _____ energy when the spring is up.*

ADVANCED

Provide Academic Language Stems to help students describe mechanical energy:

- *The pogo stick has . . .*
- *Mechanical energy is . . .*

Assess

1. **Explain What happens to the springs on a pogo stick?** (When a person jumps on the pogo stick, the springs are pushed together and gain stored energy. Then the springs go back to their original size, and the stored energy changes to energy of motion.)

2. **Apply Think of another toy that uses mechanical energy. What causes its stored energy to change to energy of motion?** (Possible answers: a top that spins when a person pulls a string to start its motion; a bicycle that moves forward when a person pushes the pedals to make the gears move; a toy that moves forward when a person puts it on a ramp and it rolls downhill)

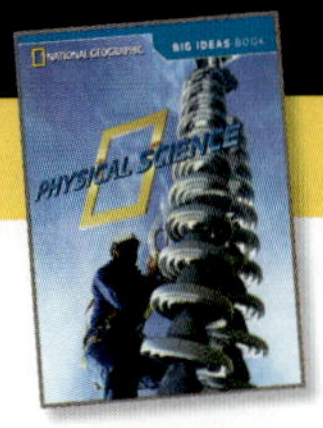

NATIONAL GEOGRAPHIC

BECOME AN EXPERT

There are many toys that use electricity. Electric train sets are fun to set up. You can hook the tracks together in a design. When you're finished, the train will move along the track. Electricity flows through wires to the tracks. Electricity causes the train to move.

Some toys use heat. Heat is the flow of energy from a warmer object to a cooler object. This toy oven heats up and bakes little cakes. The heat from the warm oven flows to the cooler cake batter. The heat from the oven cooks the batter. When you take the cakes out of the oven, the heat flows from the cakes into the air. This is how they cool off. Now they are ready to eat.

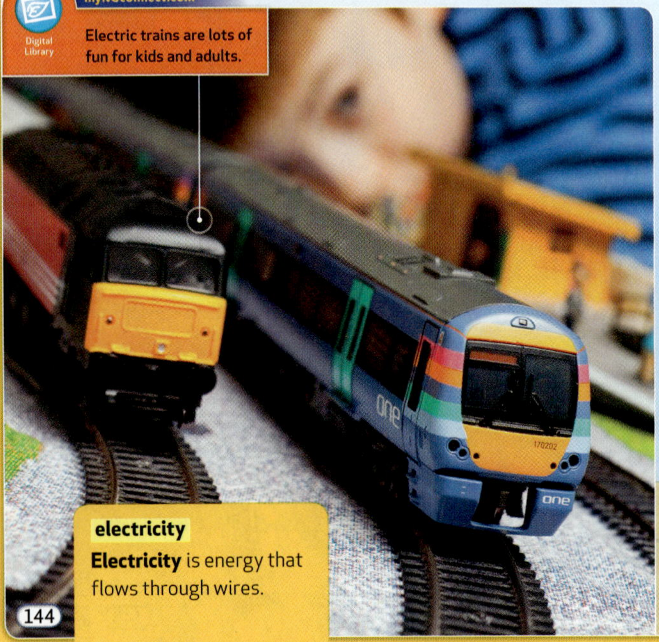

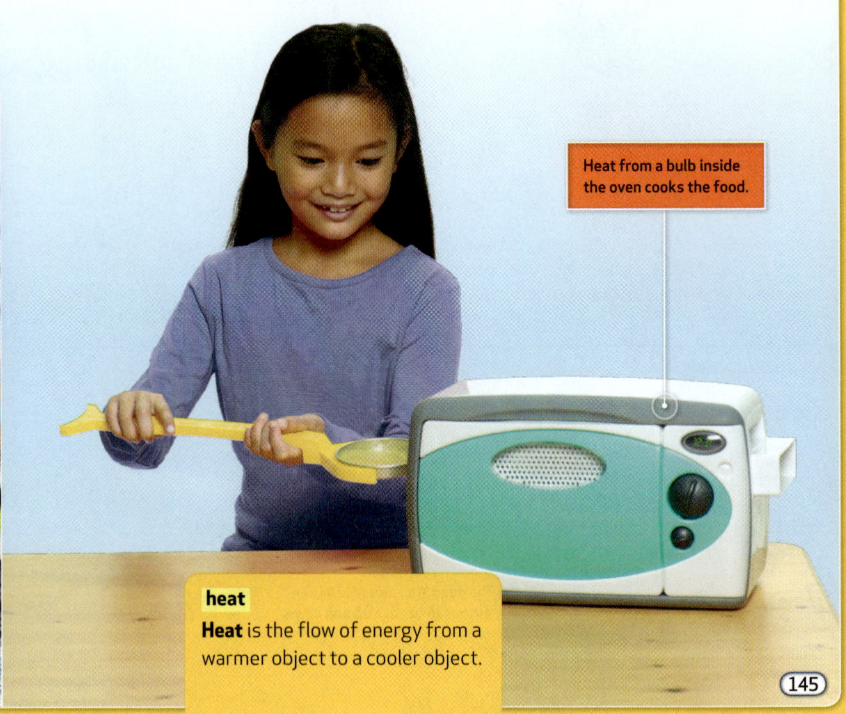

electricity
Electricity is energy that flows through wires.

144

heat
Heat is the flow of energy from a warmer object to a cooler object.

145

PROGRAM RESOURCES
- **Digital Library** at **myNGconnect.com**

Access Science Content

Explain How Toys Use Electricity and Heat

- Have students read page 144 and study the photo. Ask: **What kind of energy does this toy train use?** (electricity) **How does the electricity get to the toy train?** (It flows through wires to the tracks.)

- Have students read page 145 and study the photo. Ask: **What kind of energy does this toy oven use?** (heat energy) **What is the source of the heat?** (a light bulb) **How does the toy oven bake the cake?** (The light bulb heats the oven, and then the heat from the oven flows to the cooler cake batter and bakes the batter.)

▶ Monitor and Fix Up

Ask students if they understand how the electric trains and the toy oven work. Discuss possible fix-up strategies. For example, students may look back to pages 126–133 to review **electricity** and **heat** energy.

Digital Library

myNGconnect.com

Have students use the **Digital Library** to find photos that show toys using energy to work.

Integrated Technology

Digital Booklet Students can use the photos to create a digital booklet on energy and toys.

Notetaking

Have students record in their own words how a toy train uses **electricity** and how a toy oven uses **heat** energy. Suggest that they draw and label a diagram to illustrate their descriptions.

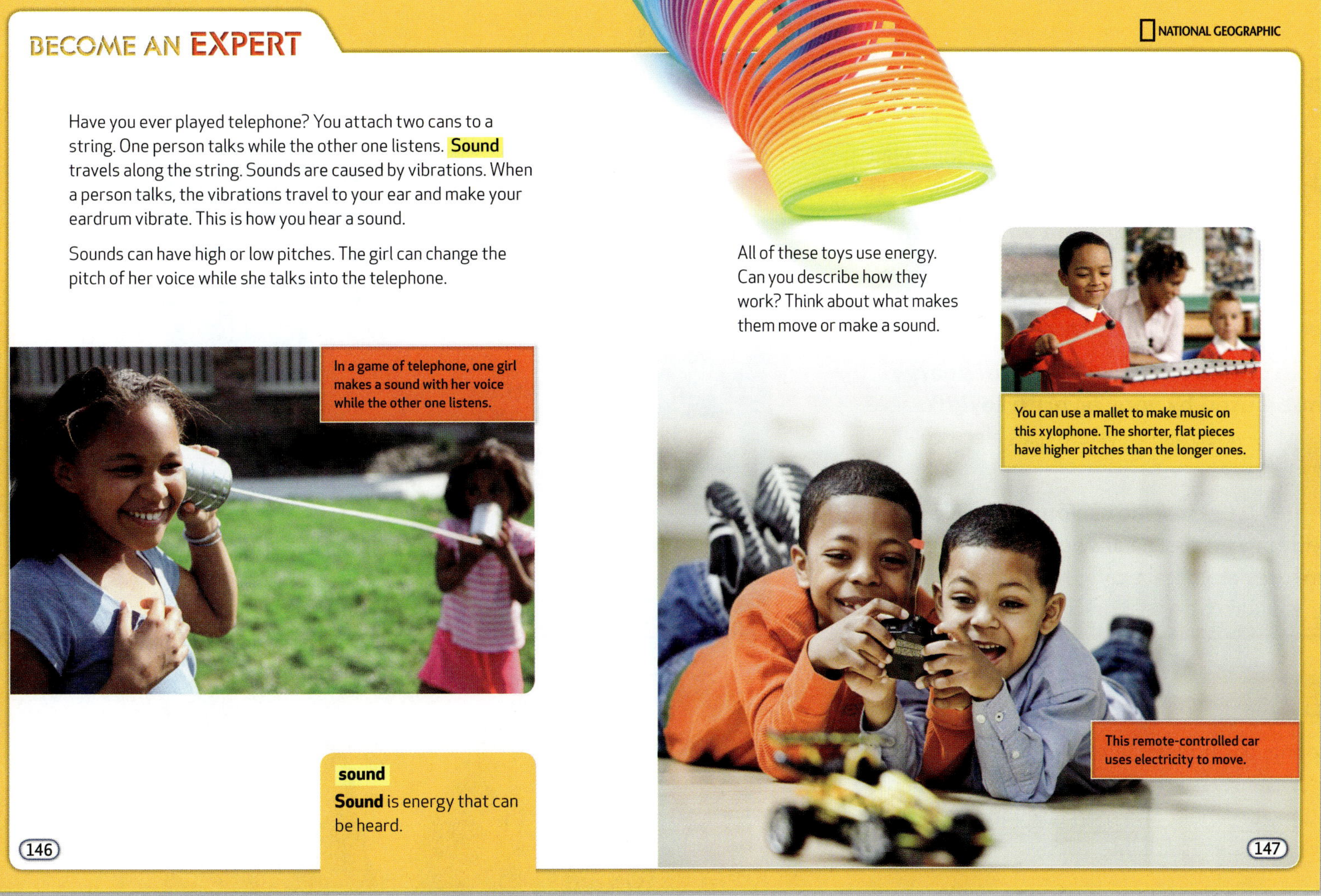

Access Science Content

Explain How Toys Produce Sound Energy

- Have students read pages 146–147. Remind students that **sound** is energy you hear. Ask: **What causes sound?** (vibrations) **How does the toy telephone help the girls talk to each other?** (The sound of the speaker's voice causes vibrations that travel along the string and then make the listener's ear drum vibrate; she hears the vibrations as sound.)

- Have students study the photos and captions on page 147. Help students pronounce *xylophone*. Say: **The boy is making music. What kind of energy does the xylophone produce?** (sound) **How does it make sounds?** (hitting the keys with a mallet causes the keys to vibrate) **Why do the shorter, flat pieces have a higher pitch?** (They vibrate faster.)

- Say: **The springlike toy at the top of the page has stored energy when the coils are pressed together. What happens when the coils move apart?** (The spring moves.) Explain: **The energy stored in the spring becomes the energy of motion. Its mechanical energy is equal to its stored energy plus its energy of motion.**

- Say: **The racecar is also moving. It needs energy to make it go.** Ask: **What gives the car energy to move?** (electricity)

Assess

1. **Explain** **What happens to the heat from a cake after you take it out of the oven?** (The heat flows from the cake to the air and warms the air.)

2. **Infer** **How is mechanical energy involved in the sound produced by a xylophone?** (When a person hits the xylophone key with a mallet to make it vibrate, the person is using mechanical energy.)

Share and Compare

Turn and Talk

Ask students to turn to partners and talk about what they learned about energy. Prompt students by asking:

1. **Recall What is energy and what are some of its basic forms?** (Energy is the ability to do work or cause a change. Some forms of energy are mechanical, sound, electrical, light, and heat.)

2. **Explain Which of these forms of energy might be used by your favorite toy?** (Possible answer: My video game system uses electrical energy.)

3. **Summarize How does your favorite toy use energy?** (Possible answer: My video game system uses electrical energy. Electricity causes the system to turn on and work.)

Read

Ask students to choose the two pages they think are most interesting. Ask them to read the words and any captions or labels, and then talk with a partner about why the pages are interesting.

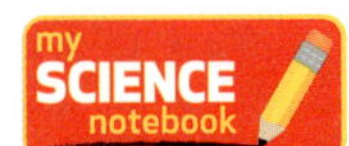 ## Write

Have students write about the different ways toys need energy to do work. Ask them to compare what they wrote with a classmate. Ask if partners recalled the same ways that toys use energy and whether their statements about the Big Idea were similar.

 ## Draw

Have students draw and label a picture of their favorite toy. Ask students to share their drawing with a classmate and describe how their toy uses energy to do work.

❭ Sum Up

Tell students that to sum up a text helps them pull together the ideas of the text. Remind students that they summed up the Become an Expert lesson in the Write section of Share and Compare. Have students take turns reading the conclusions they wrote to a partner. Invite them to compare and contrast their conclusions.

Read Informational Text

Explore on Your Own books provide additional opportunities for your students to:

• deepen their science content knowledge even more as they focus on one Big Idea.

• independently apply multiple reading comprehension strategies as they read.

After applying the strategies throughout the chapter, students will independently read Explore on Your Own books. Facsimiles of the Explore on Your Own pages are shown on pages T148a–T148h.

Pioneer

Pathfinder

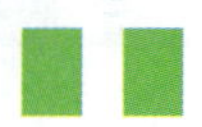

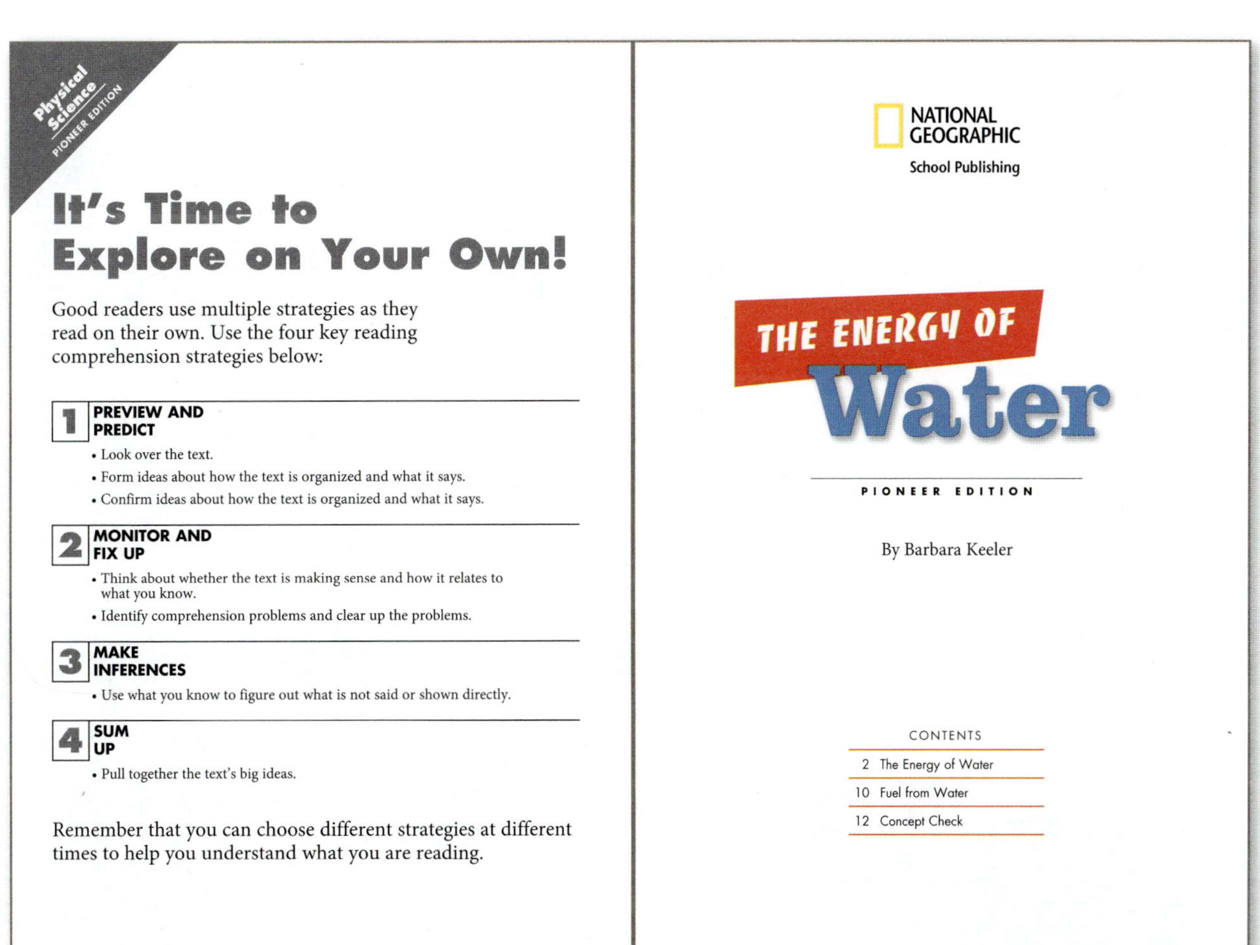

It's Time to Explore on Your Own!

Good readers use multiple strategies as they read on their own. Use the four key reading comprehension strategies below:

1 PREVIEW AND PREDICT
- Look over the text.
- Form ideas about how the text is organized and what it says.
- Confirm ideas about how the text is organized and what it says.

2 MONITOR AND FIX UP
- Think about whether the text is making sense and how it relates to what you know.
- Identify comprehension problems and clear up the problems.

3 MAKE INFERENCES
- Use what you know to figure out what is not said or shown directly.

4 SUM UP
- Pull together the text's big ideas.

Remember that you can choose different strategies at different times to help you understand what you are reading.

NATIONAL GEOGRAPHIC
School Publishing

THE ENERGY OF Water

PIONEER EDITION

By Barbara Keeler

CONTENTS

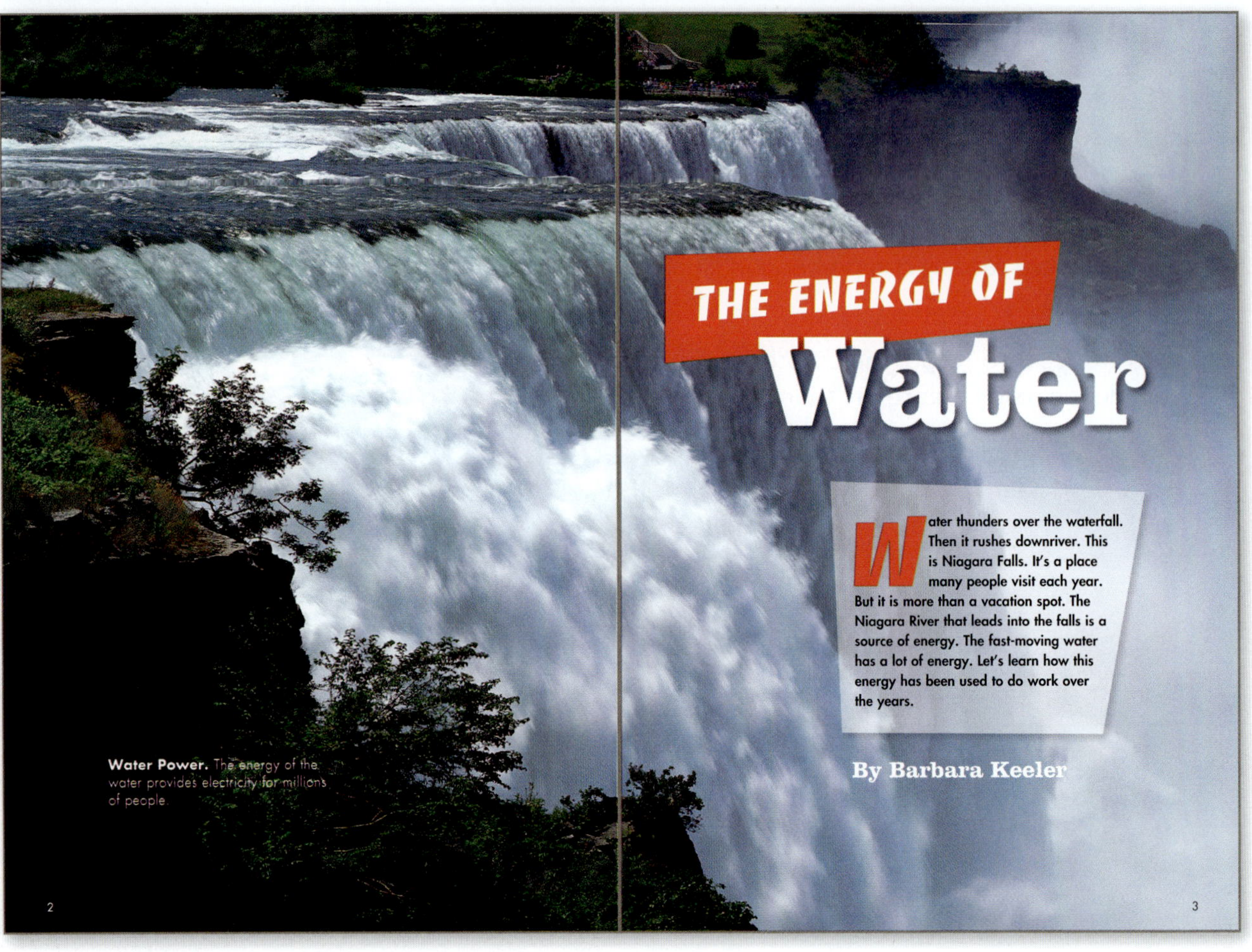

THE ENERGY OF
Water

Water thunders over the waterfall. Then it rushes downriver. This is Niagara Falls. It's a place many people visit each year. But it is more than a vacation spot. The Niagara River that leads into the falls is a source of energy. The fast-moving water has a lot of energy. Let's learn how this energy has been used to do work over the years.

By Barbara Keeler

Water Power. The energy of the water provides electricity for millions of people.

Waterwheels

People used the energy from flowing water to do work for thousands of years. The waterwheel helps use water's energy. A waterwheel is a wheel turned by water. It has paddles or cups around its edge. Moving water flows against the paddles or cups. The moving water turns the wheel. When the wheel turns, energy is created. People use this energy to do work. Waterwheels were often used to grind corn or grain. They were also used to cut wood, spin cotton, and do many other things.

Working Wheel. This shows how an early waterwheel turned a millstone at a mill. A millstone was used to grind grain.

Big Wheel. This vertical waterwheel in Syria is thousands of years old. It still works today.

The World Changes

Water power and waterwheels helped change the way people worked.

A man named Richard Arkwright invented a **machine** that made strong yarn. Yarn is cotton thread that looks like string. It is used to weave cloth. Arkwright used the energy from waterwheels to run his machines. The waterwheel passed its energy to the machine.

People began to put many machines in one building called a factory. This helped them to do more work.

More Power. Millraces can give a waterwheel more power. They flow more water to a waterwheel.

People's Lives Changed

Using the energy of moving water changed people's lives. In the early 1800s cotton cloth-making machines came to the United States. The machines used waterwheels for power. Soon there were many cloth-making mills in New England.

The new mills and factories changed the way people lived. People moved from farms to mill towns to work in the factories. More people had factory jobs instead of farm jobs.

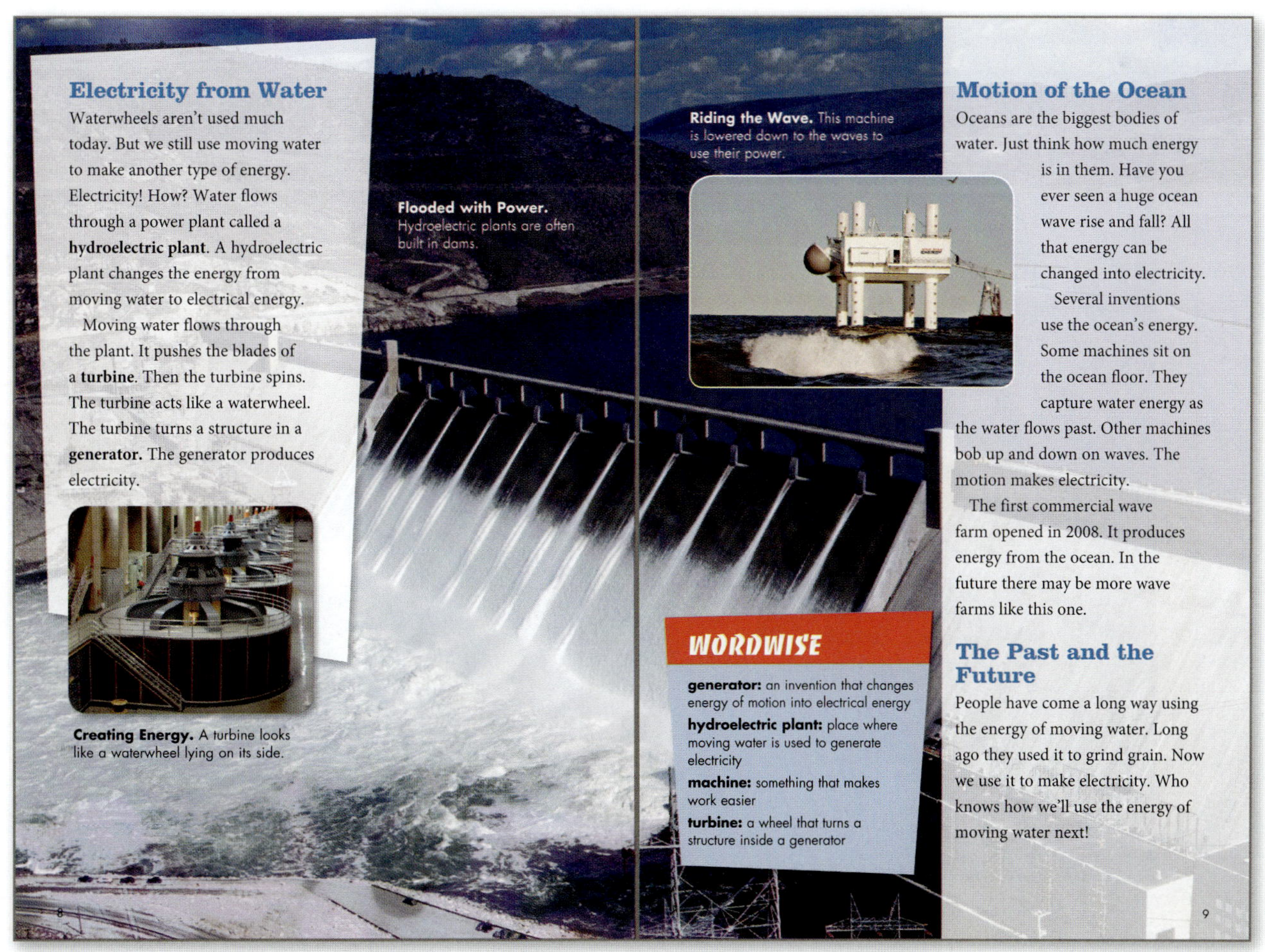

Electricity from Water

Waterwheels aren't used much today. But we still use moving water to make another type of energy. Electricity! How? Water flows through a power plant called a **hydroelectric plant**. A hydroelectric plant changes the energy from moving water to electrical energy.

Moving water flows through the plant. It pushes the blades of a **turbine**. Then the turbine spins. The turbine acts like a waterwheel. The turbine turns a structure in a **generator.** The generator produces electricity.

Creating Energy. A turbine looks like a waterwheel lying on its side.

Flooded with Power. Hydroelectric plants are often built in dams.

Riding the Wave. This machine is lowered down to the waves to use their power.

Motion of the Ocean

Oceans are the biggest bodies of water. Just think how much energy is in them. Have you ever seen a huge ocean wave rise and fall? All that energy can be changed into electricity.

Several inventions use the ocean's energy. Some machines sit on the ocean floor. They capture water energy as the water flows past. Other machines bob up and down on waves. The motion makes electricity.

The first commercial wave farm opened in 2008. It produces energy from the ocean. In the future there may be more wave farms like this one.

WORDWISE

generator: an invention that changes energy of motion into electrical energy

hydroelectric plant: place where moving water is used to generate electricity

machine: something that makes work easier

turbine: a wheel that turns a structure inside a generator

The Past and the Future

People have come a long way using the energy of moving water. Long ago they used it to grind grain. Now we use it to make electricity. Who knows how we'll use the energy of moving water next!

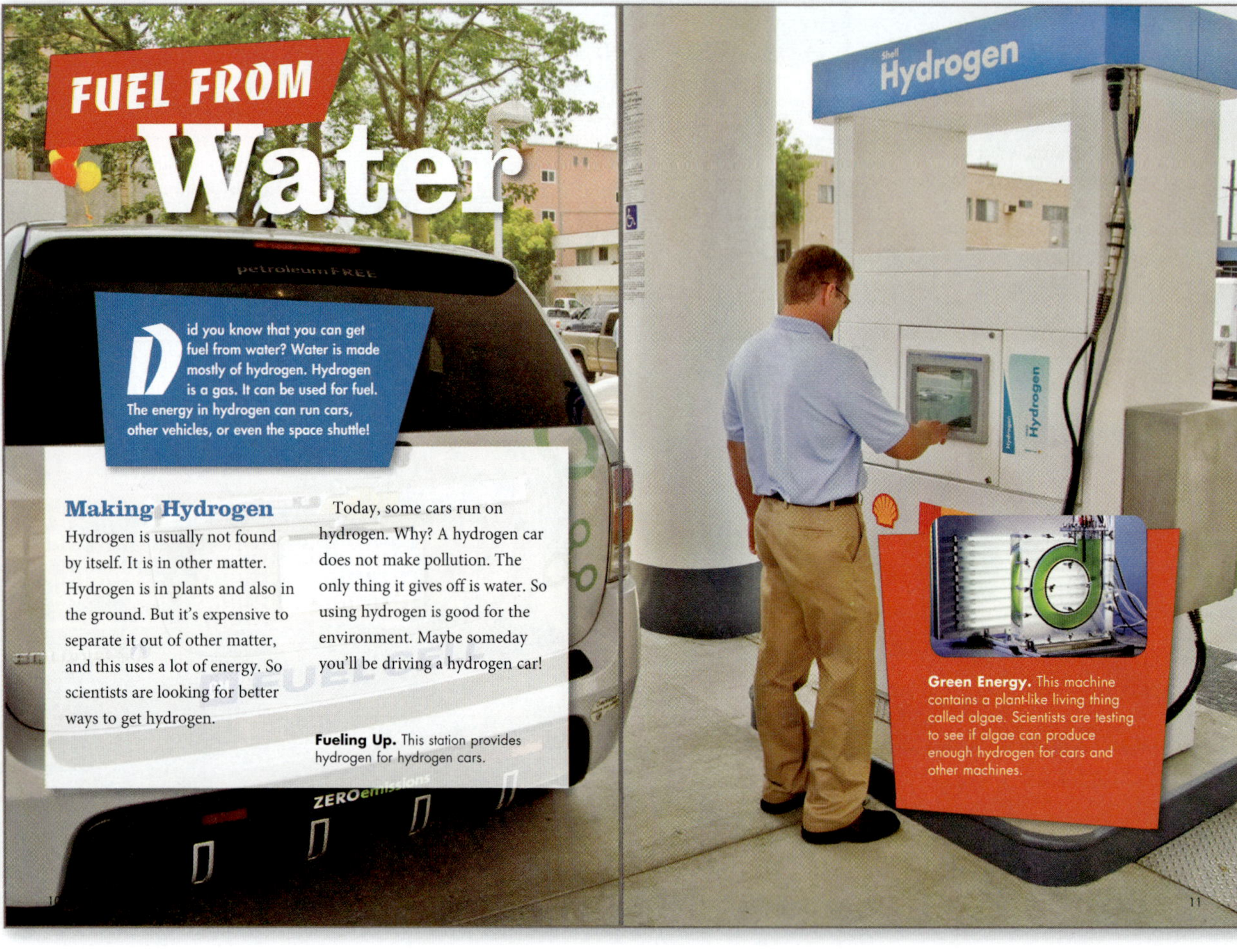

FUEL FROM Water

Did you know that you can get fuel from water? Water is made mostly of hydrogen. Hydrogen is a gas. It can be used for fuel. The energy in hydrogen can run cars, other vehicles, or even the space shuttle!

Making Hydrogen

Hydrogen is usually not found by itself. It is in other matter. Hydrogen is in plants and also in the ground. But it's expensive to separate it out of other matter, and this uses a lot of energy. So scientists are looking for better ways to get hydrogen.

Today, some cars run on hydrogen. Why? A hydrogen car does not make pollution. The only thing it gives off is water. So using hydrogen is good for the environment. Maybe someday you'll be driving a hydrogen car!

Fueling Up. This station provides hydrogen for hydrogen cars.

Green Energy. This machine contains a plant-like living thing called algae. Scientists are testing to see if algae can produce enough hydrogen for cars and other machines.

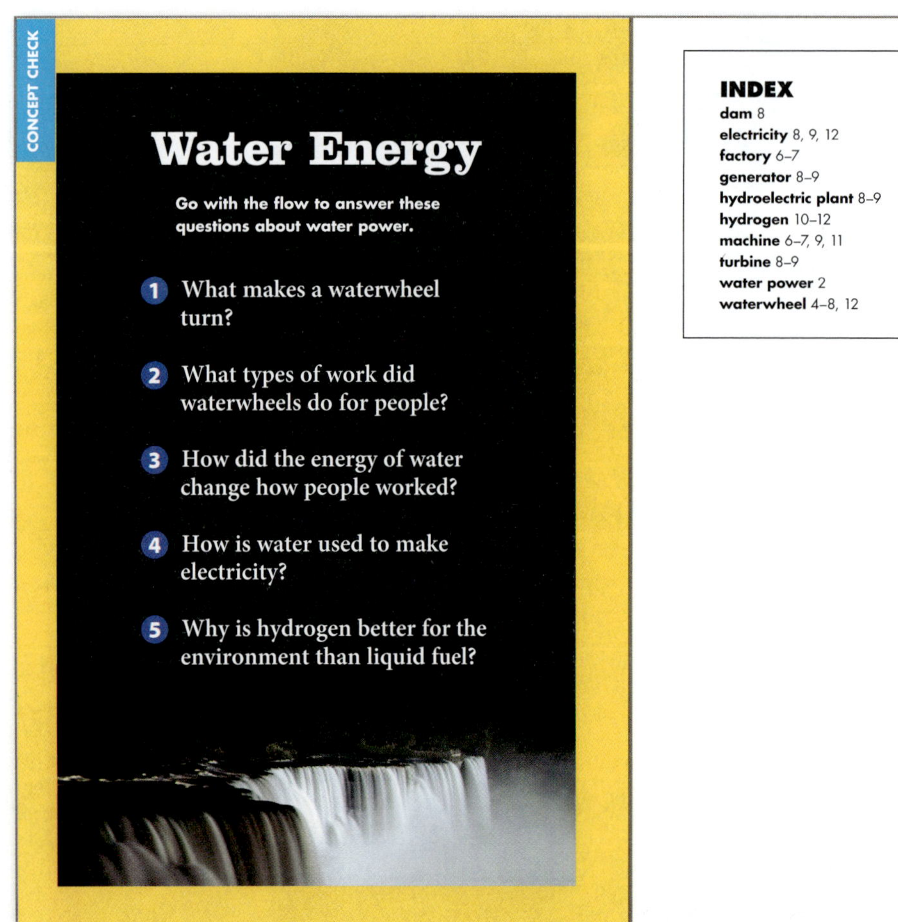

Water Energy

Go with the flow to answer these questions about water power.

1. What makes a waterwheel turn?

2. What types of work did waterwheels do for people?

3. How did the energy of water change how people worked?

4. How is water used to make electricity?

5. Why is hydrogen better for the environment than liquid fuel?

CONCEPT CHECK

Acknowledgments
Grateful acknowledgment is given to the authors, artists, photographers, museums, publishers, and agents for permission to reprint copyrighted material. Every effort has been made to secure the appropriate permission. If any omissions have been made or if corrections are required, please contact the Publisher.

Photographic Credits
Cover Creatas/Jupiterimages; 2-3 Creatas/Jupiterimages; 4-5 Styve Reineck/Shutterstock; 6 Robert Morris/Alamy Images; 8 Harris Shiffman/Shutterstock, Bettman/Corbis; 9 Henning Bagger/epa/Corbis; 10-11 Spencer Grant/PhotoEdit; 11 Pascal Goetgheluck/Photo Researchers Inc.; 12 DigitalStock/Corbis.

Illustrator Credits
5 Precision Graphics; 6 Precision Graphics; 6-7 The Granger Collection New York.

Neither the Publisher nor the authors shall be liable for any damage that may be caused or sustained as result from conducting any of the activities in this publication without specifically following instructions, undertaking the activities without proper supervision, or failing to comply with the cautions contained herein.

Program Authors
Malcolm B. Butler, Ph.D., Associate Professor of Science Education, University of South Florida, St. Petersburg, Florida; Judith Sweeney Lederman, Ph.D., Director of Teacher Education and Associate Professor of Science Education, Department of Mathematics and Science Education, Illinois Institute of Technology, Chicago, Illinois; Randy Bell, Ph.D., Associate Professor of Science Education, University of Virginia, Charlottesville, Virginia; Kathy Cabe Trundle, Ph.D., Associate Professor of Early Childhood Science Education, The Ohio State University, Columbus, Ohio; David W. Moore, Ph.D., Professor of Education, College of Teacher Education and Leadership, Arizona State University, Tempe, Arizona

The National Geographic Society
John M. Fahey, Jr., President & Chief Executive Officer
Gilbert M. Grosvenor, Chairman of the Board

Copyright © 2011 The Hampton-Brown Company, Inc., a wholly owned subsidiary of the National Geographic Society, publishing under the imprints National Geographic School Publishing and Hampton-Brown.

All rights reserved. No part of this book may be reproduced or transmitted in any form or by any means, electronic or mechanical, including photocopying, recording, or by an information storage and retrieval system, without permission in writing from the Publisher.

National Geographic and the Yellow Border are registered trademarks of the National Geographic Society.

National Geographic School Publishing
Hampton-Brown
www.NGSP.com

Printed in the USA.
RR Donnelley, Johnson City, TN

ISBN-13: 978-0-7362-7726-6

10 11 12 13 14 15 16 17 18 19
10 9 8 7 6 5 4 3 2 1

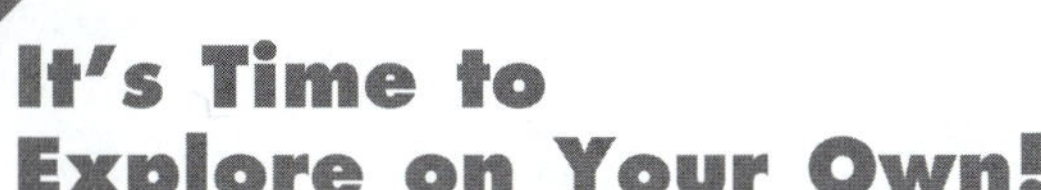

It's Time to Explore on Your Own!

Good readers use multiple strategies as they read on their own. Use the four key reading comprehension strategies below:

1 PREVIEW AND PREDICT
- Look over the text.
- Form ideas about how the text is organized and what it says.
- Confirm ideas about how the text is organized and what it says.

2 MONITOR AND FIX UP
- Think about whether the text is making sense and how it relates to what you know.
- Identify comprehension problems and clear up the problems.

3 MAKE INFERENCES
- Use what you know to figure out what is not said or shown directly.

4 SUM UP
- Pull together the text's big ideas.

Remember that you can choose different strategies at different times to help you understand what you are reading.

By Barbara Keeler

CONTENTS

Waterwheels

Flowing water is a powerful force. People have used its energy to do work for thousands of years. The waterwheel was invented to help use water's energy. A waterwheel is a wheel turned by water. Moving water flows against cups or paddles on the edge of the wheel. The moving water makes the wheel turn. The turning wheel creates energy. People use this energy to do work.

Waterwheels have been used to do work for a long time. Back in 1621, the settlers of Jamestown, Virginia, built their first mill to grind corn into flour. What made the mill work? Flowing water from the James River turned a waterwheel. As the wheel turned, it spun a heavy round stone inside the mill. The stone ground corn or grain into flour. Over time, waterwheels were used to cut wood, spin cotton, and do many other things.

Working Wheel. This shows how an early waterwheel turned a millstone at a mill. A millstone was used to grind grain.

Big Wheel. This vertical waterwheel is thousands of years old. It still works today.

The World Changes

Over time, people used waterwheels and gears to do even more work. Water power helped change the way people worked. In England a man named Richard Arkwright invented a **machine** to make a strong yarn. Yarn is cotton thread that looks like string. It is used to weave cloth. Arkwright used the energy from waterwheels to run the machines.

People began to put many machines in one building called a factory. They could make more yarn and more cloth with more machines. The factories needed a lot of energy to run the machines. People used several waterwheels at once to supply more energy.

More Power. Millraces were sometimes used to give waterwheels more power. A millrace is a man-made water channel used to flow more water to a waterwheel.

People's Lives Changed

Using the energy of moving water changed the way people lived. In the early 1800s Francis Cabot Lowell brought cotton cloth-making machines to the United States. The machines used waterwheels for power. Soon these machines and waterwheels powered clothmaking mills all over New England.

The new mills and factories changed the way people lived. People moved from farms to mill towns to work in the factories. More people had factory jobs instead of farm jobs. More people lived in cities than ever before.

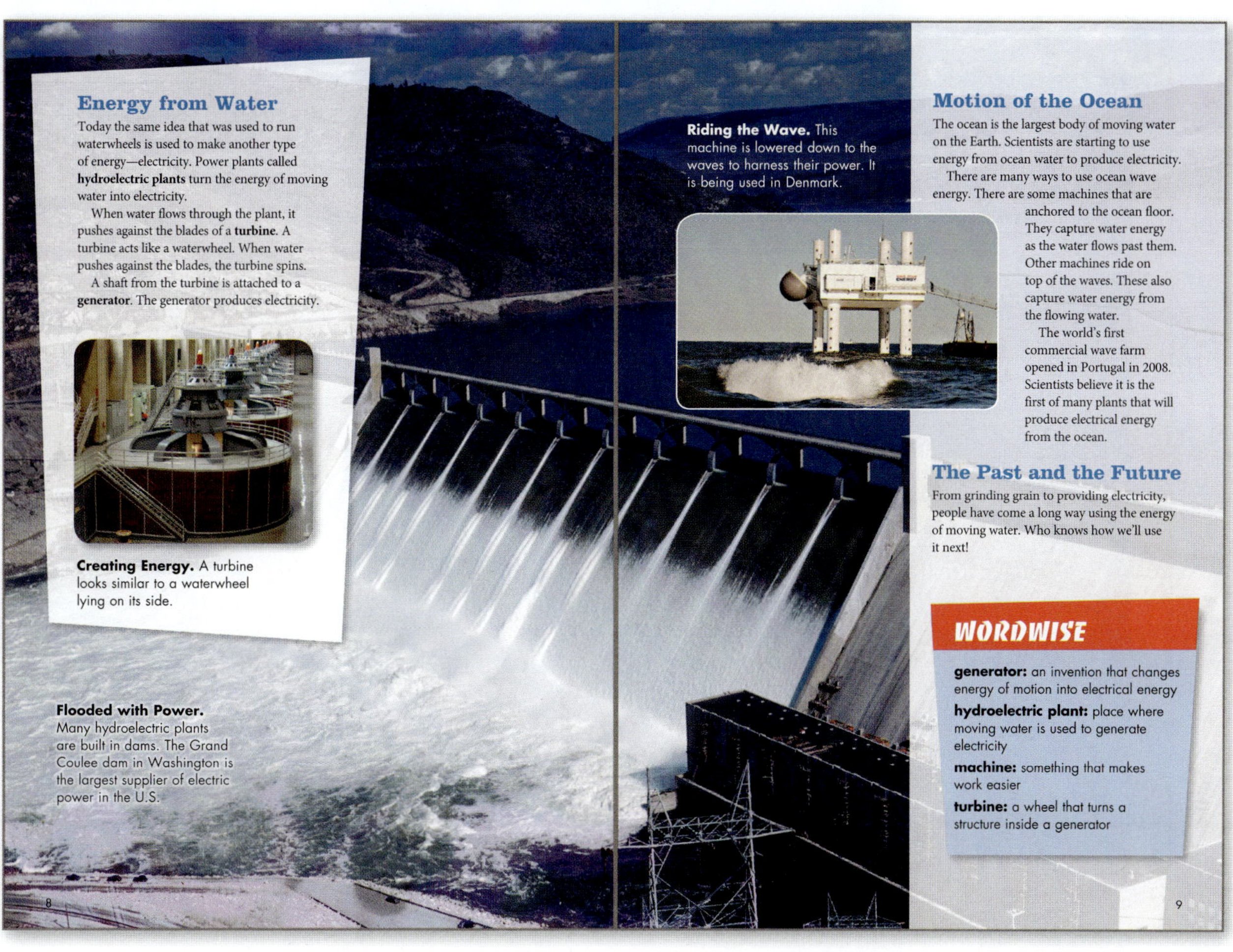

Energy from Water

Today the same idea that was used to run waterwheels is used to make another type of energy—electricity. Power plants called **hydroelectric plants** turn the energy of moving water into electricity.

When water flows through the plant, it pushes against the blades of a **turbine**. A turbine acts like a waterwheel. When water pushes against the blades, the turbine spins.

A shaft from the turbine is attached to a **generator**. The generator produces electricity.

Creating Energy. A turbine looks similar to a waterwheel lying on its side.

Flooded with Power. Many hydroelectric plants are built in dams. The Grand Coulee dam in Washington is the largest supplier of electric power in the U.S.

Riding the Wave. This machine is lowered down to the waves to harness their power. It is being used in Denmark.

Motion of the Ocean

The ocean is the largest body of moving water on the Earth. Scientists are starting to use energy from ocean water to produce electricity.

There are many ways to use ocean wave energy. There are some machines that are anchored to the ocean floor. They capture water energy as the water flows past them. Other machines ride on top of the waves. These also capture water energy from the flowing water.

The world's first commercial wave farm opened in Portugal in 2008. Scientists believe it is the first of many plants that will produce electrical energy from the ocean.

The Past and the Future

From grinding grain to providing electricity, people have come a long way using the energy of moving water. Who knows how we'll use it next!

generator: an invention that changes energy of motion into electrical energy

hydroelectric plant: place where moving water is used to generate electricity

machine: something that makes work easier

turbine: a wheel that turns a structure inside a generator

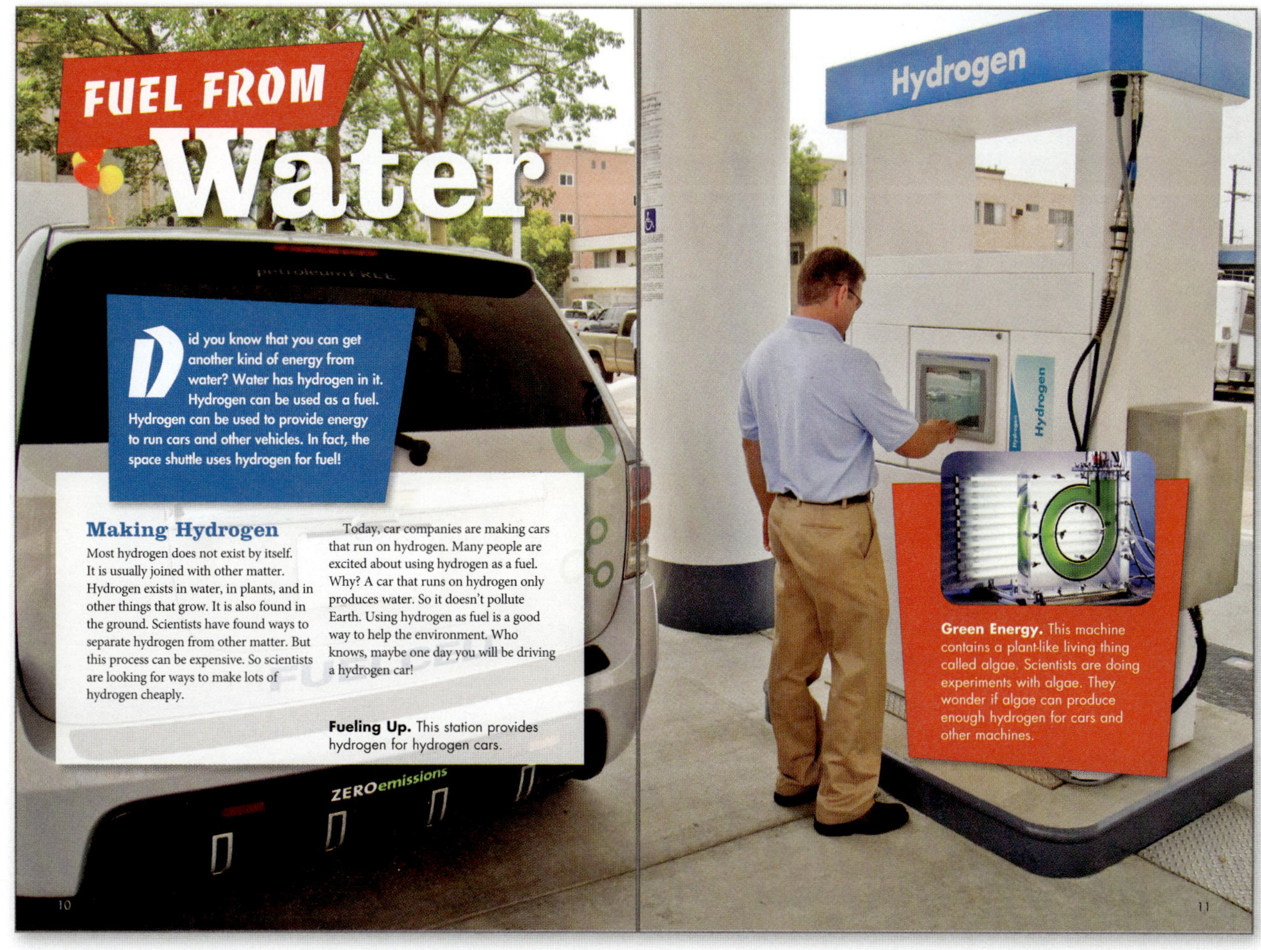

Did you know that you can get another kind of energy from water? Water has hydrogen in it. Hydrogen can be used as a fuel. Hydrogen can be used to provide energy to run cars and other vehicles. In fact, the space shuttle uses hydrogen for fuel!

Making Hydrogen

Most hydrogen does not exist by itself. It is usually joined with other matter. Hydrogen exists in water, in plants, and in other things that grow. It is also found in the ground. Scientists have found ways to separate hydrogen from other matter. But this process can be expensive. So scientists are looking for ways to make lots of hydrogen cheaply.

Today, car companies are making cars that run on hydrogen. Many people are excited about using hydrogen as a fuel. Why? A car that runs on hydrogen only produces water. So it doesn't pollute Earth. Using hydrogen as fuel is a good way to help the environment. Who knows, maybe one day you will be driving a hydrogen car!

Fueling Up. This station provides hydrogen for hydrogen cars.

Green Energy. This machine contains a plant-like living thing called algae. Scientists are doing experiments with algae. They wonder if algae can produce enough hydrogen for cars and other machines.

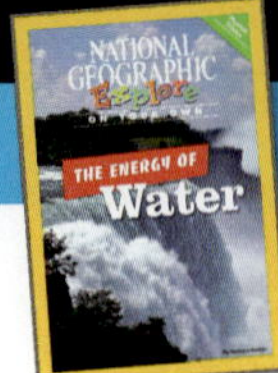

Water Energy

Go with the flow to answer these questions about water power.

1. What makes a waterwheel turn?

2. List some types of work waterwheels did for people.

3. How did the energy of water change how people worked?

4. How is water used to make electricity?

5. Why is hydrogen better for the environment than liquid fuel?

Acknowledgments
Grateful acknowledgment is given to the authors, artists, photographers, museums, publishers, and agents for permission to reprint copyrighted material. Every effort has been made to secure the appropriate permission. If any omissions have been made or if corrections are required, please contact the Publisher.

Photographic Credits
Cover Creatas/Jupiterimages; 2-3 Creatas/Jupiterimages; 4-5 Styve Reineck/Shutterstock; 6 Robert Morris/Alamy Images; 8 Harris Shiffman/Shutterstock, Bettman/Corbis; 9 Henning Bagger/epa/Corbis; 10-11 Spencer Grant/PhotoEdit; 11 Pascal Goetgheluck/Photo Researchers Inc.; 12 DigitalStock/Corbis.

Illustrator Credits
5 Precision Graphics; 6 Precision Graphics; 6-7 The Granger Collection New York.

Neither the Publisher nor the authors shall be liable for any damage that may be caused or sustained or result from conducting any of the activities in this publication without specifically following instructions, undertaking the activities without proper supervision, or failing to comply with the cautions contained herein.

Program Authors
Malcolm B. Butler, Ph.D., Associate Professor of Science Education, University of South Florida, St. Petersburg, Florida; Judith Sweeney Lederman, Ph.D., Director of Teacher Education and Associate Professor of Science Education, Department of Mathematics and Science Education, Illinois Institute of Technology, Chicago, Illinois; Randy Bell, Ph.D., Associate Professor of Science Education, University of Virginia, Charlottesville, Virginia; Kathy Cabe Trundle, Ph.D., Associate Professor of Early Childhood Science Education, The Ohio State University, Columbus, Ohio; David W. Moore, Ph.D., Professor of Education, College of Teacher Education and Leadership, Arizona State University, Tempe, Arizona

The National Geographic Society
John M. Fahey, Jr., President & Chief Executive Officer
Gilbert M. Grosvenor, Chairman of the Board

Copyright © 2011 The Hampton-Brown Company, Inc., a wholly owned subsidiary of the National Geographic Society, publishing under the imprints National Geographic School Publishing and Hampton-Brown.

All rights reserved. No part of this book may be reproduced or transmitted in any form or by any means, electronic or mechanical, including photocopying, recording, or by an information storage and retrieval system, without permission in writing from the Publisher.

National Geographic and the Yellow Border are registered trademarks of the National Geographic Society.

National Geographic School Publishing
Hampton-Brown
www.NGSP.com

Printed in the USA.
RR Donnelley, Johnson City, TN

ISBN-13: 978-0-7362-7729-7

10 11 12 13 14 15 16 17 18 19
10 9 8 7 6 5 4 3 2 1

What is Light?

After reading Chapter 5, you will be able to:

- Identify and describe light energy. Explain that we need light energy to see.
  SOURCES OF LIGHT, REFLECTION, REFRACTION, ABSORPTION, SHADOWS

- Identify sources of light energy. Recognize that light sources often give off heat as well.
  SOURCES OF LIGHT

- Recognize that light travels in a straight line. **REFLECTION, REFRACTION, SHADOWS**

- Explain that light reflects off some surfaces. **REFLECTION**

- Describe the effect of refraction when light travels from one material to another.
  REFRACTION

- Classify objects as ones that absorb light and those that reflect light. Recognize that objects heat up, depending on how much light they absorb. **ABSORPTION**

- Explain why placing an object in the path of light causes a shadow. **SHADOWS**

- **Snap!** Explain how light reflects off some surfaces. **REFLECTION**

149

Chapter 5 Contents

What is Light?

Preview and Predict

Tell students that they can preview a text by looking over it to get an idea of what the text is about and how it is organized. Tell students that predicting is forming ideas about what they will read. Students should confirm their ideas with others.

Have students preview and make predictions about the chapter. Remind them to use section heads, vocabulary words, pictures, and their background knowledge to make predictions.

CHAPTER 5 ▫ What Is Light?

LESSON	PACING	OBJECTIVES
Directed Inquiry **1** *Investigate Light and Heat* pages T149e–T149h **Science Inquiry and Writing Book** pages 202–205	**20** minutes	Investigate through Directed Inquiry (answer a question; make and compare observations; collect and record data and observations; generate explanations and conclusions based on evidence; share findings; ask questions based on observations to increase understanding). Investigate, observe, and explain that a lamp gives off energy in the forms of light and heat. Use appropriate scientific tools and measurements, such as a thermometer (Fahrenheit and Celsius) to solve problems about the natural world. Conduct investigations in which inferences are made and conclusions are drawn.
2 **Big Idea Question and Vocabulary** pages T150-T151–T152-T153 **Sources of Light** pages T154–T155	**30** minutes	Recognize light as a form of energy. Identify sources of light energy.
3 **Reflection** pages T156–T159 **Refraction** pages T160–T161	**30** minutes	Explain that light reflects off of some surfaces. Recognize that light travels in a straight line. Recognize that we need light energy to see. Explain that light reflects off objects and then enters our eyes. Describe the effect of refraction when light travels from one material to another.
4 **Absorption** pages T162–T163 **Shadows** pages T164–T165	**25** minutes	Classify objects that absorb light and objects that reflect light. Recognize that objects heat up, depending on how much light they absorb. Recognize that light travels in a straight line. Explain why placing an object in a path of light causes a shadow. Identify three things needed to produce a shadow (light source, object, and surface).
5 NATIONAL GEOGRAPHIC **Light Pollution in the Sky** pages T166–T167	**20** minutes	Identify sources of light energy. Explain that light reflects off of some surfaces.
Guided Inquiry **6** *Investigate Light and Objects* pages T167a–T167d **Science Inquiry and Writing Book** pages 206–209	**30** minutes	Investigate through Guided Inquiry (answer a question; make and compare observations; collect and record data and observations; generate explanations and conclusions based on evidence; share findings; ask questions based on observations to increase understanding; adjust explanations based on findings and new ideas). Demonstrate that light can be reflected by, absorbed by, and can pass through some objects. Keep records that describe observations, carefully distinguish actual observations from ideas and speculations, and are understandable weeks and months later. Evaluate the reasonableness of an explanation.

VOCABULARY	RESOURCES	ASSESSMENT
measure **conclude**	Science Inquiry and Writing Book: *Physical Science* Science Inquiry Kit: *Physical Science* Directed Inquiry: Learning Masters 223–226	Inquiry Rubric: Assessment Handbook, page 215 Inquiry Self-Reflection: Assessment Handbook, page 225 Reflect and Assess, page T149h
light	Vocabulary: Learning Master 227 Big Ideas Book: *Physical Science*	Assess, page T155
reflection **refraction**		Before You Move On, page T159 Assess, page T161
absorption	Science Inquiry and Writing Book: *Physical Science* Extend Learning: Learning Master 228	Before You Move On, page T163 Assess, page T165
	Share and Compare: Learning Master 229	Assess, page T167
compare	Science Inquiry and Writing Book: *Physical Science* Science Inquiry Kit: *Physical Science* Guided Inquiry: Learning Masters 230–233	Inquiry Rubric: Assessment Handbook, page 215 Inquiry Self-Reflection: Assessment Handbook, page 226 Reflect and Assess, page T167d

TECHNOLOGY RESOURCES

STUDENT RESOURCES

myNGconnect.com

- Student eEdition
- Big Ideas Book
- Science Inquiry and Writing Book
- Explore on Your Own Books
- Read with Me
- Vocabulary Games
- Digital Library
- Enrichment Activities

National Geographic Kids

National Geographic Explorer!

TEACHER RESOURCES

myNGconnect.com

- Teacher eEdition
- Teacher's Edition
- Science Inquiry and Writing Book
- Explore on Your Own Books
- Online Lesson Planner
- National Geographic Unit Launch Videos
- Assessment Handbook
- Presentation Tool
- Digital Library

NGSP ExamView CD-ROM

▶▶▶

IF TIME IS SHORT...
FAST FORWARD.

CHAPTER 5 ▫ What Is Light?

LESSON	PACING	OBJECTIVES
7 Conclusion and Review pages T168–T169	**15** minutes	
8 NATIONAL GEOGRAPHIC **PHYSICAL SCIENCE EXPERT** *Inventor* pages T170–T171 NATIONAL GEOGRAPHIC **BECOME AN EXPERT** *Laser: A Special Kind of Light* pages T172-T173–T180	**35** minutes	Recognize that laser light is a form of light energy. Recognize light as a form of energy.
Open Inquiry **9** *Do Your Own Investigation* pages T180j–T180k **Science Inquiry and Writing Book** pages 210–219 **Write Like a Scientist** pages T180l–T180m **Investigation Model** pages T180n–T180q	**50** minutes	Investigate through Open Inquiry (generate questions to investigate; plan, make, and compare investigations; collect and record data and observations; generate explanations and conclusions based on evidence or observations; share findings; ask questions based on observations to increase understanding; adjust interpretations based on findings and new ideas). Investigate through Write Like a Scientist (choose a question; gather materials to use; make a hypothesis or prediction; identify, manipulate, and control variables if needed; make and carry out a plan for an investigation; collect and record data in charts, tables, and graphs; analyze data; explain and share results; tell a conclusion; think of another question). Identify sound as a form of energy and relate fast and slow vibrations to variations in pitch. Indicate materials to be used and steps to follow to conduct an investigation, and describe how data will be recorded. Explore and solve problems generated from school, home, and community situations, using concrete objects and materials.
10 Think Like a Scientist *How Scientists Work: Using Observations to Evaluate Explanations* pages T180r–T180u **Science Inquiry and Writing Book** pages 220–223	**25** minutes	Explain why keeping accurate and detailed records is important when doing science. Read and interpret data in simple tables produced by others. Determine the reasonableness of estimates and measurements. Use simple logical reasoning to develop conclusions, recognizing patterns and relationships in the environment. Question explanations heard from others, identifying similarities and differences, seeking clarification, and comparing them with their own observations and understandings.

FAST FORWARD ▶▶▶
ACCELERATED PACING GUIDE

DAY 1 🕐 **20** minutes

Directed Inquiry

Investigate Light and Heat, page T149e

DAY 2 🕐 **35** minutes

NATIONAL GEOGRAPHIC **PHYSICAL SCIENCE EXPERT** *Inventor,* page T170

NATIONAL GEOGRAPHIC **BECOME AN EXPERT** *Laser: A Special Kind of Light,* page T172–T173

VOCABULARY	RESOURCES	ASSESSMENT
		Assess the Big Idea, page T169
		Chapter 5 Test: Assessment Handbook, pages 106–109
		NGSP ExamView CD-ROM
		Assess, pages T171, T174–T175, T178–T179
question	Science Inquiry and Writing Book: *Physical Science*	Inquiry Rubric: Assessment Handbook, page 216
	Science Inquiry Kit: *Physical Science*	Inquiry Self-Reflection: Assessment Handbook, page 227
	Open Inquiry: Learning Masters 234–237	Reflect and Assess, pages T180k, T180q
	Investigation Model: Learning Masters 238–239	
	Science Inquiry and Writing Book: *Physical Science*	Assess, page T180u
	Think Like a Scientist: Learning Masters 240–241	

TECHNOLOGY RESOURCES

STUDENT RESOURCES

myNGconnect.com

- **Student eEdition**
- Big Ideas Book
- Science Inquiry and Writing Book
- Explore on Your Own Books
- **Read with Me**
- **Vocabulary Games**
- **Digital Library**
- **Enrichment Activities**

National Geographic Kids

National Geographic Explorer!

TEACHER RESOURCES

myNGconnect.com

- **Teacher eEdition**
- Teacher's Edition
- Science Inquiry and Writing Book
- Explore on Your Own Books
- Online Lesson Planner
- National Geographic Unit Launch Videos
- Assessment Handbook
- **Presentation Tool**
- **Digital Library**

NGSP ExamView CD-ROM

DAY 3 30 minutes

Guided Inquiry

Investigate Light and Objects, page T167a

DAY 4  50 minutes

Open Inquiry

Do Your Own Investigation, page T180j

Objectives

Students will be able to:

- Investigate through Directed Inquiry (answer a question; make and compare observations; collect and record data and observations; generate explanations and conclusions based on evidence; share findings; ask questions based on observations to increase understanding).
- Investigate, observe, and explain that a lamp gives off energy in the forms of light and heat.
- Use appropriate scientific tools and measurements, such as a thermometer (Fahrenheit and Celsius) to solve problems about the natural world.
- Conduct investigations in which inferences are made and conclusions are drawn.

Science Process Vocabulary

measure, conclude

PROGRAM RESOURCES

- Science Inquiry and Writing Book: *Physical Science*
- Science Inquiry and Writing Book **eEdition** at ⊘ **myNGconnect.com**
- **Inquiry eHelp** at ⊘ **myNGconnect.com**
- Science Inquiry Kit: *Physical Science*
- Learning Masters Book, pages 223–226, or at ⊘ **myNGconnect.com**
- Inquiry Rubric: Assessment Handbook, page 215, or at ⊘ **myNGconnect.com**
- Inquiry Self-Reflection: Assessment Handbook, page 225, or at ⊘ **myNGconnect.com**

MATERIALS

Kit materials are listed in italics.

thermometer; lamp; *stopwatch*

❶ Introduce

Tap Prior Knowledge

- Ask: **Would you rather be outside on a sunny day or a cloudy day? Why?** (a sunny day, because it might feel warmer) Lead a discussion about how sunlight gives us light and heat on Earth.

Investigate Light and Heat

What happens to an object's temperature when light shines on it?

Science Process Vocabulary   Materials

measure verb

When you **measure,** you find out how much or how many.

conclude verb

When you **conclude,** you use information, or data, from an investigation to come up with a decision or answer.

thermometer

lamp

stopwatch

202

MANAGING THE INVESTIGATION

Time

20 minutes

Groups

Small groups of 4

Advance Preparation

- Make sure the lamps you use have low-wattage bulbs so students do not burn their fingers.
- Set up and plug in the lamps for each group.
- Allow space between each lamp so each group can access a section of the table that is not heated by the lamp.

What to Do

1. Put the thermometer on your desk. Place the lamp over the thermometer. Do not turn on the lamp. Wait 5 minutes. Then record the temperature on the thermometer in your science notebook as the Start temperature.

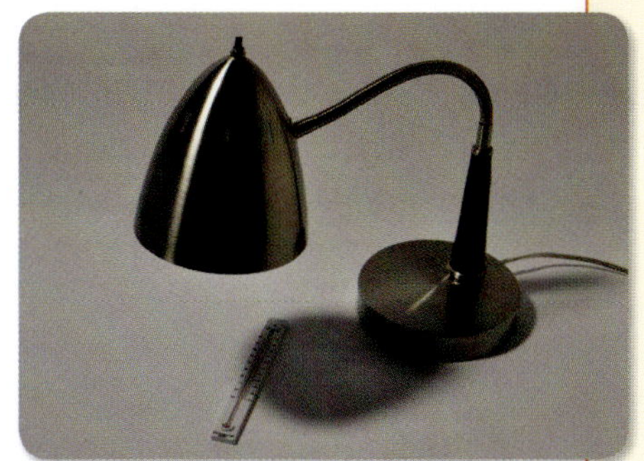

2. **Predict** what will happen to the temperature if you turn on the lamp. Record your prediction.

3. Turn on the lamp. Wait 5 minutes. **Measure** and record the temperature on the thermometer.

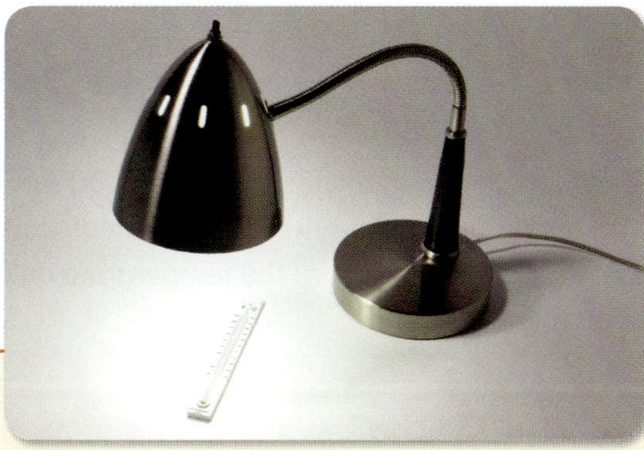

203

Teaching Tips

- If students have trouble reading the thermometer, have them line up a card or folded piece of paper with the top of the red liquid.
- When students read the thermometers, tell them to round to the nearest degree.

What to Expect

- Students should observe that when light shines on an object, the object's temperature increases.

Introduce, continued

Connect to the Big Idea

- Review the Big Idea Question, *What is light?* Explain to students that this inquiry will demonstrate that sources of light also give off heat energy.

- Have students open their Science Inquiry and Writing Books to page 202. Read the Question and invite students to share ideas about light and heat.

❷ Build Vocabulary

Science Process Words: measure, conclude

Use this routine to introduce the words.

1. **Pronounce the Word** Say **measure**. Have students repeat it in syllables.

2. **Explain Its Meaning** Choral read the sentence. Ask students for another word or phrase that means the same as **measure**. (find out how much or how many)

3. **Encourage Elaboration** Ask: **What are some things you can measure?** (how long something is, temperature)

Repeat for the word **conclude**. To encourage elaboration, ask: **What do scientists use to make a conclusion?** (observations and analyzed data)

❸ Guide the Investigation

- Distribute materials. Read the inquiry steps on pages 203–204 together with students. Move from group to group and clarify steps if necessary.

- In step 3, remind students to focus the center of the light over the bulb of the thermometer to obtain a more accurate reading.

- Throughout the investigation ask students to refer to the temperature in both Fahrenheit and Celsius. Ask students to compare how the two are related to each other. Remind students that the Celsius scale is used by scientists.

- Tell students not to touch the thermometer bulb. Ask: **How might touching the thermometer bulb change your results?** (Touching the thermometer with your hand might interfere with the temperature reading.)

Physical Science **T149f**

Guide the Investigation, continued

- In step 5, ask students to think about what they already observed to make an inference. Ask: **Do you think the temperature will increase or decrease over the next 5 minutes?** (decrease) Repeat the question before step 6.

> ⚠ **Safety**First
>
> **Be sure the bulb does not exceed the required wattage for the lamp. Tell students not to touch the lamp.**

❹ Explain and Conclude

- Guide students as they record their data in their tables. Have students share their observations.

- Students should then analyze the data and look for patterns. If students have difficulty seeing the pattern, make a simple graph on the board with time on the horizontal axis (*x*-axis) and temperature on the vertical axis (*y*-axis). Students should see that the temperature goes up when the lamp is on and down when the lamp is off.

- Ask students to compare their predictions with the results. Ask: **How did your understanding change?**

- Have students practice giving their conclusions in complete sentences. Have them use the words *increased* and *decreased* to describe how the temperature changed over time.

- Ask: **If I placed a different object under the lamp, what would happen to it?** (The object's temperature would increase.) Make sure students understand that light and heat are two forms of energy. Sources of light energy often give off heat energy, too. Ask students to think of another source of light and heat energy besides the sun and a light bulb. (fire)

Answers

1. Answers will vary, but students may have predicted that the light would increase the temperature of the thermometer.

What to Do, continued

4 Predict what will happen to the temperature if you leave the lamp on for 5 more minutes. Record your prediction. Then wait 5 minutes. Record the temperature on the thermometer.

5 Predict what will happen to the temperature if you turn the lamp off. Record your prediction. Turn off the lamp. Move the thermometer to a place on your desk that has not been heated by the lamp. Wait 5 minutes. Record the temperature on the thermometer.

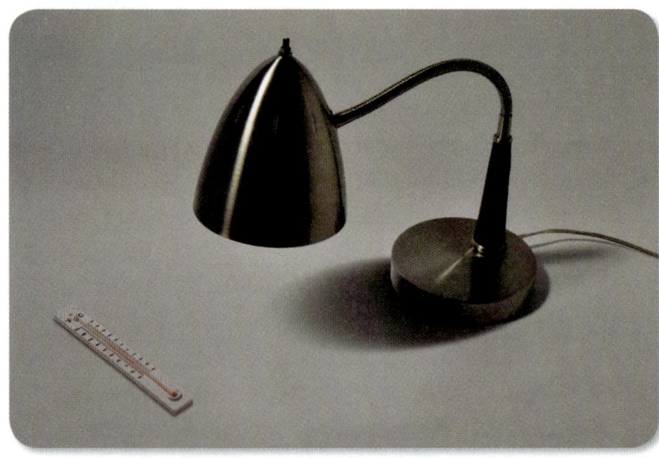

6 Predict what will happen to the temperature if you leave the lamp off for 5 more minutes. Record your prediction. Wait 5 more minutes. Then, record the temperature on the thermometer again.

204

NATIONAL GEOGRAPHIC **Raise Your SciQ!**

Energy and Heat There are many everyday examples of the connection between energy and heat. For example, if you are cold you can run around to warm up. This uses more energy and heats your body. The longer you run, the warmer you feel. The same is true for light energy. Think of how a car interior heats up on a hot summer day compared to a cloudy summer day. If the sunlight increases, the heat also increases.

Record

Write in your science notebook.
Use a table like this one.

Light and Temperature

	Temperature (°C)	Predictions
Start		What will happen to the temperature if you turn on the lamp?

Explain and Conclude

1. Did your **observations** support your **predictions?** Explain.
2. What happened to the temperature as the lamp shined on the thermometer longer? What happened after you turned off the lamp?
3. What can you **conclude** about what can happen to the temperature of an object when light shines on it?

Think of Another Question

What else would you like to find out about what happens to an object's temperature when light shines on it? How could you find an answer to this new question?

In some restaurants, lights are used to keep food warm.

205

Inquiry Rubric at ⊙ myNGconnect.com	Scale			
The student used a thermometer to **measure** temperature.	4	3	2	1
The student collected and recorded temperature **data** in a table.	4	3	2	1
The student made **predictions** and checked if the **observations** supported the predictions.	4	3	2	1
The student came up with a **conclusion** about what happens to the temperature of an object when light shines on it.	4	3	2	1
The student **shared** and **compared** data and conclusions with other groups.	4	3	2	1
Overall Score	4	3	2	1

Explain and Conclude, continued

2. Possible answer: The temperature of the thermometer kept going up when the lamp was turned on. It kept going down when the lamp was turned off.
3. The temperature of an object goes up when light shines on it.

❺ Find Out More

Think of Another Question

- Students should use their observations to generate new questions that arise from their investigations or research. Record student questions for possible future investigations. Discuss what students could do to find answers to the new questions.

❻ Reflect and Assess

- To assess student work with the Inquiry Rubric shown below, see Assessment Handbook, page 215, or go online at ⊙ **myNGconnect.com**
- Have students use the Inquiry Self-Reflection on Assessment Handbook, page 225, or at ⊙ **myNGconnect.com**

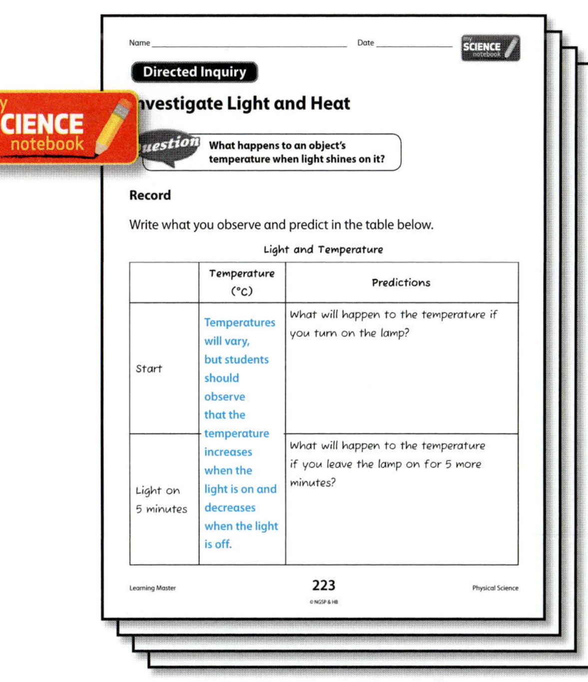

	Temperature (°C)	Predictions
Start	Temperatures will vary, but students should observe that the temperature increases when the light is on and decreases when the light is off.	What will happen to the temperature if you turn on the lamp?
Light on 5 minutes		What will happen to the temperature if you leave the lamp on for 5 more minutes?

Learning Masters 223–226
or at ⊙ **myNGconnect.com**

Objectives
Students will be able to:
- Recognize light as a form of energy.

Science Academic Vocabulary
light, reflection, refraction, absorption

PROGRAM RESOURCES
- Big Ideas Book: *Physical Science*
- Big Ideas Book: *Physical Science* **eEdition**
  at **myNGconnect.com**
- **Vocabulary Games** at **myNGconnect.com**
- **Digital Library** at **myNGconnect.com**
- **Enrichment Activities** at **myNGconnect.com**
- **Read with Me** at **myNGconnect.com**
- Learning Masters Book, page 227, or at
  myNGconnect.com

❶ Introduce

Tap Prior Knowledge
- Have students identify all sources of light in the classroom. They may point out natural light coming through windows, electric lights, the light coming from computer screens, and so on.

❷ Focus on the Big Idea

Big Idea Question

- Read the Big Idea Question aloud, and have students echo it.
- Preview pages 154–155, 156–159, 160–161, 162–163, and 164–165, linking the headings with the Big Idea Question.

Differentiated Instruction

ELL Vocabulary Support

BEGINNING	INTERMEDIATE	ADVANCED
Involve students in a chant that helps them remember the word *light. Without light, there would be no sight.* Repeat the activity for the other three words.	Provide Academic Language Frames to help students describe the properties of light. • In _____, light bounces off a surface. • In _____, light bends. • In _____, a material takes in light.	Provide Academic Language Stems to help students describe the properties of light. • In reflection, light … • In refraction, light … • In absorption, light …

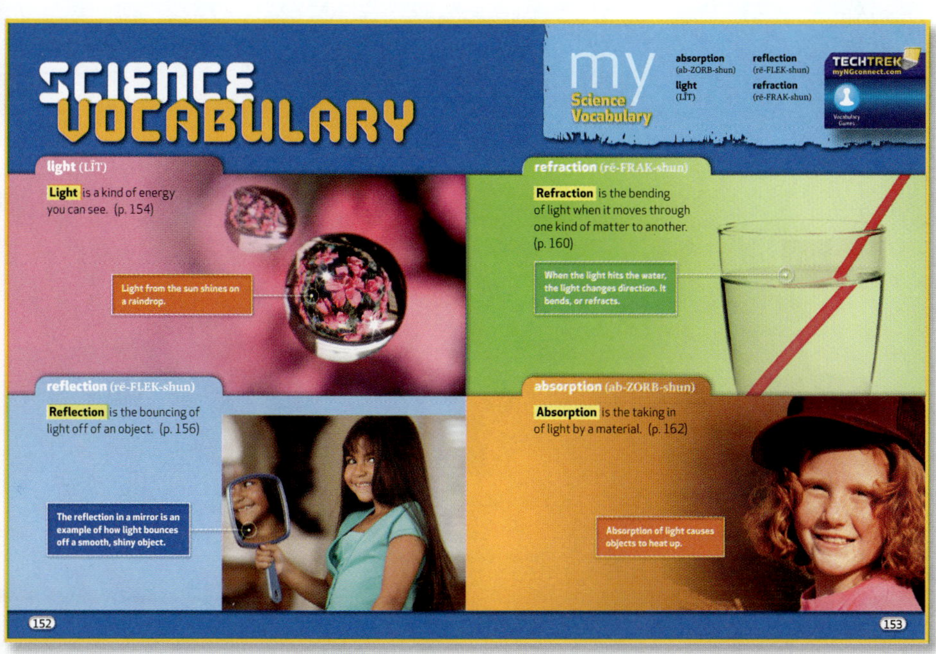

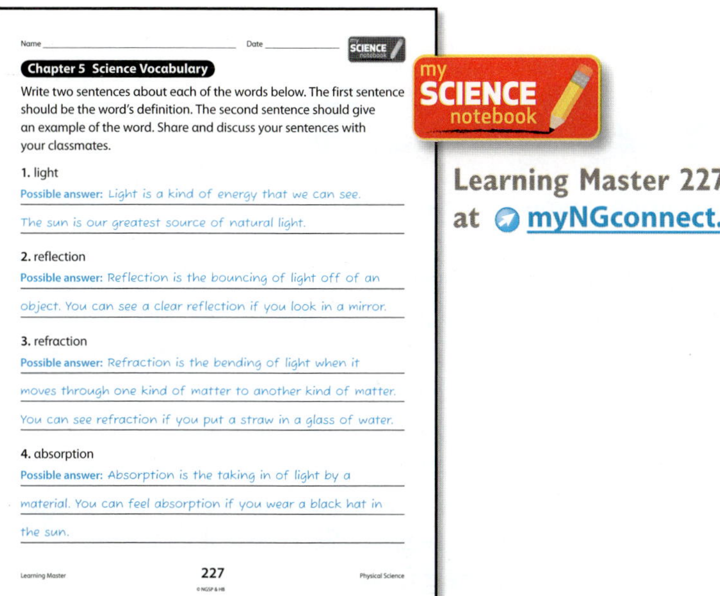

my SCIENCE notebook

Learning Master 227 or at 🔵 myNGconnect.com

Focus on the Big Idea, continued

• With student input, post a chart that displays the chapter headings. Have students orally share what they expect to find in each section. Then read page 150 aloud.

❸ Teach Vocabulary

Have students look at pages 152–153, and use this routine to teach each word. For example:

1. **Pronounce the Word** Say **reflection** and have students repeat it.

2. **Explain Its Meaning** Read the word, definition, and sample sentence, and use the photo to explain the word's meaning. Point out the girl's **reflection** in the mirror: **When you look into a mirror, light bounces off the mirror and goes to your eyes, which makes it possible for you to see your reflection.**

3. **Encourage Elaboration** Have students look at what else the photo shows. Ask: **How does the image of the girl's face appear in the mirror?** (Light bounces off her face and travels straight to the mirror, which reflects it.)

Repeat for the words **light, refraction,** and **absorption,** using the following Elaboration Prompts:

• **Light is a type of what?** (energy)

• **What happens during refraction?** (Light bends as it moves through one kind of matter to another.)

• **During absorption, what happens to light?** (It is taken in by a material.)

Objectives

Students will be able to:

- Identify sources of light energy.

Science Academic Vocabulary

light

❶ Introduce

Tap Prior Knowledge

- Ask students to share what they know about **light** sources.

- Have students study the photos on pages 154–155 and describe the **light** source in each picture. Record their answers.

Set a Purpose and Read

- Read the heading. Then tell students that they will read about the different sources of **light**.

- Have students read pages 154–155.

❷ Teach

Academic Vocabulary: *light*

- Point out the word **light**. Have a student read the definition. Say: **Light** energy from the sun is the main source of **light** energy on Earth.

Discuss the Sun as a Source of **Light** and Other Forms of Energy

- Draw students' attention to the picture of the sun on page 154. Say: **The sun is providing light energy, which you can see. What can you feel when the sun is out?** (heat) **The sun gives off heat energy too. It also provides energy to all living things on Earth. Living things could not grow without the sun's energy.**

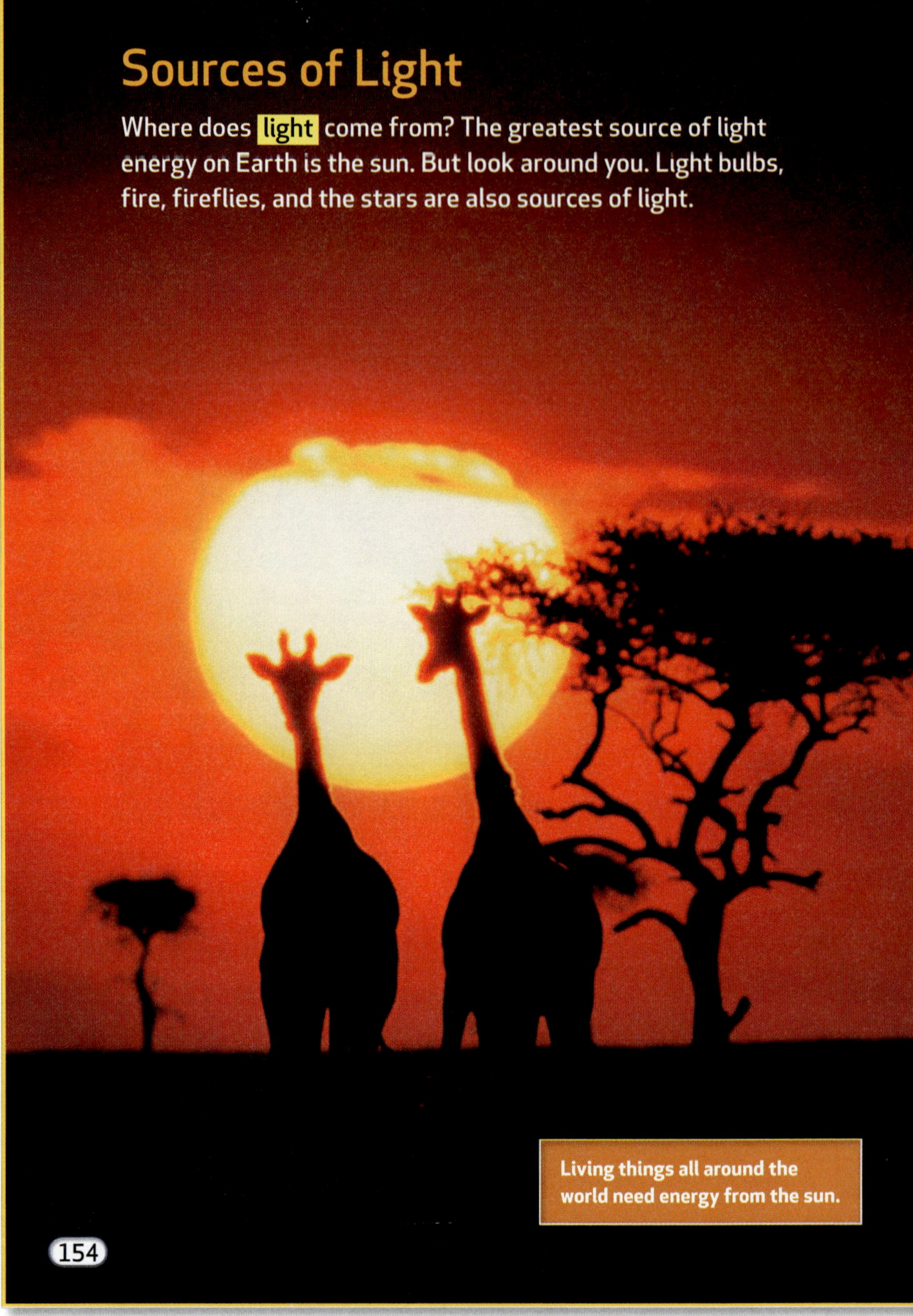

Sources of Light

Where does **light** come from? The greatest source of light energy on Earth is the sun. But look around you. Light bulbs, fire, fireflies, and the stars are also sources of light.

154

Raise Your SciQ!

Bioluminescence Some living things produce light. This is called *bioluminescence*. One of best-known forms of bioluminescence is found in fireflies. But many other creatures, including other insects, spiders, and even fungi, possess bioluminescent abilities. Like fireflies, many land-based living things only light up at night. It is estimated that 90% of deep sea marine life produce bioluminescence. They remain lit all the time, perhaps because they are in darkness all the time. Most deep sea bioluminescence is blue or green, which can be seen through seawater most easily. Scientists believe different living things use bioluminescence for different reasons. Some use it to attract mates, while others use it to attract food.

Many sources of light also give off heat. When you sit in the sun, you can see the light. You can also feel its heat. You can see and feel light and heat from other light sources, too. Think about what happens when you turn on a lamp. The light bulb in the lamp gives off light. But be careful. Never touch a lit light bulb. It also gives off heat.

Campers sit by a campfire on a cool, dark night. The campfire helps them see what is around them. It also keeps the campers warm.

Light from the sun is natural light. Light from a light bulb is artificial light.

Before You Move On

1. Name some sources of light.
2. Why is the sun important?
3. **Analyze** How is the sun like a light bulb? How is it different?

155

Observe That Other Light Sources Also Give Off Heat

- Have students look at the photo of the campfire. Ask: **Have you ever been near a campfire or a fire in a fireplace?** (Some students will likely say yes.) Have students describe the fire, and write the adjectives they use on the board. (bright, orange, yellow, red, warm, hot)

- Ask: **What kinds of energy did the fire give off?** (light energy, heat energy) Say: **Sources of light also often give off heat.**

❸ Assess

» Before You Move On

1. **Recall** **Name some sources of light.** (the sun, light bulbs, fire, fireflies, stars; note that the moon reflects light and is not a source of light)

2. **Explain** **Why is the sun important?** (Without the sun, there would be almost no light or heat on Earth's surface. The sun is Earth's greatest source of light, heat, and other energy.)

3. **Analyze** **How is the sun like a light bulb? How is it different?** (Many sources of light also give off heat. That is true of the sun and light bulbs. However, the sun was not made by people, and it will not burn out for billions of years, while light bulbs are made by people, and even the most long-lasting bulbs will burn out within a few years. The sun makes its own energy, while light bulbs get their energy from electricity.)

Differentiated Instruction

ELL **Language Support for Observing That Light Sources Give Off Heat**

BEGINNING	INTERMEDIATE	ADVANCED
Help students discuss light sources by asking yes/no questions, for example: **Do many light sources also give off heat? Is the sun a light source? Does a light bulb give off heat?**	Have students fill in Academic Language Frames to talk about light sources. • *The ______ is a natural light source.* • *Light bulbs are a human-made ______ source.* • *Many light sources also give off ______.*	Have students use Academic Language Stems elaborate on ideas about light sources. • *Natural ______ sources include…* • *Human-made, ______ light sources include…* • *Many light sources give off ______. For example…*

Objectives
Students will be able to:
- Explain that light reflects off of some surfaces.

Science Academic Vocabulary
reflection

PROGRAM RESOURCES
- **Digital Library** at ⊘ **myNGconnect.com**

❶ Introduce

Tap Prior Knowledge
- Ask students to look at the photo of the trees reflected in the water. Ask them to explain how they can see trees when they look at the surface of the water.

Preview and Read
- Read the heading. Have students preview the photos on pages 156–159.
- Have students read pages 156–159.

❷ Teach

Academic Vocabulary: *reflection*
- Display photos of **reflections** from various surfaces. Pronounce and write **reflection**. Say: **A reflection is the bouncing of light from an object's surface. Light travels toward a surface in a straight line, and then it bounces off the surface in a straight line.**

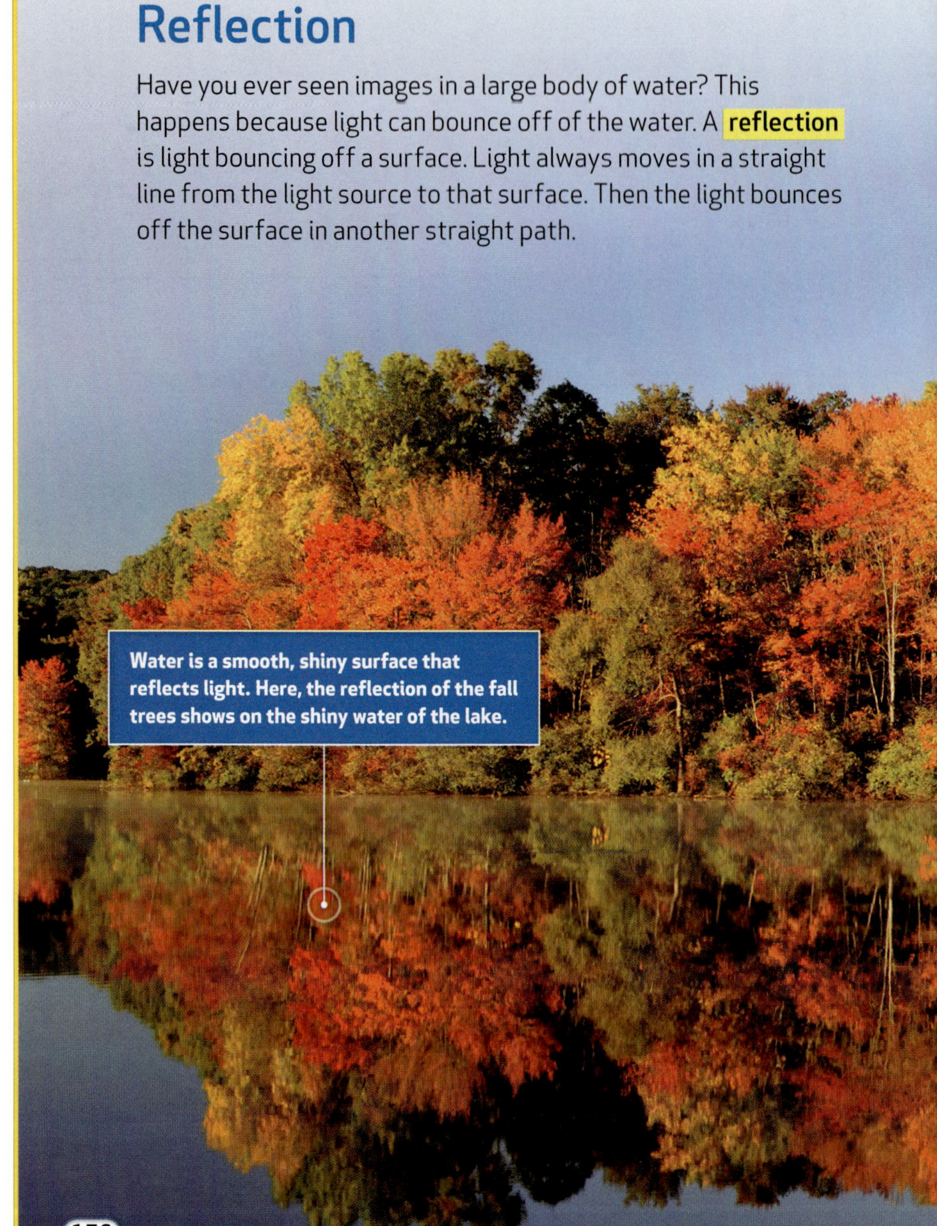

Reflection

Have you ever seen images in a large body of water? This happens because light can bounce off of the water. A **reflection** is light bouncing off a surface. Light always moves in a straight line from the light source to that surface. Then the light bounces off the surface in another straight path.

156

Science Misconceptions

Reverse Image? Some students might think that when light reflects off a mirror, it reverses the image a person sees. Students often have this misconception because when they hold printed material up to a mirror, it appears to be written backward. Have a volunteer hold up a sign in front of a mirror. Explain that the words on the sign appear backward because they *are*—the volunteer turned the sign around to present it to the mirror. Explain that light reflects straight from the mirror to the eye, and that mirrors do not reverse images.

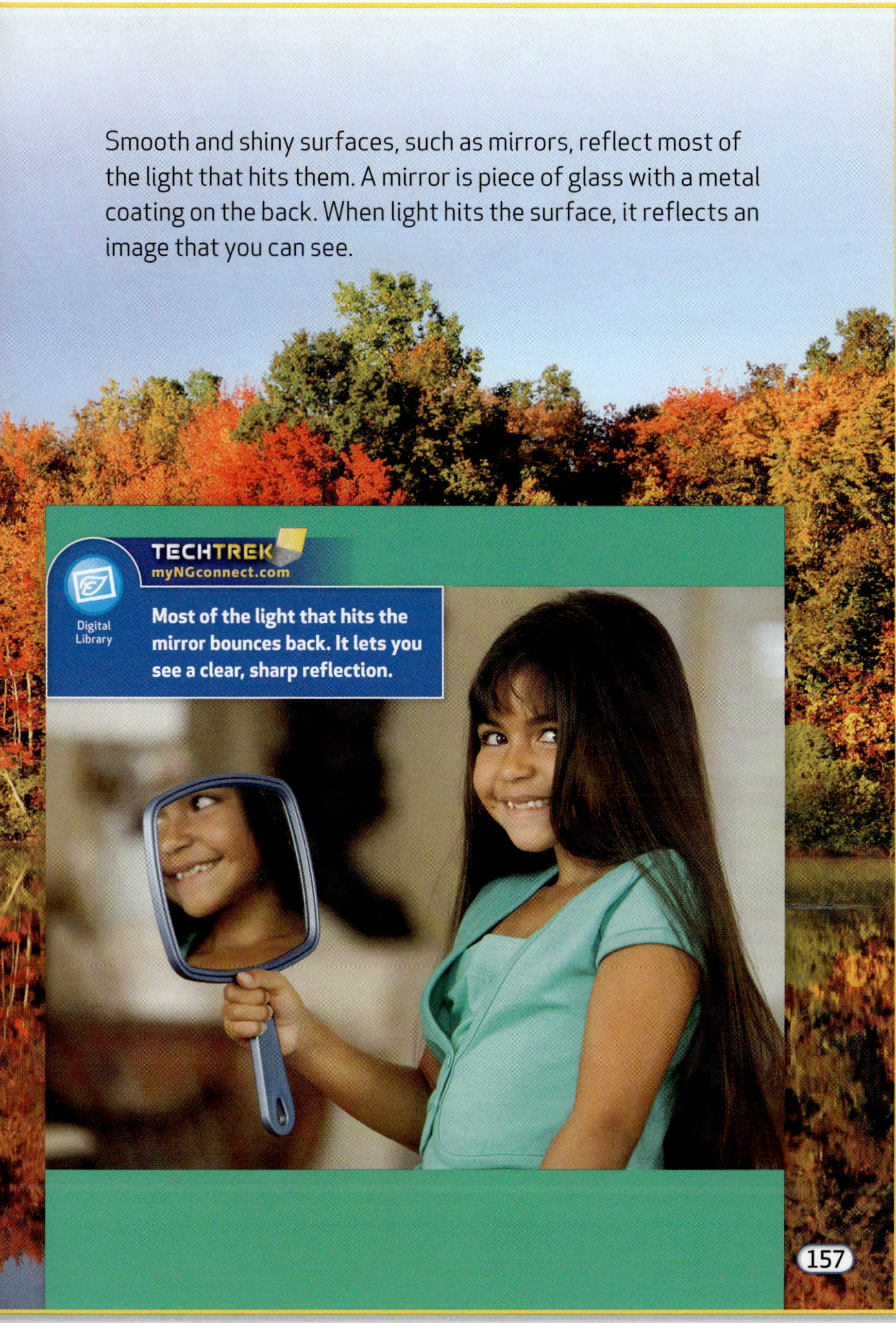

Smooth and shiny surfaces, such as mirrors, reflect most of the light that hits them. A mirror is piece of glass with a metal coating on the back. When light hits the surface, it reflects an image that you can see.

157

Teach, continued

Demonstrate How Light Is Reflected

- Have students stand in two lines facing each other. Ask each student to hold a mirror facing out at chest height. Have one student stand at the head of the lines and shine a flashlight into the mirror of the first person in one of the lines. Ask: **What happens to the light coming from the flashlight?** (It bounces off the mirror.)

- Have students angle their mirrors so that the light coming from the flashlight bounces between the mirrors down the length of the two lines. Ask: **What property or properties of light does this experiment demonstrate?** (Light moves in a straight line; light reflects from some surfaces.)

Explain Why Some Surfaces Reflect Light

- Ask students to think about where they have seen reflections. Ask: **Do some surfaces have clearer reflections?** (Students might see clearer reflections in mirrors, glass, calm water, and some shiny metals than in other surfaces.)

- Explain: **Smooth and shiny surfaces reflect most of the light that hits them, which is why they have clearer reflections.**

Digital Library

myNGconnect.com

Have students use the Digital Library to find photos of mirrors and objects reflected in mirrors.

Integrated Technology

Whiteboard Presentation Students can use the information to make a computer presentation about reflection. You can help them show the presentation on a whiteboard.

Differentiated Instruction

ELL **Language Support for Explaining Why Some Surfaces Reflect Light**

BEGINNING	INTERMEDIATE	ADVANCED
Help students explain why some surfaces reflect light by asking either/or questions, for example: **Do shiny or dull surfaces best reflect light? Do smooth or rough surfaces best reflect light?**	Name these objects, and have students identify each as a surface that would or would not reflect light well: **a mirror, a rough plaster wall, a calm pond, a wooden table**	Help students elaborate on surfaces that reflect light well by providing Academic Language Stems. • *Smooth surfaces _____ light well. For example…* • *When light hits a mirror,…*

Objectives

Students will be able to:

- Recognize that light travels in a straight line.
- Recognize that we need light energy to see.
- Explain that light reflects off objects and then enters our eyes.

Teach, continued

Explain How Reflected Light Enables Us to See Objects

- Ask: **How does reflected light allow us to see?** (Light travels from a light source to an object. Some of the light reflects off the object and enters our eyes.) Ask: **What kind of path does light take?** (It travels in a straight line.)

- Ask: **Could we see objects without light energy?** (no)

- Have students describe what they see around them in the classroom. Say: **You can see these things because of light energy. Ask: From where is the light energy coming?** (Depending on your classroom, either sunlight through a window or light fixtures in the room.)

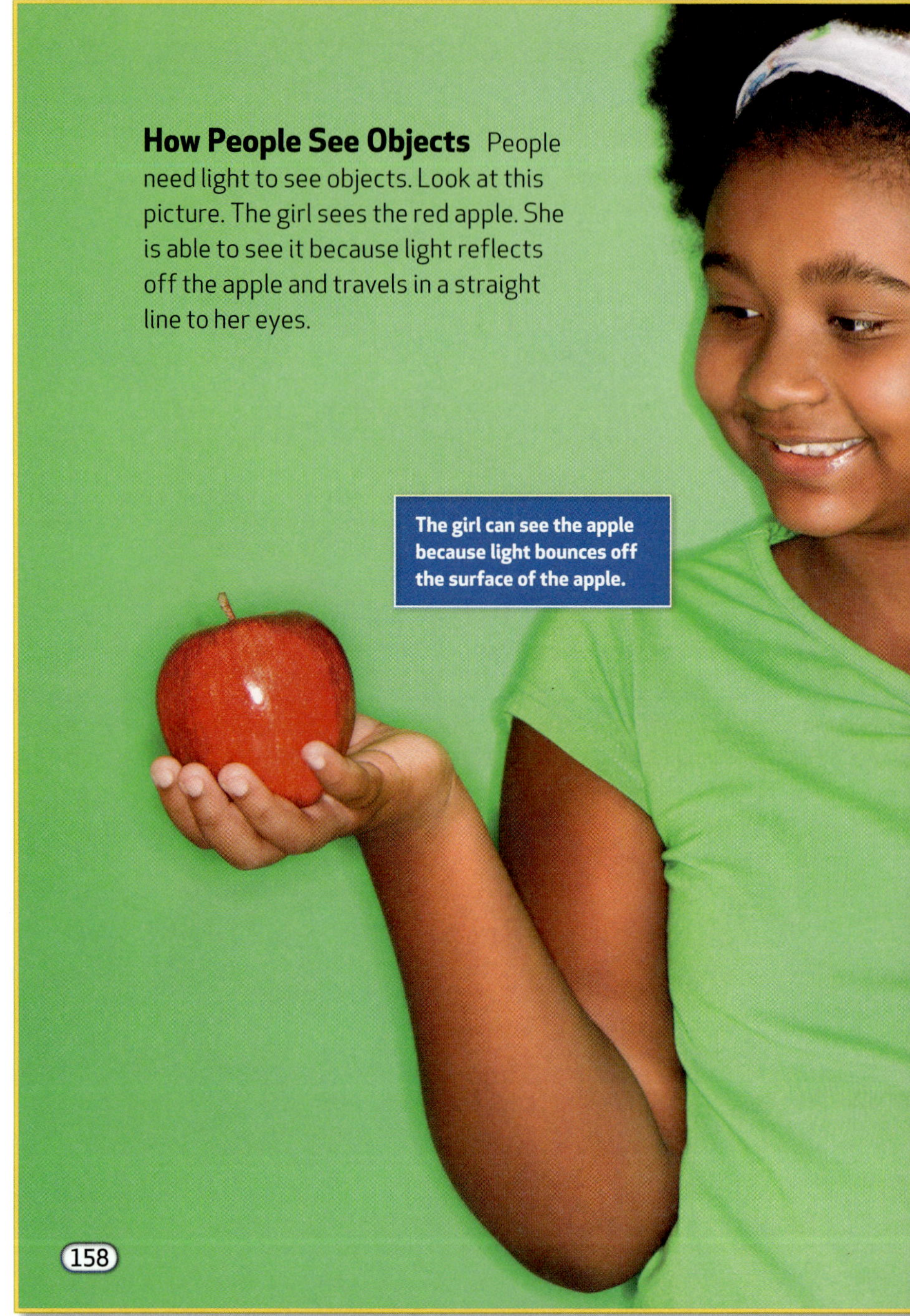

How People See Objects People need light to see objects. Look at this picture. The girl sees the red apple. She is able to see it because light reflects off the apple and travels in a straight line to her eyes.

158

Science in a Snap!

Investigating Reflection 

Materials index cards, pencils, aluminum foil, flashlights

Have students work in pairs. Distribute an index card, a pencil, a piece of aluminum foil, and a flashlight to each pair of students. Have students read and follow the instructions on page 159 of the Big Ideas Book. Students should record their observations and results in their science notebook.

What to Expect Students should observe that they have to manipulate the flashlight to get just the right angle.

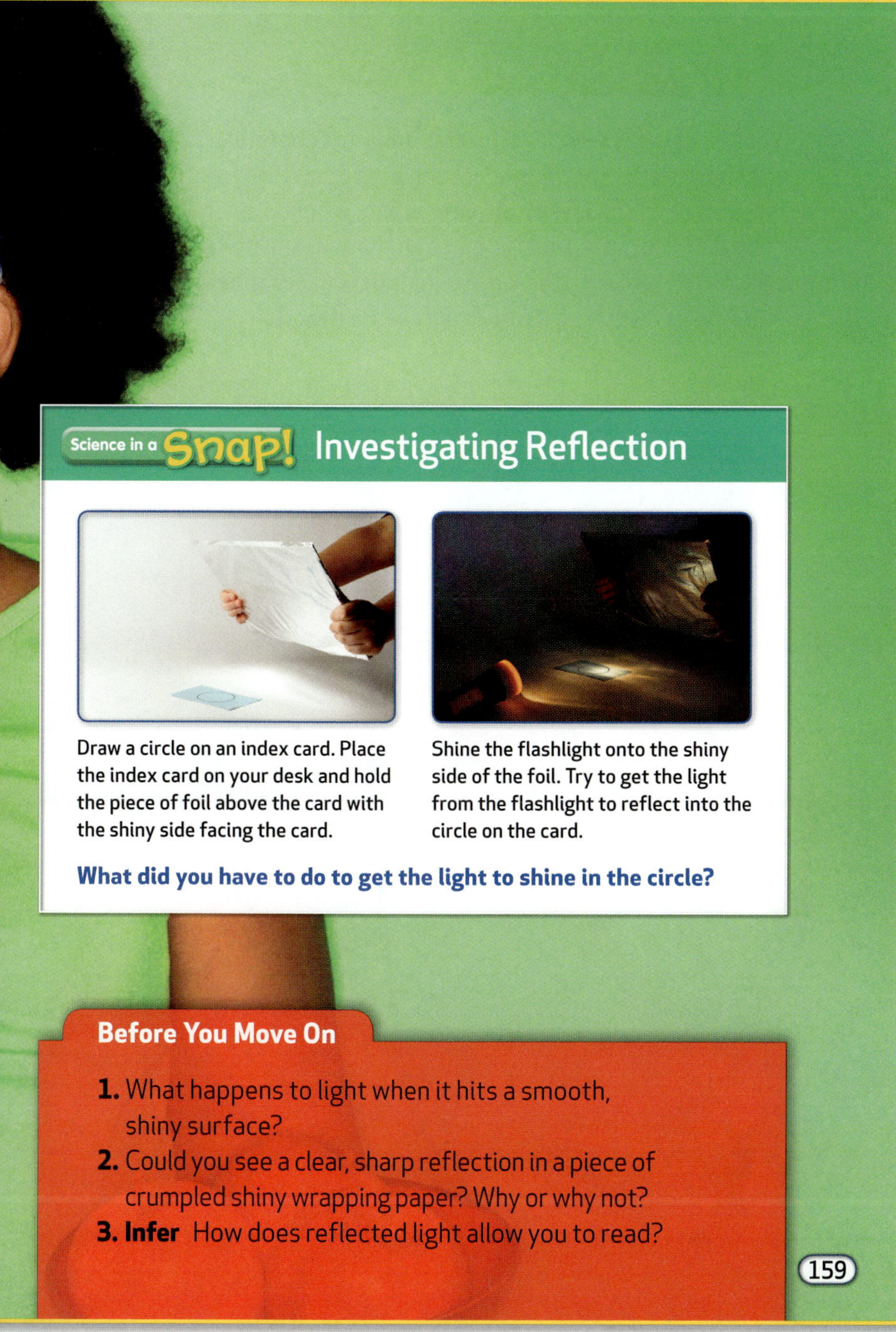

Making and Recording Observations

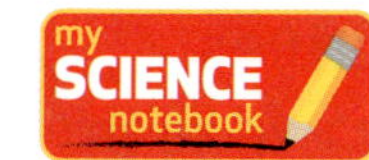

Have students think about times and places in which it is hard for them to see. For example, students might find it difficult to see at dusk. This is because there is less light that reflects from objects and enters the eye. They also might think of when they are in the shadow of an object. Students will learn about shadows in this chapter. In their science notebook, they should write about when and where it's hard to see. Have them explain why they think seeing is hard for them at those times and in those places.

» Before You Move On

1. **Describe What happens to light when it hits a smooth, shiny surface?** (Most of it bounces, or reflects, off the shiny surface.)

2. **Explain Could you see a clear, sharp reflection in a piece of crumpled shiny silver wrapping paper? Why or why not?** (No. Although the surface is shiny, it is not smooth. Light needs a smooth, shiny surface to make a clear reflection.)

3. **Infer How does reflected light allow you to read?** (Light goes from the light source to the book page. Then, it reflects from the book page to your eyes.)

Quick Questions Ask students the following questions:

- **Why did you have to manipulate the flashlight to get just the right angle before the light would reflect into the circle?** (Because light reflects off of a surface in a straight line, we had to manipulate the flashlight to get the light to reflect straight off the foil and into the circle.)

- **How does reflection make it possible for you to see in this experiment?** (Some of the light reflects off the index card to your eyes.)

LESSON 3 ▫ Refraction

Objectives

Students will be able to:

- Recognize that light travels in a straight line.
- Describe the effect of refraction when light travels from one material to another.

Science Academic Vocabulary

refraction

PROGRAM RESOURCES

- **Enrichment Activities** at ⊘ **myNGconnect.com**

❶ Introduce

Tap Prior Knowledge

- Ask students if they have ever seen a rainbow and have them describe what they observed. Ask: **What was the weather like when the rainbow appeared in the sky?**

Set a Purpose and Read

- Read the heading. Tell students that they will read about what happens when light passes through different materials instead of being reflected.
- Have students read pages 160–161.

❷ Teach

Academic Vocabulary: *refraction*

- Point out the word **refraction**. Reread the sentence to define it. Say: **refraction occurs when light passes from air into water.**

Recognize How Light Travels

- Say: **Light travels in a straight line through air.** Ask: **What everyday examples can you give of this property of light?** (a movie projector, a flashlight, car headlights)
- Ask: **What is different about the way in which light travels when it goes from air into water?** (The light can change direction, or bend, when it travels from air into water.)

Refraction

You've probably used a straw to drink water before. Did you ever notice that the straw can look bent from the side of the glass? As light passes from air to water, it bends, or changes direction. The straw shows that light bent at the place where it passed from air into the water. Light bending as it passes from one kind of matter to another is called **refraction**.

(160)

Differentiated Instruction

Extra Support

To help students understand the difference between reflection and refraction, have them draw a picture of each concept. Correct any misconceptions students might have about these concepts.

Challenge

Bring in a variety of prisms, and let students take turns shining the light from a flashlight through them. Ask: **Does each prism cause white light to break into the same colors in the same order every time?** (yes)

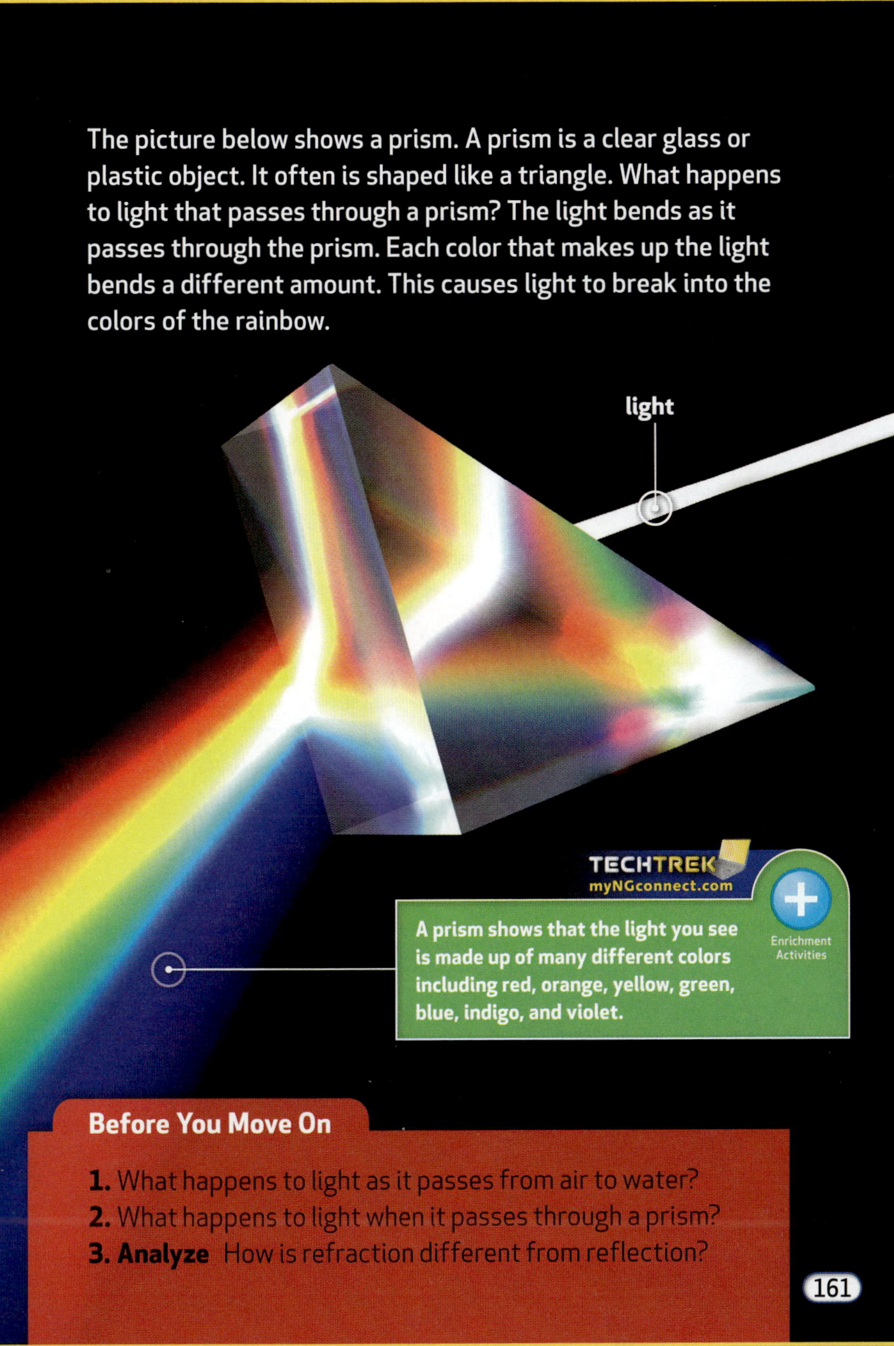

The picture below shows a prism. A prism is a clear glass or plastic object. It often is shaped like a triangle. What happens to light that passes through a prism? The light bends as it passes through the prism. Each color that makes up the light bends a different amount. This causes light to break into the colors of the rainbow.

Before You Move On

1. What happens to light as it passes from air to water?
2. What happens to light when it passes through a prism?
3. **Analyze** How is refraction different from reflection?

161

Teach, continued

Demonstrate and Describe Refraction

- Give each student a spoon and a clear glass filled halfway with water. Have students observe the spoon and then lower it into the water and observe it again. Say: **Describe what you see.** (The spoon looked straight until I put it in the water. Then it looked bent.) Ask: **What is happening?** (Refraction occurs; the light bends as it passes from the air to the water in the glass.)

- Explain: **Refraction occurs when light passes through a prism. The light bends and separates into the colors of the rainbow.** Ask: **What causes the light to bend?** (The light passes from the air into the glass prism.) Explain that a similar effect occurs when sunlight passes from the air through raindrops, which act like a prism to form a rainbow in the sky.

 **Enrichment Activities**

 **myNGconnect.com**

Have students view the photos and animations and do the **Enrichment Activities** to further their understanding of reflection and **refraction**.

Integrated Technology

Whiteboard Presentation Students can use the information to make a computer presentation about reflection and **refraction**. You can help them show the presentation on a whiteboard.

❸ Assess

» Before You Move On

1. **Recall** **What happens to light as it passes from air to water?** (When light passes from air to water, it refracts, which means it bends or changes direction.)

2. **Explain** **What happens to light when it passes through a prism?** (It bends. Each color that makes up the light bends a different amount, causing the light to break into the colors of the rainbow.)

3. **Analyze** **How is refraction different from reflection?** (When light reflects, it bounces off the surface of the object. When light refracts, it passes through a material and bends as it passes from one kind of matter to another.)

Differentiated Instruction

ELL Language Support for Describing Refraction

BEGINNING	INTERMEDIATE	ADVANCED
Help students talk about refraction by asking yes/no questions.	Provide Academic Language Frames for students to discuss refraction.	Have students use Academic Language Stems to explain refraction.
Is reflection when light bounces straight back? Is refraction when lights bends?	*Light travels in a _____ line. Light _____ when it passes from one kind of matter to another. This is called _____.*	• *Light travels…* • *In refraction, light…*

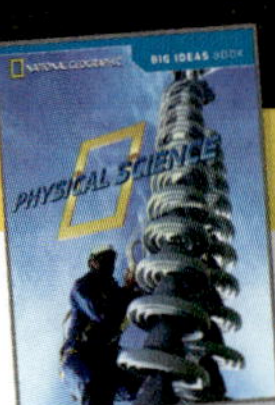

LESSON 4 □ Absorption

Objectives

Students will be able to:

- Classify objects that absorb light and objects that reflect light.
- Recognize that objects heat up, depending on how much light they absorb.

Science Academic Vocabulary

absorption

PROGRAM RESOURCES

- Science Inquiry and Writing Book: *Physical Science*
- Science Inquiry and Writing Book **eEdition** at ⊘ **myNGconnect.com**
- **Digital Library** at ⊘ **myNGconnect.com**

❶ Introduce

Tap Prior Knowledge

- Have students look at the clothes they are wearing. Ask: **Which articles of clothing will heat up faster in the sunlight?** (dark-colored ones)

Preview and Read

- Read the heading and have students preview the pictures on pages 162–163.
- Have students read pages 162–163.

❷ Teach

Academic Vocabulary: *absorption*

- Pronounce **absorption** and write it. Have a student read the definition. Say: **Dark-colored objects absorb a lot of light. Light-colored objects reflect more light than they absorb.**

Relate Heat and Absorption of Light

- Say: **The more light an object absorbs, the more it will heat up.** Ask: **Why does a black car get hotter than a white one on a sunny day?** (A dark-colored car absorbs a lot more light than a white one, so it heats up more.)

Digital Library

⊘ **myNGconnect.com**

Have students use the **Digital Library** to find photos of light- and dark-colored objects.

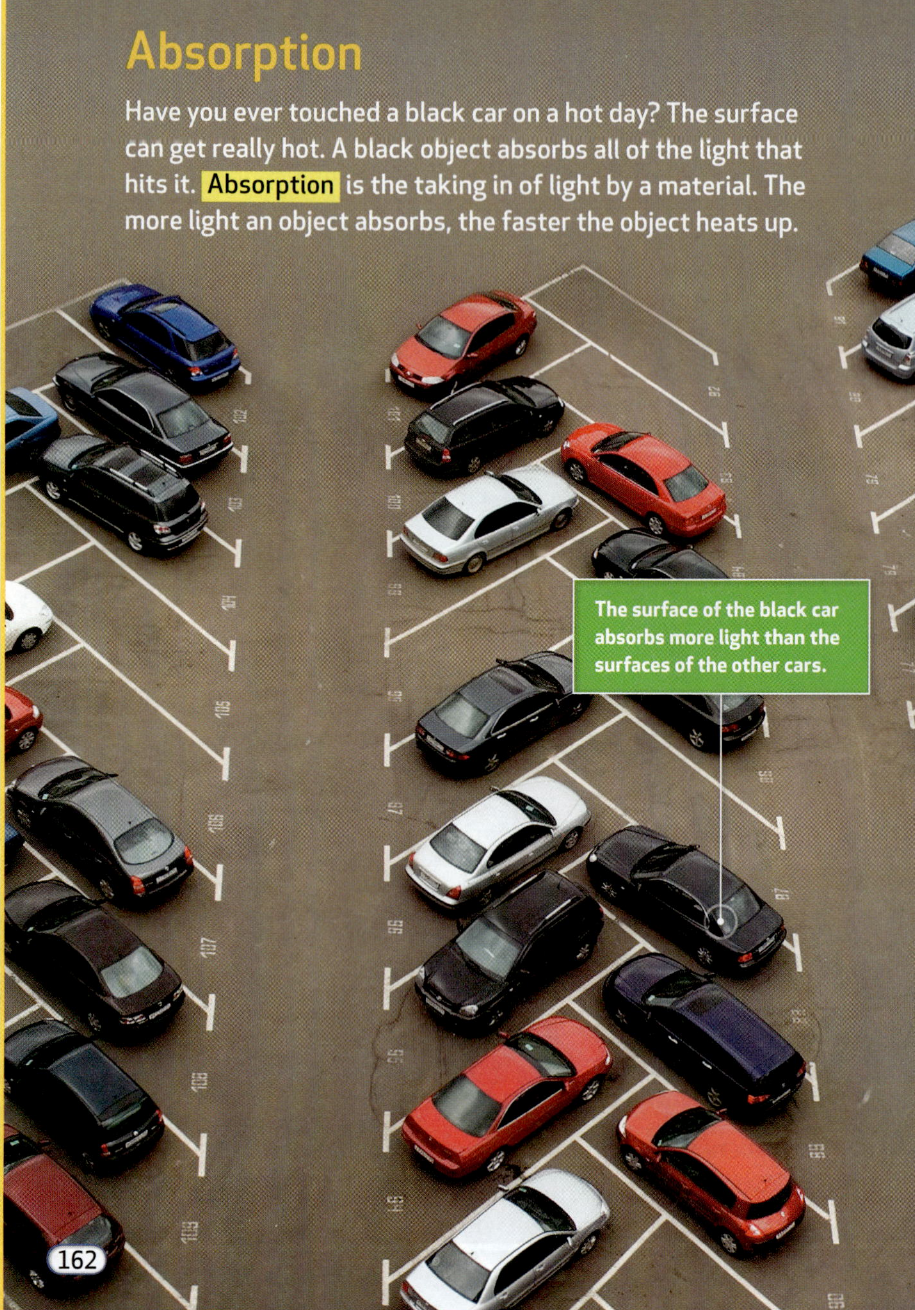

Absorption

Have you ever touched a black car on a hot day? The surface can get really hot. A black object absorbs all of the light that hits it. **Absorption** is the taking in of light by a material. The more light an object absorbs, the faster the object heats up.

Differentiated Instruction

ELL Language Support for Describing Absorption

BEGINNING	INTERMEDIATE	ADVANCED
Have students point to dark-colored articles of clothing that absorb more light and heat up more quickly. Then have students identify light-colored articles of clothing that absorb less light and heat up less quickly.	Have students use Academic Language Frames to explain absorption: • A _____-colored object absorbs most of the light that strikes it. • A _____-colored object does not absorb much light.	Have students use Academic Language Stems to elaborate on ideas about absorption. • A light-colored object … • A dark-colored object …

A white object does not absorb much light. Look at the photos. If you are out in the hot sun, which cap would you wear? If you want to stay cool, you would choose the white cap. The dark cap will get hot faster than the white cap.

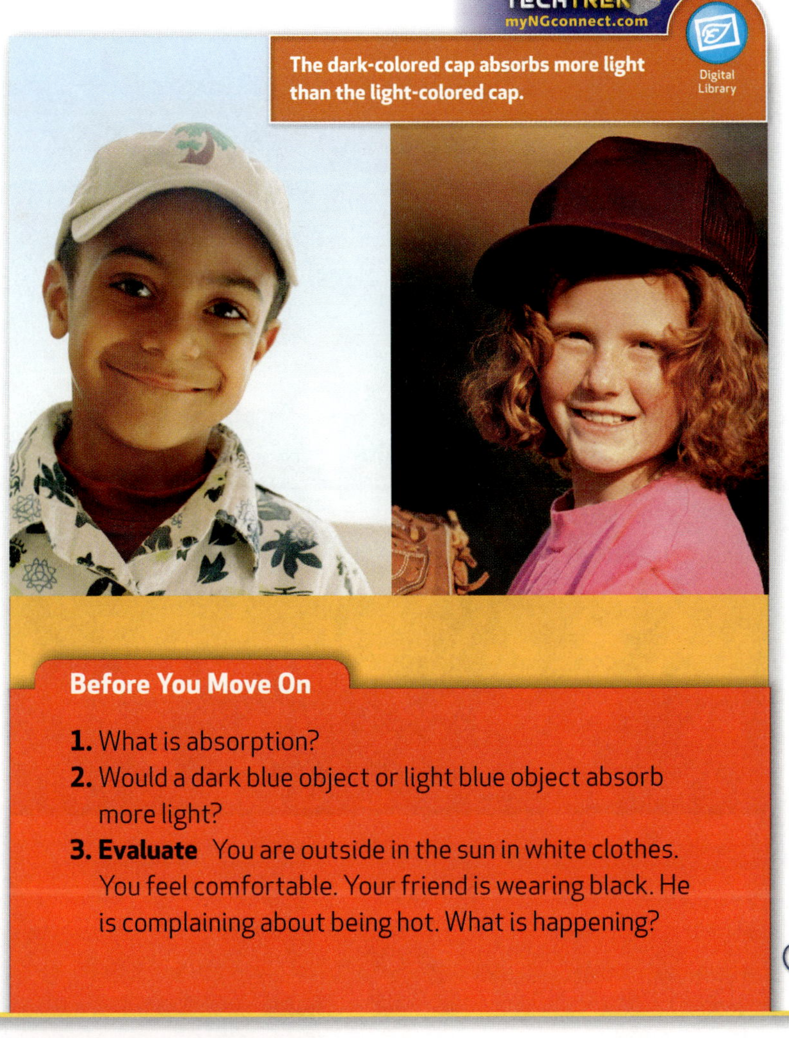

Before You Move On

1. What is absorption?
2. Would a dark blue object or light blue object absorb more light?
3. **Evaluate** You are outside in the sun in white clothes. You feel comfortable. Your friend is wearing black. He is complaining about being hot. What is happening?

(163)

Science in a Snap! 

Observe the Heat 

Science Inquiry and Writing Book: *Physical Science,* page 165

Materials white construction paper, black construction paper, foil, tissue, plastic wrap

Caution students not to touch the light bulb. Have students record their observations and results.

What to Expect Materials that absorb less and reflect more of the light that hits them (white construction paper) feel cool, while materials that absorb more and reflect less of the light (black construction paper) feel warm.

» Before You Move On

1. **Recall** What is **absorption**? (Absorption is the taking in of light by a material.)

2. **Compare** Would a dark blue object or light blue object absorb more light? (dark blue object)

3. **Evaluate** You are outside in the sun in white clothes. You feel comfortable. Your friend is wearing black. He is complaining about being hot. What is happening? (My friend's black clothes are absorbing more light than my white clothes. As a result, his clothes are heating up faster, and he feels hot.)

Extend Learning

Absorption vs. Reflection

Find Out	Think and Do	Describe and Compare
Remind students that most objects both absorb and reflect light, but darker-colored materials tend to absorb more light and lighter-colored things reflect more light. Have groups of students classify objects in the classroom as things that absorb more light and things that reflect more light.	 Have students make a chart in their science notebook to classify materials found in the classroom according to their ability to absorb and reflect light. They may want to establish three or more categories.	Have groups take turns sharing their charts with the class. Be sure they can explain their reasons for classifying each object or material as they did.

LESSON 4 □ Shadows

Objectives

Students will be able to:

- Recognize that light travels in a straight line.
- Explain why placing an object in a path of light causes a shadow.
- Identify three things needed to produce a shadow (light source, object, and surface).

PROGRAM RESOURCES

- **Digital Library** at ⊘ **myNGconnect.com**
- Learning Masters Book, page 228, or at ⊘ **myNGconnect.com**

❶ Introduce

Tap Prior Knowledge

- Ask students if they have ever noticed their shadow. Have them share their observations.

Preview and Read

- Have students preview the pictures and then read the text on pages 164–165.

❷ Teach

Explain How Shadows Form

- Say: **Light travels in a straight line.** Ask: **What happens if an object is in the path of the light?** (A shadow forms where the light cannot reach.) Using the picture on page 164, have students point out the three things needed to make a shadow.

Extend Learning

MANAGING THE INVESTIGATION

Time

🕐 20 minutes

Groups

👥 Small groups of 4

PROGRAM RESOURCES

- Learning Master 228

MATERIALS

- toy figures
- flashlights
- paper
- pencils
- a desk

Shadows

You need three things to make a shadow: a light source, an object, and an area where the light cannot reach.

You can make hand shadows by blocking the light from a flashlight. Since light always travels in a straight line, the light cannot curve around your hand. A shadow forms in the area where the light cannot reach.

(164)

Shadows Where?

Preview	What To Do
 **What causes shadows to form?** Students will demonstrate that light travels in a straight line until it strikes an object. They will observe that when light strikes an object, a shadow forms behind the object.	**1.** Arrange the toy figures on a desk. Put a flashlight in the middle of the figures. Turn the flashlight on. **2.** Draw what you see. Where are the shadows? Why did they form where they did? **3.** Move the flashlight to a new position. Draw what you see. How did the shadows change? Why did the shadows change?

Demonstrate That Light Travels in a Straight Line

- Have students study the photo on page 164. Ask: **Why is the boy's hand casting a shadow on the wall?** (because he put his hand in front of the flashlight) **Why didn't the light just go around his hand?** (because light travels in a straight line)

- Explain that outdoors, shadows change through the day as the position of the sun changes in the sky.

myNGconnect.com

Have students use the **Digital Library** to find photos of shadows.

❸ Assess

❱❱ Before You Move On

1. **List What three things do you need to make a shadow?** (light source, object to block the light source, an area where light cannot reach)

2. **Explain It's a sunny day. You stand on the sidewalk at noon. You stand in the same place in the early evening. Your shadow is different each time. Why?** (At noon, the sun is high in the sky—almost overhead. It makes my shadow look short. In the early evening, the sun is low in the sky. It makes my shadow looks long.)

3. **Infer How can you make a shadow disappear?** (Remove the object that is blocking the light source; turn off the light source.)

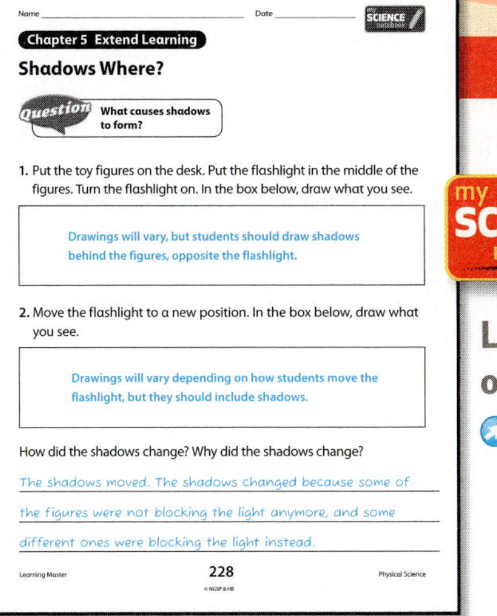

my SCIENCE notebook

Learning Master 228 or at

myNGconnect.com

Explain Results

Students should explain that:

- When the light from a flashlight shines on a toy figure, a shadow forms behind the figure, because the figure is blocking the light source.

- When you move the flashlight, the shadows change because the light source is in a different position and different toy figures are blocking the light.

- Three things are needed to produce a shadow: a light source (the flashlight), an object (the toy figures), and a surface upon which a shadow can form (the desk).

LESSON 5 □ Light Pollution in the Sky

Objectives

Students will be able to:
- Identify sources of light energy.
- Explain that light reflects off of some surfaces.

PROGRAM RESOURCES
- Learning Masters Book, page 229, or at
  ☁ **myNGconnect.com**

❶ Introduce

Tap Prior Knowledge
- Ask students to talk about the difference in night skies in the city and outside the city. Discuss why more stars are visible outside of the city.

Preview and Read
- Read the heading. Have students preview the photos on pages 166–167.
- Have students read pages 166–167.

❷ Teach

Identify Light Energy and Sources of Light Energy
- Ask: **What is light?** (It is a kind of energy we can see.) **What are some of the properties of light?** (Light always travels in a straight line; light can be reflected, refracted, or absorbed.)
- Have students think about sources of light energy that are made by people. Ask: **What sources of light could cause light pollution?** (streetlights, signs, yard lights, lights in or on houses or businesses, car headlights)

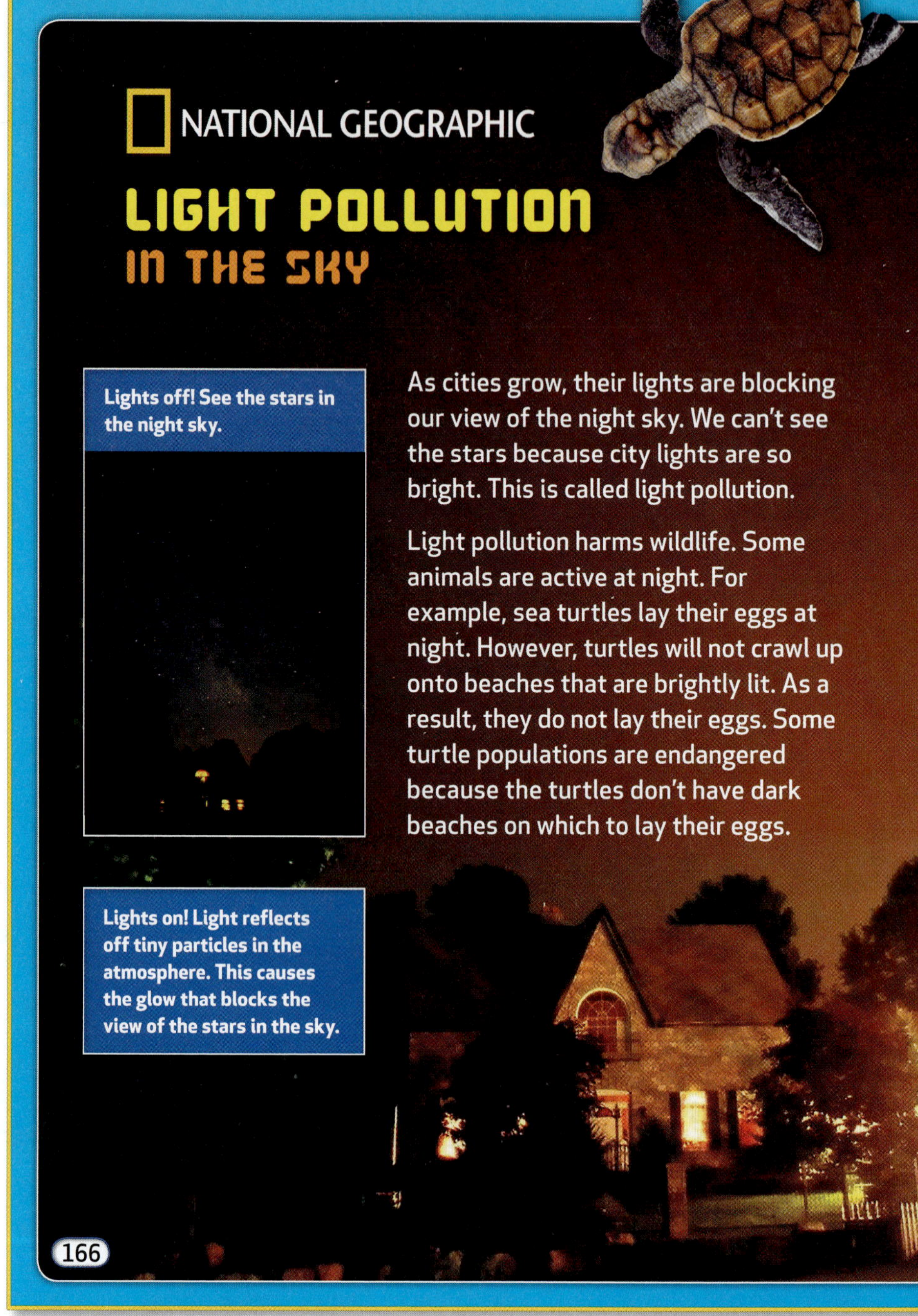

Social Studies in Science

Loggerhead Turtles in Florida Some species of turtles, such as loggerhead turtles, spend parts of their lives migrating around the Sargasso Sea. Show a map of the world and point out the Sargasso Sea. Tell students that loggerhead turtles return to the beaches of Florida to lay their eggs. Point out Florida on the map. Tell students that to avoid predators, newly hatched loggerhead turtles wait under the sand until night before they walk to the ocean for the first time. If there are lights on the beach, they become confused, walk toward the lights, and often die. Thus, light pollution affects not only mother turtles but also their offspring.

Light pollution can be reduced by using lights that are designed to keep light from shining upward and outward. Decreasing light pollution will keep the night sky dark and will benefit many living things.

167

Demonstrate the Reflection of Light

- Give students flashlights and darken the classroom. Have groups of students point their flashlights at the ceiling, the floor, or straight out from their chests. Have them observe the reflections. Ask: **How are our flashlights like city lights?** (Building lights point into the sky, streetlights point to the ground, and headlights point horizontally.)

❭ **Make Inferences**

Have students reread the paragraph on page 167. Tell students that many people want to live or build businesses such as restaurants near the ocean. Ask: **How do you think people are making beaches too bright for turtles to lay their eggs on?** (There are lights on the inside and the outside of people's houses and businesses near the ocean. Cars with their bright headlights drive on lighted roads near the ocean. All those things light up the beaches.)

❸ Assess

1. **Describe** **In some places, why is it hard to see the stars at night?** (As cities grow, lights made by people are blocking our view of the stars.)

2. **Cause and Effect** **How is light pollution affecting sea turtles?** (Sea turtles will not lay their eggs on brightly lit beaches, so the more light pollution there is, the more endangered sea turtles become.)

3. **Analyze** **How can people help reduce light pollution?** (They can point lights at the ground, so they don't shine into the sky and block the stars or shine onto the beaches and keep sea turtles from laying their eggs. People can also use fewer lights on the outside of their homes and businesses.)

Share and Compare

Reflection, Refraction, or Absorption?

Give students the Learning Master. Have them classify the examples as reflection, refraction, or absorption. Then have them write, in their own words, a definition of these three properties of light under the graphic organizer. Have partners brainstorm and discuss other examples of reflection, refraction, and absorption.

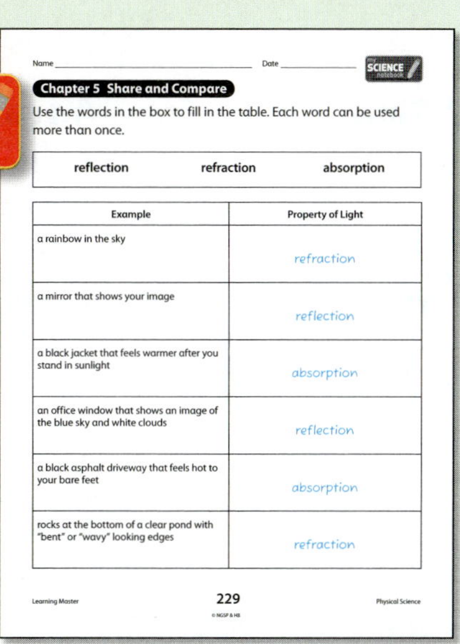

Name _______________ Date _______________

Chapter 5 Share and Compare

Use the words in the box to fill in the table. Each word can be used more than once.

reflection	refraction	absorption

Example	Property of Light
a rainbow in the sky	refraction
a mirror that shows your image	reflection
a black jacket that feels warmer after you stand in sunlight	absorption
an office window that shows an image of the blue sky and white clouds	reflection
a black asphalt driveway that feels hot to your bare feet	absorption
rocks at the bottom of a clear pond with "bent" or "wavy" looking edges	refraction

Learning Master   229   Physical Science

Learning Master 229 at 🌐 **myNGconnect.com**

Physical Science **T167**

LESSON 6 ▫ Guided Inquiry

Objectives

Students will be able to:

- Investigate through Guided Inquiry (answer a question; make and compare observations; collect and record data and observations; generate explanations and conclusions based on evidence; share findings; ask questions based on observations to increase understanding; adjust explanations based on findings and new ideas).
- Demonstrate that light can be reflected by, absorbed by, and can pass through some objects.
- Keep records that describe observations, carefully distinguish actual observations from ideas and speculations, and are understandable weeks and months later.
- Evaluate the reasonableness of an explanation.

Science Process Vocabulary *compare*

PROGRAM RESOURCES

- Science Inquiry and Writing Book: *Physical Science*
- Science Inquiry and Writing Book **eEdition** at ⊙ **myNGconnect.com**
- **Inquiry eHelp** at ⊙ **myNGconnect.com**
- Science Inquiry Kit: *Physical Science*
- Learning Masters Book, pages 230–233, or at ⊙ **myNGconnect.com**
- Inquiry Rubric: Assessment Handbook, page 215, or at ⊙ **myNGconnect.com**
- Inquiry Self-Reflection: Assessment Handbook, page 226, or at ⊙ **myNGconnect.com**

MATERIALS

Kit materials are listed in italics.

white paper; masking tape; *meterstick; flashlight; mirror; foil; black paper; cloth;* paper towel

For teacher use: scissors

❶ Introduce

Tap Prior Knowledge

- Ask students to think about the amount of light that passes through different curtains. Ask: **Do all curtains let in the same amount of light?** Discuss how different materials allow different amounts of light to pass through them.

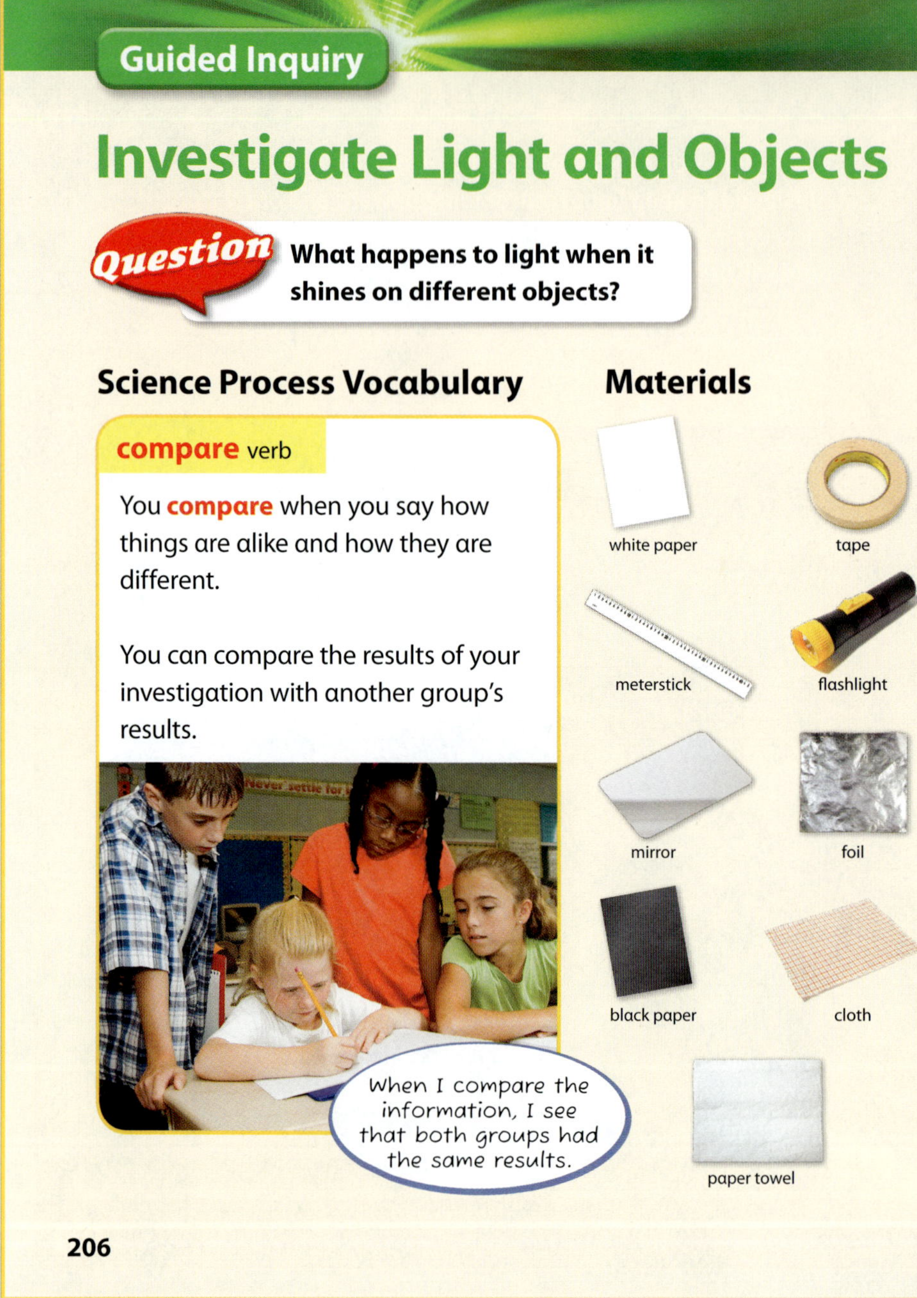

MANAGING THE INVESTIGATION

Time

30 minutes

Groups

Small groups of 4

Advance Preparation

- Cut a piece of foil and a piece of cloth for each group. Each piece should be 8½ × 11 in., the same size as the piece of paper.
- Make sure all flashlights work properly. Replace batteries if necessary.

What to Do

1 Tape a piece of white paper to the wall at eye level. **Measure** a distance of 50 cm from the paper. Put a piece of tape at that distance.

2 Stand at the piece of tape. Turn on the flashlight. Shine the flashlight on the white paper. **Observe** the light on the paper. Record your observations in your science notebook. Turn off the flashlight.

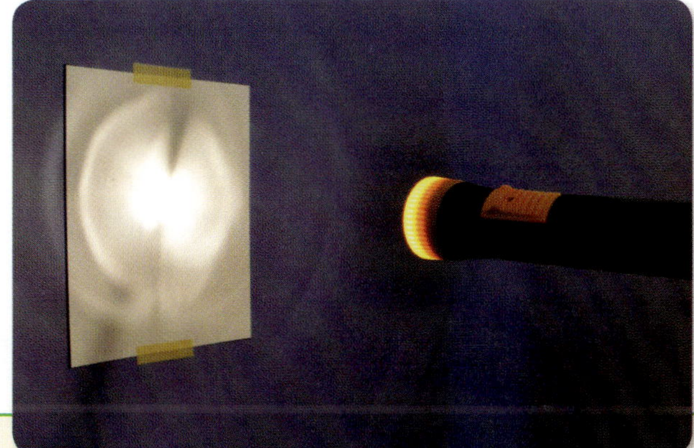

207

Teaching Tips

• Turn off overhead lights and close the blinds. Space out the groups so the light from neighboring groups does not affect the results.

• Make sure students can distinguish between different degrees of light and dark.

What to Expect

• Students will observe that the mirror and the foil reflected the most light. The cloth, the paper towel, and the black paper absorbed light and reflected some light. Some light passed through the cloth and the paper towel, but none passed through the black paper.

Introduce, continued

Connect to the Big Idea

• Review the Big Idea Question, *What is light?* Explain to students that this inquiry will demonstrate what happens when light shines on different objects.

• Have students open their Science Inquiry and Writing Books to page 206. Read the Question and invite students to share ideas about the energy of light.

❷ Build Vocabulary

Science Process Word: compare

Use this routine to introduce the word.

1. **Pronounce the Word** Say **compare**. Have students repeat it in syllables.

2. **Explain Its Meaning** Choral read the sentence. Ask students for another word or phrase that means the same as **compare**. (tell how objects or events are alike and different)

3. **Encourage Elaboration** Ask: **How would you compare the sun to a flashlight?** (They both give off light and produce heat. The sun is much larger, hotter, and brighter than a flashlight.)

ELL Use Cognates

Spanish speakers may know the word **comparar**. Use it to help them access the English word **compare**.

❸ Guide the Investigation

• Distribute materials. Read the steps on pages 207–208 together with students.

• In step 2, tell students that they will observe how the light appears on the white paper and compare it with how it appears on other materials.

• Students should think about what they observed as they write their predictions.

• In step 3, students will choose three of the following: mirror, foil, black paper, cloth, or paper towel. Make sure each material is tested by at least one group.

• Remind students to keep the distances consistent. Ask: **Why is it important to keep the distances consistent?** (Changing distances could affect how the light is absorbed and reflected on the white paper and on the objects.)

❹ Explain and Conclude

- Guide students as they record their observations in tables and write their conclusions. Ask students to compare predictions and results and tell how their understanding has changed.

- Review what students have learned about light. Lead them to make inferences. Ask: **What is a reflection?** (light bouncing off a surface) **Which objects reflected light?** (All of the materials reflected light.) **How do you know?** (I could see some light from the flashlight on all of the materials.) Then ask: **How could you tell which objects let light pass through them?** (I could still see light on the wall when I shined the flashlight toward them.) Tell students to evaluate their answers to check if they are reasonable.

- Ask students to review their observations. Discuss that clear records, based on observation, are important so they can be used later. Have students use their observations to evaluate the reasonableness of their answers.

- Make sure students understand that light reflects off objects, travels through space, and enters our eyes. That is how we see the objects.

- Discuss objects such as clothing, curtains, and napkins. These objects absorb some light, but also reflect some light or we would not see them. Invite students to name things that reflect light in clear images, such as water or glass in buildings.

Answers

1. Possible answer: I could see some light on the white paper when I shined the flashlight at the cloth. I did not see any light on the white paper when I shined the flashlight at the black paper and the mirror.

2. Possible answer: I could see some light on the black paper and the cloth, but I could see the most light on the mirror and the foil. Some of the light was absorbed by the cloth and the black paper. Some light could pass through the cloth. Most of the light bounced off the mirror and the foil.

3. Possible answer: Shiny objects reflect the most light. Other objects may reflect some light, absorb some light, or let some light pass through.

What to Do, continued

3 Select 3 objects. **Predict** what will happen to the light on the white paper when you put each object between the flashlight and the paper. Record your predictions.

4 Have a partner hold 1 object in front of you. Hold the flashlight behind the object, pointing it toward the white paper.

5 Turn on the flashlight. Observe the amount of light on the white paper and the object. Record your observations. Repeat with the other 2 objects.

208

NATIONAL GEOGRAPHIC **Raise Your SciQ!**

Is Light a Form of Matter? Students may assume that because they can see light, it must be matter. The traditional definition of matter says that matter is any material with mass. However, light does not have mass. It is energy that sometimes acts like a wave and sometimes acts like a particle, depending on how it is measured. Like other waves, light can be reflected, refracted (bent), or absorbed. Like particles, light travels in a straight line and can bounce off of objects.

Record

Write in your science notebook.
Use a table like this one.

Light on Paper			
Material Between Flashlight and Paper	Predictions	Observations	
	On the white paper	On the white paper	On the object
None			

Explain and Conclude

1. **Compare** how bright the light on the white paper was when you shined the flashlight at the 3 different objects.

2. Were you able to see the same amount of light on all 3 objects in step 5? Why do you think that is so?

3. Based on the results of your **investigation,** what can you **infer** about what happens to light when it shines on different objects?

Think of Another Question

What else would you like to find out about what happens to light when it shines on different objects? How could you find an answer to this new question?

Inquiry Rubric at ⊘ **myNGconnect.com**	Scale			
The student **observed** the light on the white paper.	4	3	2	1
The student recorded observations in a table.	4	3	2	1
The student **compared** how light interacted with different objects.	4	3	2	1
The student made **inferences** about how light is reflected by, absorbed by, or can pass through different objects.	4	3	2	1
The student **shared** observations and **conclusions** with other students.	4	3	2	1
Overall Score	4	3	2	1

❺ Find Out More

Think of Another Question

- Students should use observations made from this investigation to generate other questions that could be studied using readily available materials. Have them use the question stems, *What happens when…, What would happen if…?* or *How can I/we…?* to form testable questions.

- Record student questions for possible future investigations.

- Discuss what students could do to find answers.

❻ Reflect and Assess

- To assess student work with the Inquiry Rubric shown below, see Assessment Handbook, page 215, or go online to ⊘ **myNGconnect.com**

- Score each item separately and then decide on one overall score.

- Have students use the Inquiry Self-Reflection on Assessment Handbook, page 226, or at ⊘ **myNGconnect.com**

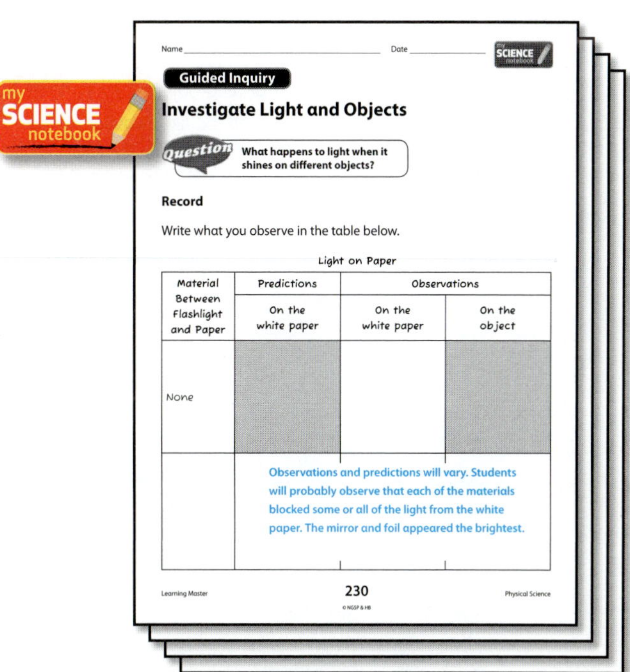

Learning Masters 230–233
or at ⊘ **myNGconnect.com**

Physical Science **T167d**

PROGRAM RESOURCES
- Chapter 5 Test, Assessment Handbook, pages 106–109, or at 🌐 **myNGconnect.com**
- NGSP ExamView CD-ROM

❶ Sum Up the Big Idea

- Display the chart from page T152–T153. Read the Big Idea Question. Then add new information to the chart.

- Ask: **What did you find out in each section? Is this what you expected to find?**

What is Light?

Sources of Light
Joshua: The sun is the greatest source of light energy on Earth, but there are other sources of light. Many sources of light also give off heat.

Reflection
Mika: A reflection is light bouncing off a surface. Light travels in straight lines.

Refraction
Carla: Refraction is the bending of light when it moves through one kind of matter to another.

Absorption
Jamaal: Absorption is the taking in of light by a material. The more light an object absorbs, the faster the object heats up.

Shadows
Winona: Light cannot curve around objects, so a shadow forms in the area that the light cannot reach.

Conclusion

Light is energy you can see. Light can be reflected, refracted, or absorbed, depending on what object it hits.

Big Idea Light energy travels in a straight line and comes from sources that give off light and often give off heat.

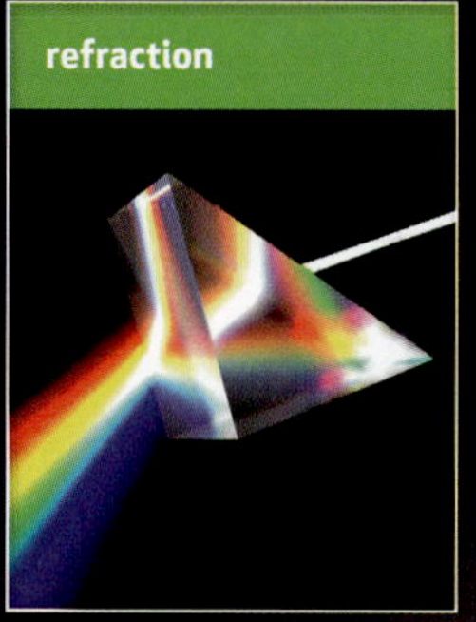

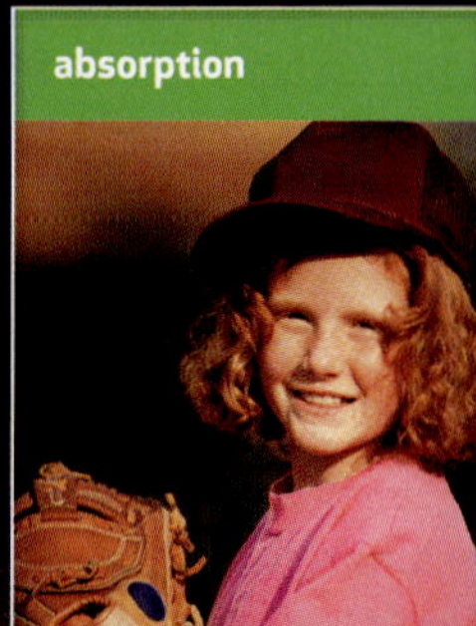

Vocabulary Review

Match the following terms with the correct definition.

A. light	**1.** The bouncing of light off of an object
B. reflection	**2.** The taking in of light by a material
C. refraction	**3.** A kind of energy you can see
D. absorption	**4.** The bending of light when it moves through one kind of matter to another

Review Academic Vocabulary

Academic Vocabulary Tell students that scientists use special words when describing their work to others. For example:

light   reflection   refraction   absorption

Have students write the list in their science notebook and use each word in a sentence that tells about the properties of light. They should then share their sentences with a partner.

Big Idea Review

1. Cause and Effect What happens when light hits an object?

2. Recall How do people see objects?

3. Analyze How can a mirror help you see what's behind you?

4. Explain Why does light cause some materials to heat up faster than others?

5. Predict What will happen when an object blocks the path of light?

6. Apply Imagine putting a pencil into a glass of water. Why might the pencil look like it bends?

Write About Refraction

Explain What is happening in this photo? How is the mist from the waterfall acting as a prism?

❷ Discuss the Big Idea

Notetaking

Have students write in their science notebook to show what they know about the Big Idea. Have them:

1. Write about how light enables people to see.

2. Explain how light acts when it passes from one kind of matter to another.

3. Describe how light can be reflected, refracted, and absorbed.

❸ Assess the Big Idea

Vocabulary Review

1. B 2. D 3. A 4. C

Big Idea Review

1. Some light will be reflected, especially if the surface is smooth and shiny. Some light may be absorbed and cause the object to heat up, especially if the object is a dark color.

2. Light reflects from the surface of an object and travels to a person's eyes.

3. Light reflects from objects behind you and goes to the mirror. That light then reflects from the mirror and goes to your eyes.

4. The object that absorbs more light heats up faster than an object that does not absorb as much light.

5. The object will make a shadow because light travels in a straight line and does not go around corners.

6. As light passes from air to water, it bends, or changes direction. The pencil will look bent at the place where the light passes from one kind of matter to another.

Write About Refraction

Light is passing from the air to the water that is falling. When the light passes into the water, it changes direction. This means the light bends or refracts. The water acts like a prism. This mean that different colors in the light bend, or refract, different amounts, and separate from one another. The bent light creates the appearance of a rainbow.

Assess Student Progress Chapter 5 Test

Have students complete their Chapter 5 test to assess their progress in this chapter.

Chapter 5 Test, Assessment Handbook, pages 106–109, or at 🌐 **myNGconnect.com**

or NGSP ExamView CD-ROM

Physical Science **T169**

LESSON 8 □ Physical Science Expert

Objectives

Students will be able to:

- Recognize that laser light is a form of light energy.

PROGRAM RESOURCES

- Big Ideas Book: *Physical Science*
- Big Ideas Book: *Physical Science* **eEdition** at ⊙ **myNGconnect.com**
- **Digital Library** at ⊙ **myNGconnect.com**

❶ Introduce

Tap Prior Knowledge

- Ask: **Where have you seen laser light?** (laser pointers, laser light shows, maybe in a TV show about surgical or industrial use of lasers) **What uses are there for lasers?** (CD and DVD players, correcting eyesight, cutting metal)

Preview and Read

- Read the heading. Have students preview the photos on pages 170–171. Ask them to suggest ideas about what kind of work a light expert does.
- Have students read pages 170–171.

❷ Teach

Identify and Describe Light

- Ask: **What is light?** (It is a kind of energy we can see.) **What are some sources of light?** (the sun, the stars, light bulbs, fire)

Classify Lasers as Light Energy

- Say: **A laser focuses light of a single color in one thin beam.** Ask: **What type of energy is laser light?** (It is a form of light energy.) **How do you think laser light travels?** (in a straight line) **How did you reach this conclusion?** (Laser light is a form of light energy, and all light travels in a straight line until it strikes an object or travels from one medium to another.)
- Say: **Dr. Bath invented a new device for using lasers.** Ask: **What discovery did she make?** (She discovered that she could help people see better by using the laser to remove cataracts.)

Inventor: Dr. Patricia Bath

Imagine inventing a device that could help people to see better! Dr. Patricia Bath did just that. She discovered that laser light could help.

Dr. Bath is an ophthalmologist. An ophthalmologist is a doctor who does eye surgery and treats diseases of the eye. Dr. Bath invented the laserphaco. This machine is used to remove cataracts. A cataract is a cloudy film over part of the eye. Cataracts can cause blurred vision.

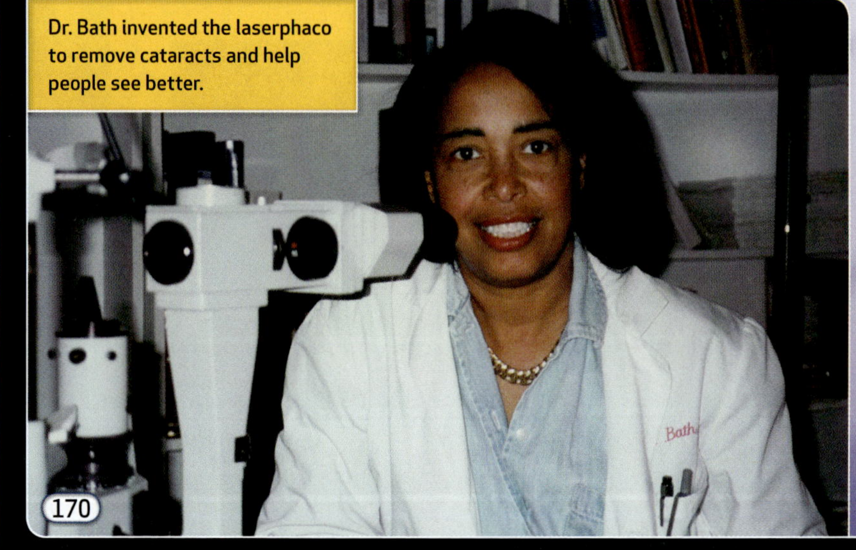

Dr. Bath invented the laserphaco to remove cataracts and help people see better.

170

NATIONAL GEOGRAPHIC Raise Your SciQ!

Laser Eye Surgery Doctors can use lasers to correct a variety of vision problems. One surgical procedure is called LASIK (laser-assisted in situ keratomileusis). In this procedure, doctors use a laser to reshape a person's cornea (the clear cover of the eye) by making tiny cuts in it. Sometimes a misshapen cornea will focus images either in front of or behind the retina, or the back of the eye. When a laser reshapes the cornea, it is then able to focus images on the retina, where they are supposed to be. People who undergo LASIK surgery usually no longer need to wear glasses or contact lenses.

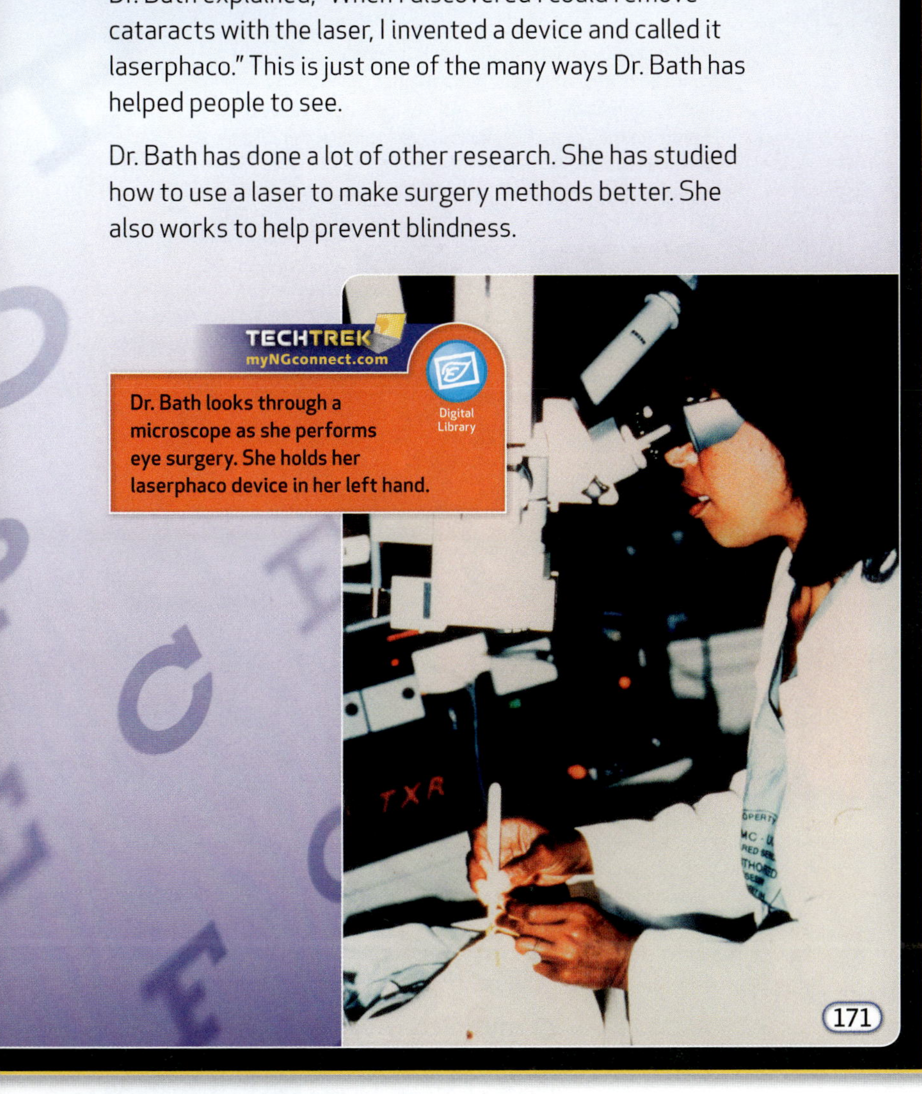

Dr. Bath explained, "When I discovered I could remove cataracts with the laser, I invented a device and called it laserphaco." This is just one of the many ways Dr. Bath has helped people to see.

Dr. Bath has done a lot of other research. She has studied how to use a laser to make surgery methods better. She also works to help prevent blindness.

Dr. Bath looks through a microscope as she performs eye surgery. She holds her laserphaco device in her left hand.

171

Differentiated Instruction

ELL ### Language Support for Classifying Lasers as Light Energy

BEGINNING

Help students use words related to lasers by asking yes/no questions:

Are lasers a form of light energy? Do lasers travel in a straight line?

INTERMEDIATE

Have students describe laser light using Academic Language Frames.

- *Lasers are a form of ______ energy.*
- *Lasers travel in a ______ line.*

ADVANCED

Have students elaborate on lasers by using Academic Language Stems.

- *Lasers are . . .*
- *Laser beams travel in . . .*

Analyze Word Parts

- Write the words *ophthalmology* and *ophthalmologist* on the board. Underline the root *ophthalmo-*. Say: **This word root comes from a Greek word that means "eye."** Underline the suffix *-logy*. Say: **This suffix comes from a Greek word that means, "the study of (something)."** Underline the suffix *-ist*. Say: **This suffix comes from a Greek word that means, "an expert in." So an *ophthalmologist* is a person who studies and is an expert on the eye.**

Digital Library

myNGconnect.com

Have students us the **Digital Library** to find photos relating to ophthalmology.

Integrated Technology

Computer Presentation Students can use the information to make a computer presentation about ophthalmology. Help students share their work.

Find Out More

- Ask: **Is this a career you would find interesting? Why or why not?** (Accept all answers, positive or negative.)

- Encourage interested students to research careers in ophthalmology. Have them find out about the technology used in this career. Also have them find out about other discoveries ophthalmologists have made.

❸ Assess

1. **Recall** What does an ophthalmologist do? (This kind of doctor does eye surgery and treats diseases of the eye.)

2. **Explain** How does Dr. Bath's invention use laser light? (Dr. Bath's laserphaco uses laser light to remove cataracts from people's eyes.)

3. **Draw Conclusions** Why might being an ophthalmologist be a rewarding job? (Possible answer: You would get to help people take care of their eyes. Helping people makes me feel good.)

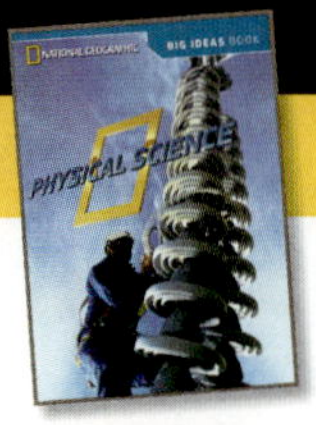

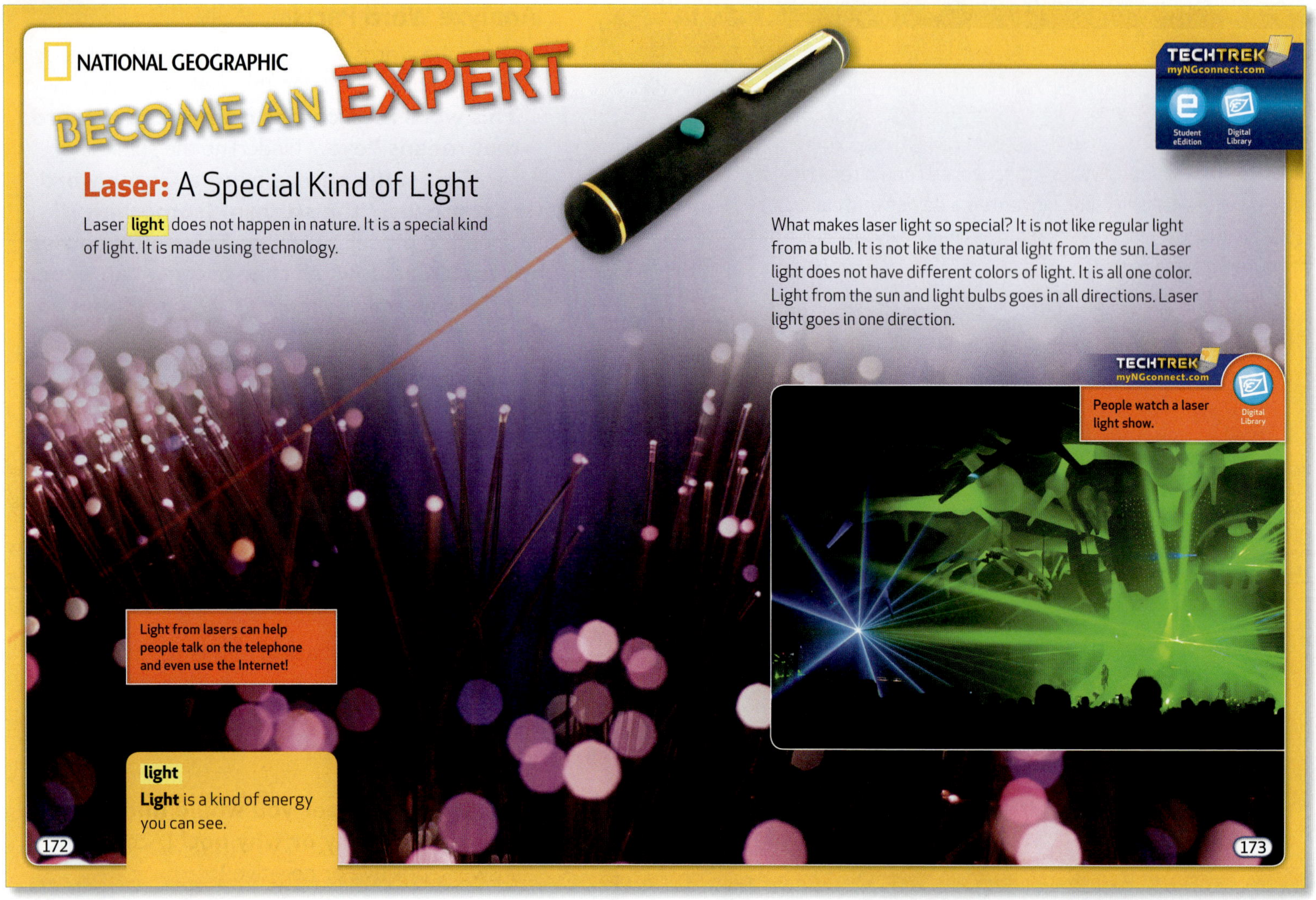

NATIONAL GEOGRAPHIC

BECOME AN EXPERT

Laser: A Special Kind of Light

Laser **light** does not happen in nature. It is a special kind of light. It is made using technology.

What makes laser light so special? It is not like regular light from a bulb. It is not like the natural light from the sun. Laser light does not have different colors of light. It is all one color. Light from the sun and light bulbs goes in all directions. Laser light goes in one direction.

People watch a laser light show.

Light from lasers can help people talk on the telephone and even use the Internet!

light
Light is a kind of energy you can see.

172

173

PROGRAM RESOURCES
- Big Ideas Book: *Physical Science*
- Big Ideas Book: *Physical Science* | eEdition | at ⊘ **myNGconnect.com**
- | Digital Library | at ⊘ **myNGconnect.com**

Access Science Content

Identify Laser **Light** as a Form of **Light** Energy
- Have students read page 172. Point out the photo of the laser pointer. Say: **Laser light is a special form of light energy.** Ask: **Is it the same or different from natural light that comes from the sun or human-made light that comes from a light bulb?** (different)
- Have students study the illustrations and read the captions on page 173. Ask: **How is laser light different from natural light?** (Natural light comes from a source in nature, such as the sun, the stars, or very hot objects that give off light energy. Laser light does not occur in nature; it is made only by people using technology. Laser light is all one color and goes in one direction away from its source, while natural light has many colors and can go out in all directions from a source.)

Digital Library

⊘ **myNGconnect.com**

Have students us the | Digital Library | to find photos of technology that makes use of lasers.

Integrated Technology

Digital Booklet Students can use the information to make a digital booklet about lasers and the ways in which people use them.

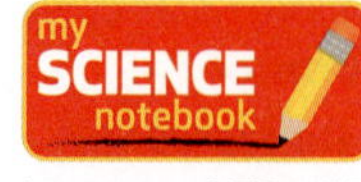

Notetaking

As students read pages 172–173, ask them to write down the word *laser*. Ask them to consult a dictionary and write the definition of this word in their science notebook.

BECOME AN EXPERT

The First Laser

The first laser was made in 1960. A ruby crystal was used. It was placed between two mirrors. The mirrors helped light reflect back and forth. This made a very powerful laser light. The ruby crystal showed the **reflection** of red light. The result was a ruby-colored laser light.

What Makes Lasers Useful?

We have learned a lot about lasers since 1960. We know laser light is powerful. A laser light beam can travel great distances—even to the moon! It is also easy to control. It can be pointed at the exact spot where it is needed. These traits make lasers very useful. Lasers are used in medicine and construction. They are also used in entertainment and in outer space. Lasers are everywhere.

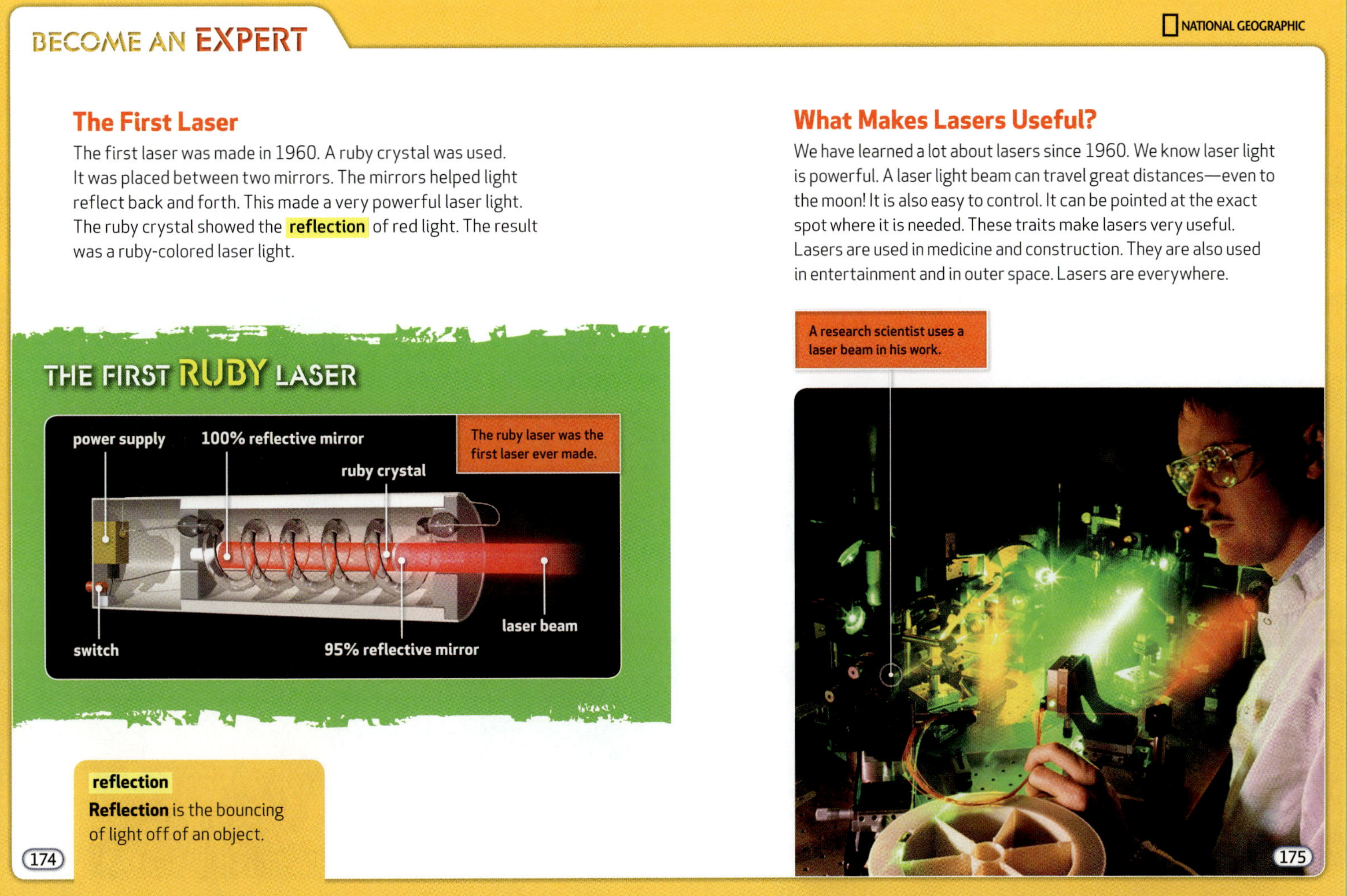

A research scientist uses a laser beam in his work.

reflection
Reflection is the bouncing of light off of an object.

174

175

Access Science Content

Describe the First Laser and Explain Why Lasers Are Useful

- Have students read page 174. Point out the word **reflection**. Say: **Remember that words that are highlighted are important words to understand.** Ask students to describe in their own words what **reflection** means.

- Ask: **Have you ever seen a laser?** Have students study the diagram on page 318. Ask: **How many mirrors are there?** (two) Ask: **What was the purpose of the mirrors?** (They helped the light reflect back and forth and boosted the laser's power.)

- Have students read page 175. Say, **When lasers were first invented, people called them "a solution looking for a problem." That is, scientists did not know what to use lasers for. But over the years, scientists have found uses for lasers.** Ask: **What do people use lasers for today?** (We use them in medicine, construction, entertainment, and in outer space.)

Assess

1. **Identify** What properties of laser light make lasers useful to people? (Laser light is powerful, can travel great distances in one direction, and is easy to control.)

2. **Contrast** What are the differences between natural light and laser light? (Natural light has light of different colors and scatters in all directions from its source. Laser light is one color and goes in one direction from the source, so that the laser source is bright even from far away.)

BECOME AN EXPERT

NATIONAL GEOGRAPHIC

How Eye Doctors Use Lasers

Do you know people who wear glasses? If they take off their glasses, things look out of focus or blurry. This happens because some people's eyes do not refract light correctly. **Refraction** is the bending of light. Eye doctors can use lasers to change the shape of the eye. After laser eye surgery, the eye can correctly refract the light. Then things are in focus and not blurry anymore.

Lasers Shape Metal

A powerful laser can cut through strong metals including steel. The metal **absorbs**, or takes in, the light energy. This makes the metal heat up until it melts. Lasers cut metal precisely. They are sometimes used to engrave small pieces of metal.

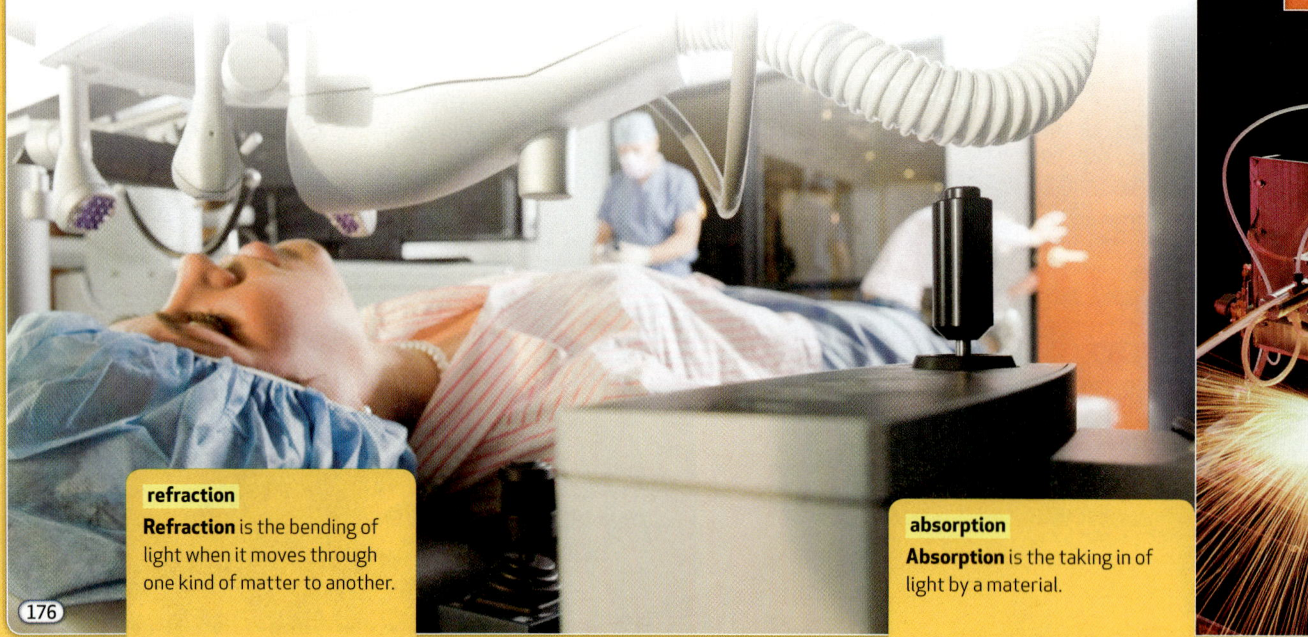

refraction
Refraction is the bending of light when it moves through one kind of matter to another.

176

absorption
Absorption is the taking in of light by a material.

177

Access Science Content

Explain How Lasers Are Used in Medicine

- Have students read pages 176–177. Explain: **Some people cannot see clearly because their eyes do not refract, or bend, light properly.** Ask: **How do doctors use lasers to correct problems with refraction?** (They use lasers to change the shape of the eye so it bends the light correctly.) Point out that a laser is also strong enough to cut steel because the metal can **absorb** the laser's powerful light energy.

Differentiated Instruction

ELL ## Language Support for Explaining How Lasers are Used in Medicine

BEGINNING

Ask students yes/no questions to help them with the uses of lasers. For example:

Can doctors use lasers to change the shape of the eye?

INTERMEDIATE

Have students use verbs to complete the Academic Language Frame:

During eye surgery, a laser can ______ the shape of the eye.

ADVANCED

Have student use Academic Language Stems to compare and contrast the uses of lasers.

Lasers can be used for...

Notetaking

Have students study the text and captions on pages 176–177. Have them list the words *reflection,* **refraction,** and **absorption** in their science notebook. Then have them write a sentence for each word telling how a laser can be used to demonstrate each of those properties of light.

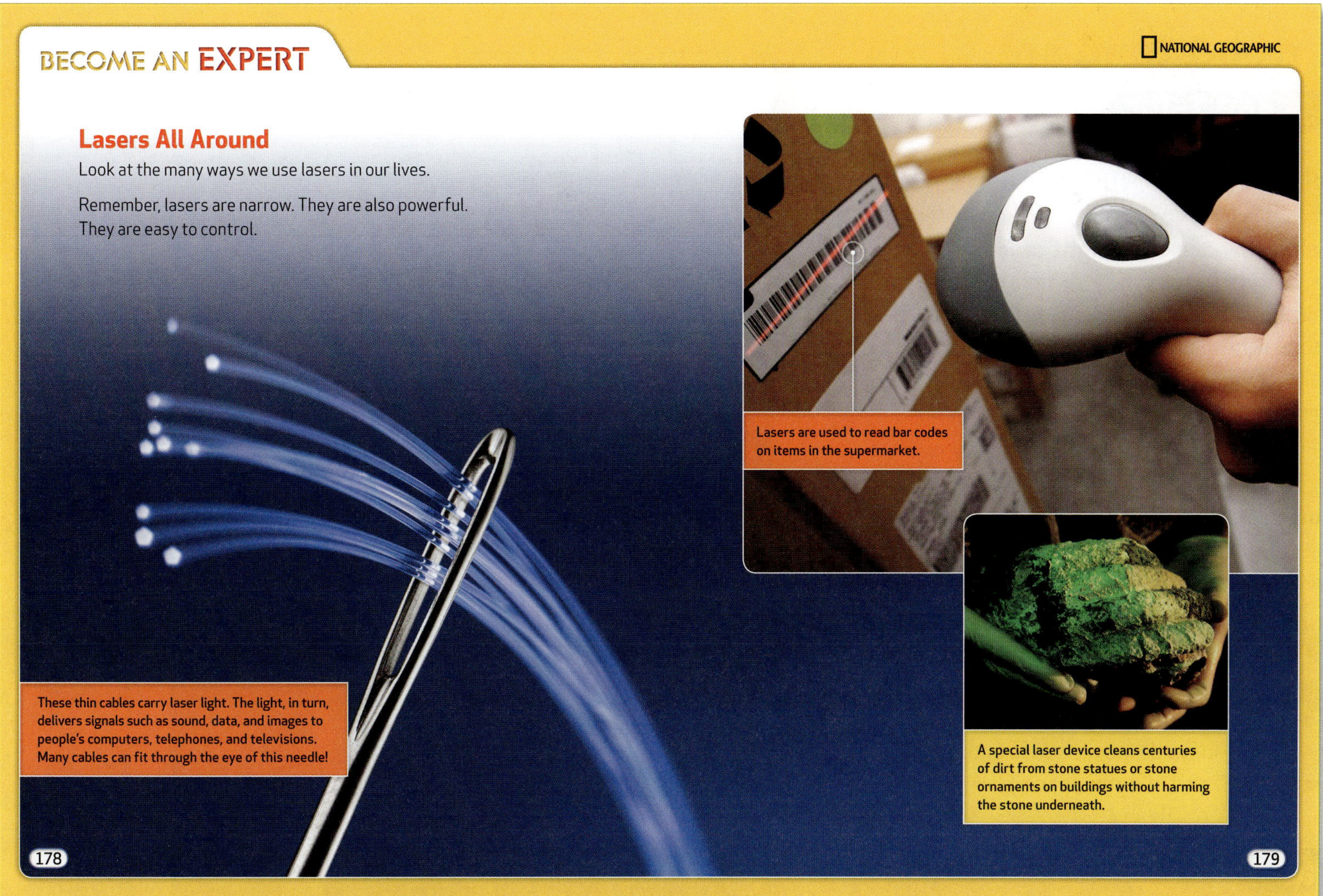

Lasers All Around

Look at the many ways we use lasers in our lives.

Remember, lasers are narrow. They are also powerful. They are easy to control.

These thin cables carry laser light. The light, in turn, delivers signals such as sound, data, and images to people's computers, telephones, and televisions. Many cables can fit through the eye of this needle!

Lasers are used to read bar codes on items in the supermarket.

A special laser device cleans centuries of dirt from stone statues or stone ornaments on buildings without harming the stone underneath.

178

179

Access Science Content

Describe How People Use Lasers

- Have students read the text on page 178 and study the photos and captions on pages 178–179. Explain that the thin cables in the photo are called fiber optic cables. Say: **Lasers are part of people's everyday lives.** Ask: **How do people use lasers to communicate?** (People use fiber optic cables to carry laser light, which delivers sound, data, and images to people's computers, telephones, and televisions.)

- Say: **Do you know any uses for lasers that are not shown on these pages?** (Possible answer: There are lasers in CD and DVD players.)

❯ Monitor and Fix Up

Ask students if anything they read was confusing. Students may have difficulty recognizing the laser as a form of light energy. Review how lasers are different from natural light sources and help students summarize similarities and differences in a compare and contrast chart. Then have them draw and label a picture showing one common use of lasers.

Assess

1. **Recall** What do fiber optic cables carry? (laser light)

2. **Predict** What uses do you think scientists will find for lasers in the future? (Accept all answers given.)

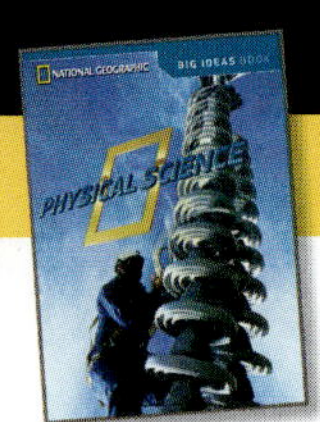

Share and Compare

Turn and Talk

Ask students to turn to partners and talk about what they learned about lasers. Prompt students by asking:

1. **Recall** **How are laser light and sunlight alike, and how are they different?** (They are both light sources that have many uses. Sunlight is natural light that radiates in all directions and contains all the colors; laser light is made by people and machines, contains one color, and moves in one direction from its source.)

2. **Describe** **What did scientists not know about lasers when they were invented?** (They did not know the many ways that laser light could be used.)

3. **Summarize** **How have lasers changed our lives?** (Possible answer: They have made things possible in medicine that were not possible before, such as correcting people's vision. Lasers are used everywhere—in medicine, in construction, in entertainment, and in outer space.)

Read

Ask students to select the two pages in this section that they find the most interesting. Have them reread the pages and then read them aloud to a partner or small group. Then have students tell what it was they found interesting about the pages they chose.

Write 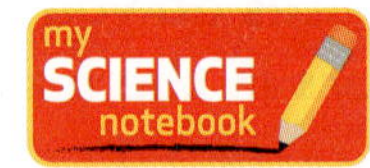

Have students write a conclusion that summarizes what they learned about lasers. Ask them to compare what they wrote with a classmate. Ask how their conclusion and Big Idea differed from their classmate's conclusion and Big Idea.

Draw 

Have students draw a picture that shows how a laser is used. Have them include labels in the picture. Ask students to share their drawing with a classmate. Have them explain how their drawings are alike and how they are different.

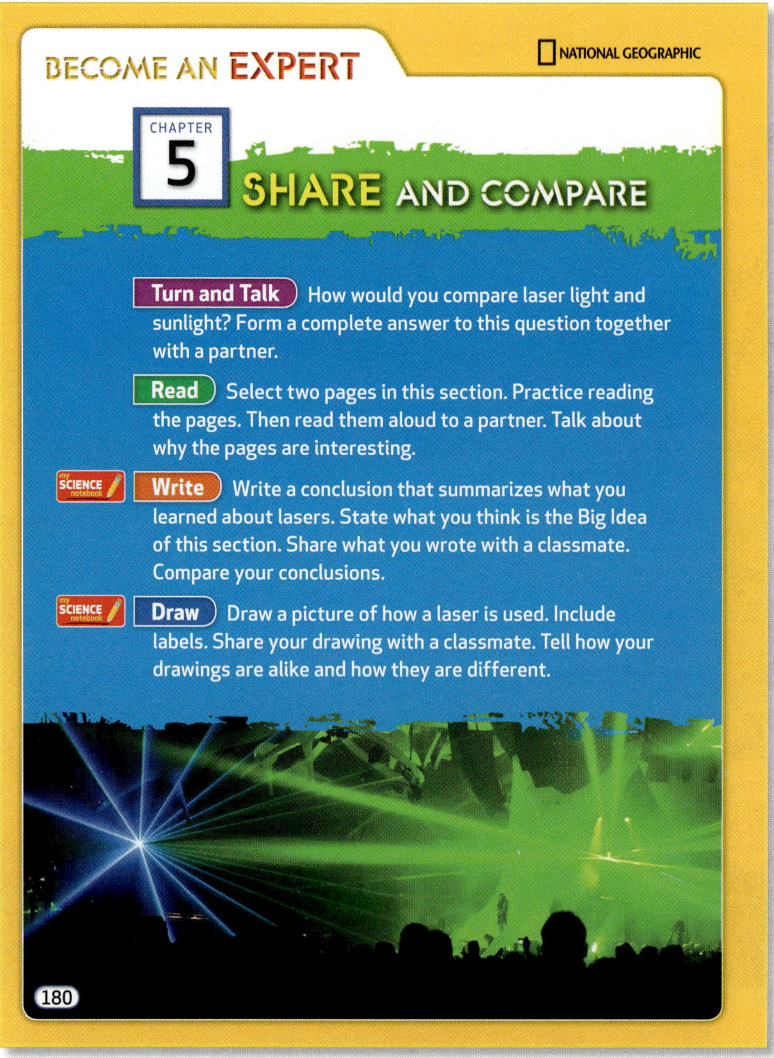

❭ Sum Up

Tell students that to sum up a text helps them pull together the ideas of the text. Remind students that they summed up the Become an Expert lesson in the Write section of Share and Compare. Have students take turns reading the conclusions they wrote to a partner. Invite them to compare and contrast their conclusions.

Read Informational Text

Explore on Your Own books provide additional opportunities for your students to:

• deepen their science content knowledge even more as they focus on one Big Idea.

• independently apply multiple reading comprehension strategies as they read.

After applying the strategies throughout the chapter, students will independently read Explore on Your Own books. Facsimiles of the Explore on Your Own pages are shown on pages T180a–T180h.

Pioneer

Pathfinder

It's Time to Explore on Your Own!

Good readers use multiple strategies as they read on their own. Use the four key reading comprehension strategies below:

1 PREVIEW AND PREDICT
• Look over the text.
• Form ideas about how the text is organized and what it says.
• Confirm ideas about how the text is organized and what it says.

2 MONITOR AND FIX UP
• Think about whether the text is making sense and how it relates to what you know.
• Identify comprehension problems and clear up the problems.

3 MAKE INFERENCES
• Use what you know to figure out what is not said or shown directly.

4 SUM UP
• Pull together the text's big ideas.

Remember that you can choose different strategies at different times to help you understand what you are reading.

NATIONAL GEOGRAPHIC
School Publishing

Scope This Out

PIONEER EDITION

By Juan Quintana

CONTENTS

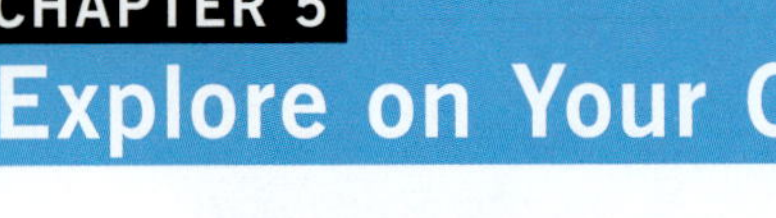

The Hubble Space Telescope travels around the Earth. It sends back amazing pictures of space. These help us learn more about our universe.

Telescopes have been used for about 400 years. They let people see things that are very far away.

But the story of telescopes really started with eyeglasses. Read on to scope it out.

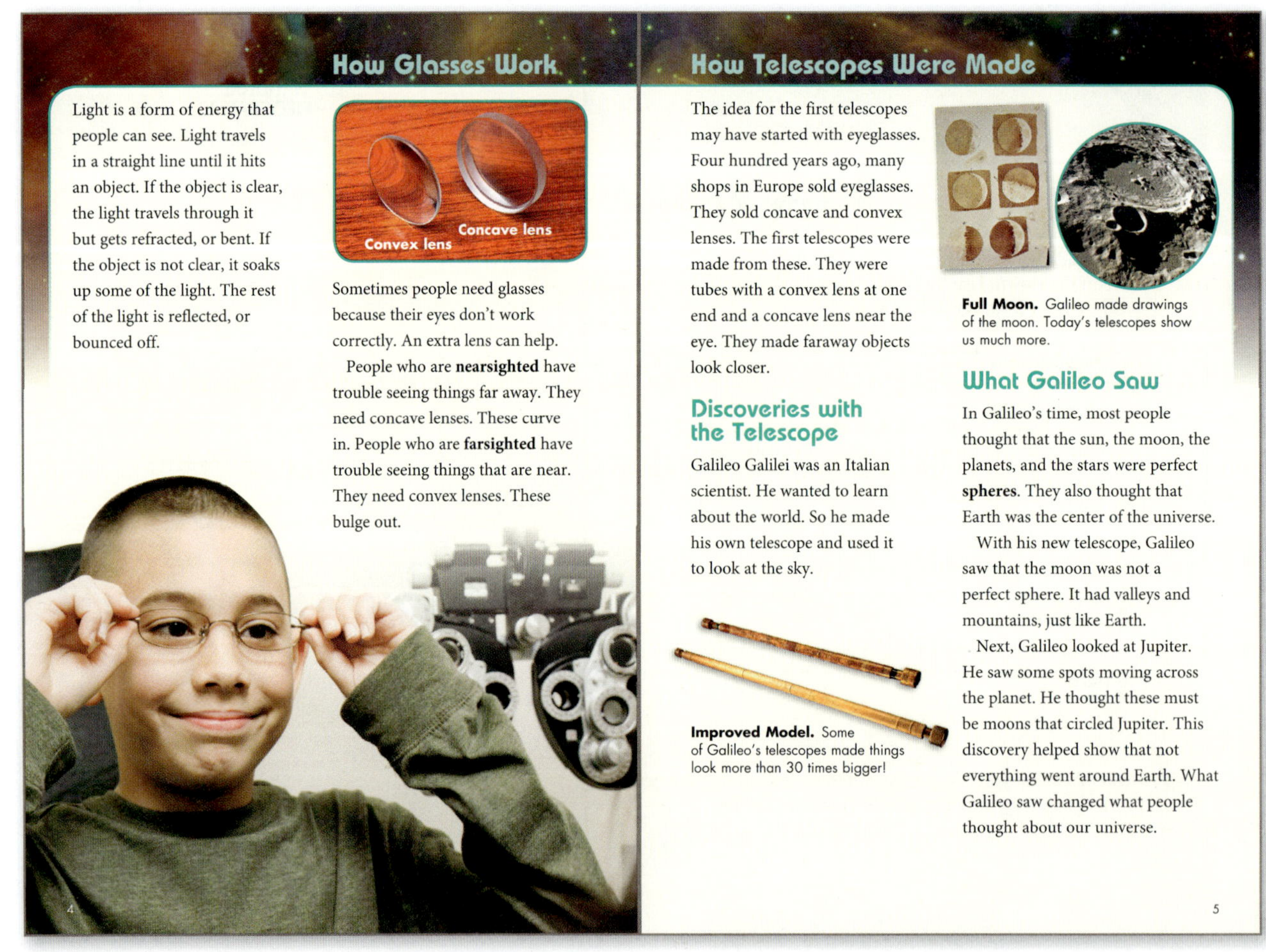

How Glasses Work

Light is a form of energy that people can see. Light travels in a straight line until it hits an object. If the object is clear, the light travels through it but gets refracted, or bent. If the object is not clear, it soaks up some of the light. The rest of the light is reflected, or bounced off.

Sometimes people need glasses because their eyes don't work correctly. An extra lens can help.

People who are **nearsighted** have trouble seeing things far away. They need concave lenses. These curve in. People who are **farsighted** have trouble seeing things that are near. They need convex lenses. These bulge out.

How Telescopes Were Made

The idea for the first telescopes may have started with eyeglasses. Four hundred years ago, many shops in Europe sold eyeglasses. They sold concave and convex lenses. The first telescopes were made from these. They were tubes with a convex lens at one end and a concave lens near the eye. They made faraway objects look closer.

Full Moon. Galileo made drawings of the moon. Today's telescopes show us much more.

Discoveries with the Telescope

Galileo Galilei was an Italian scientist. He wanted to learn about the world. So he made his own telescope and used it to look at the sky.

Improved Model. Some of Galileo's telescopes made things look more than 30 times bigger!

What Galileo Saw

In Galileo's time, most people thought that the sun, the moon, the planets, and the stars were perfect **spheres**. They also thought that Earth was the center of the universe.

With his new telescope, Galileo saw that the moon was not a perfect sphere. It had valleys and mountains, just like Earth.

Next, Galileo looked at Jupiter. He saw some spots moving across the planet. He thought these must be moons that circled Jupiter. This discovery helped show that not everything went around Earth. What Galileo saw changed what people thought about our universe.

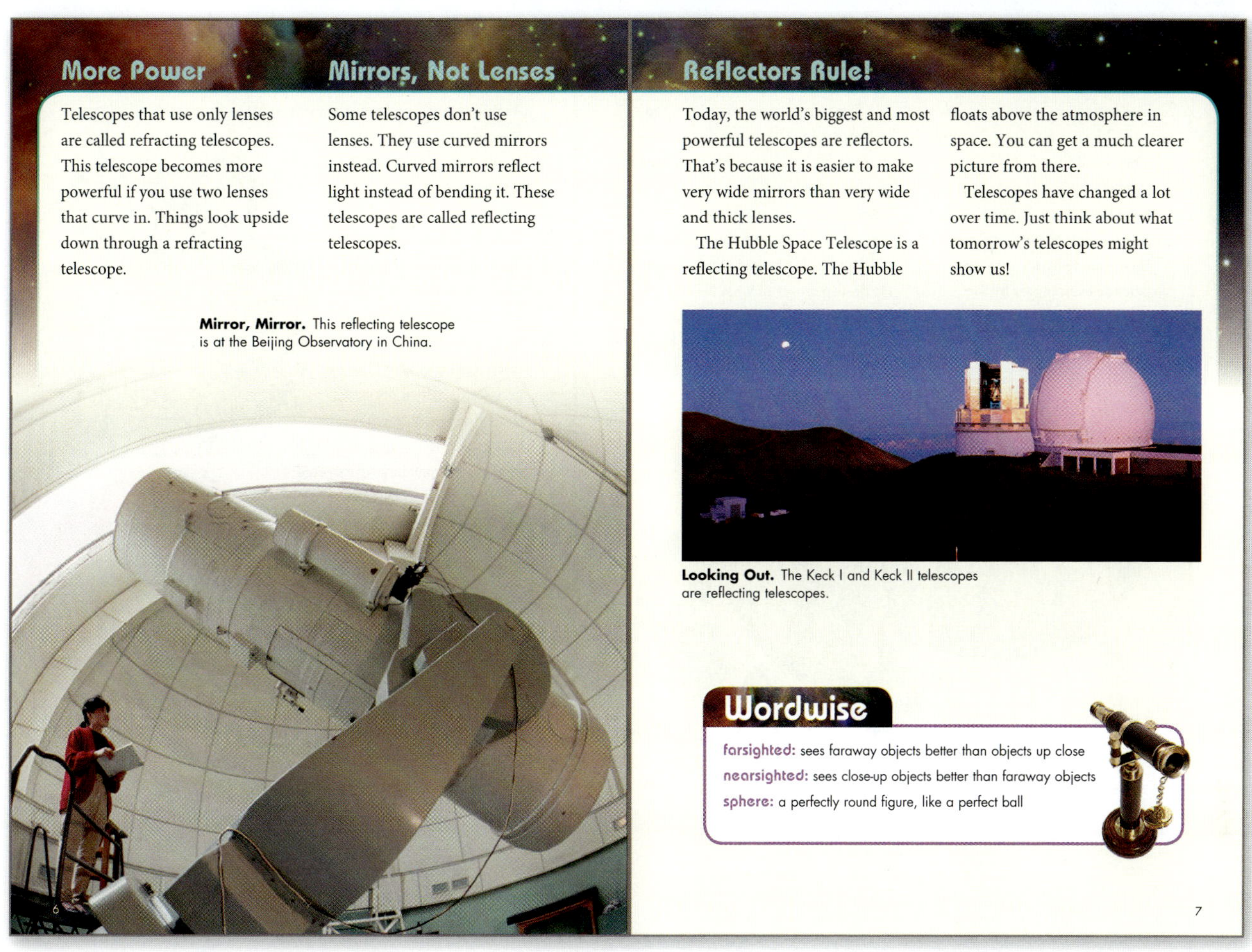

More Power

Telescopes that use only lenses are called refracting telescopes. This telescope becomes more powerful if you use two lenses that curve in. Things look upside down through a refracting telescope.

Mirror, Mirror. This reflecting telescope is at the Beijing Observatory in China.

Mirrors, Not Lenses

Some telescopes don't use lenses. They use curved mirrors instead. Curved mirrors reflect light instead of bending it. These telescopes are called reflecting telescopes.

Reflectors Rule!

Today, the world's biggest and most powerful telescopes are reflectors. That's because it is easier to make very wide mirrors than very wide and thick lenses.

The Hubble Space Telescope is a reflecting telescope. The Hubble floats above the atmosphere in space. You can get a much clearer picture from there.

Telescopes have changed a lot over time. Just think about what tomorrow's telescopes might show us!

Looking Out. The Keck I and Keck II telescopes are reflecting telescopes.

Wordwise

farsighted: sees faraway objects better than objects up close

nearsighted: sees close-up objects better than faraway objects

sphere: a perfectly round figure, like a perfect ball

7

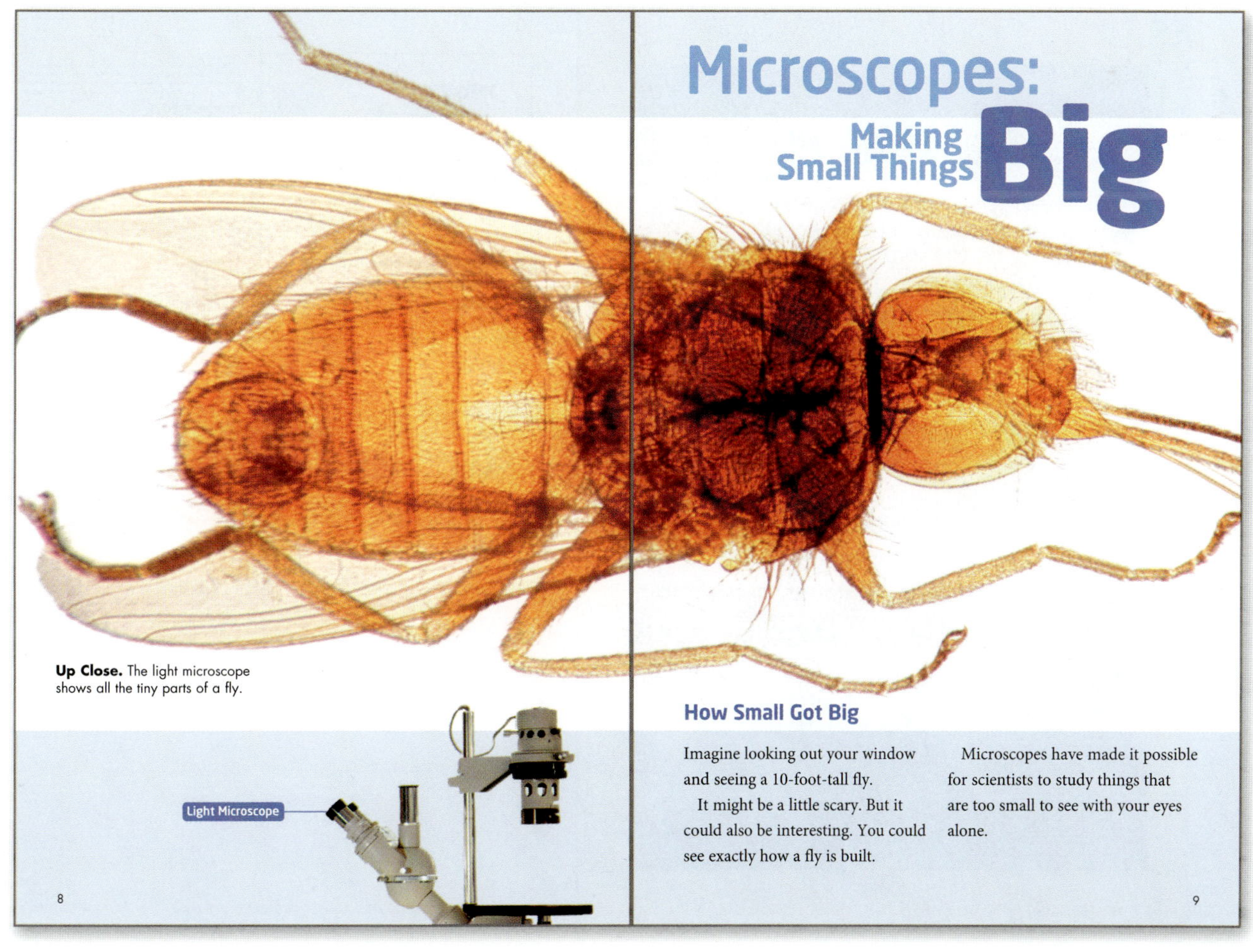

Up Close. The light microscope shows all the tiny parts of a fly.

How Small Got Big

Imagine looking out your window and seeing a 10-foot-tall fly.

It might be a little scary. But it could also be interesting. You could see exactly how a fly is built.

Microscopes have made it possible for scientists to study things that are too small to see with your eyes alone.

8

9

Physical Science **T180c**

The First Microscopes

The lenses in some microscopes work like a refracting telescope. They bend light rays together to make a bigger image. They have two lenses that make small images look bigger. The bigger image made by the one lens is made even bigger by the other lens.

Portrait of a Flea

The first microscopes were invented 400 years ago. They were known as "flea glasses." Some people used them to look at fleas and other insects.

Robert Hooke was a scientist. He used the microscope to look at fleas, too. But he also looked at other things. He drew pictures of what he saw.

Hooke's drawings appeared in a book called *Micrographia*. This book made people interested in things too small to see.

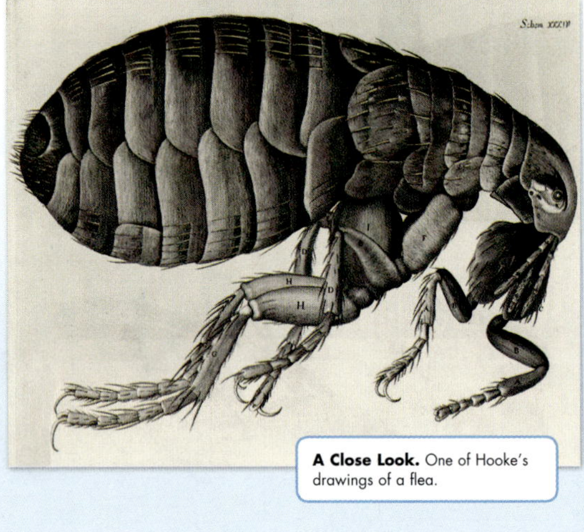

A Close Look. One of Hooke's drawings of a flea.

Microscopes and Health

As microscopes improved scientists were able to see tiny animals and plants. Scientists called these *microorganisms*.

Some microorganisms cause diseases. These "bad" microorganisms are called *germs*. Scientists wanted to control diseases caused by germs. Medicines were discovered that killed the germs and helped prevent some diseases. Remember that a scientist using the microscope may have discovered the medicine you take when you are sick.

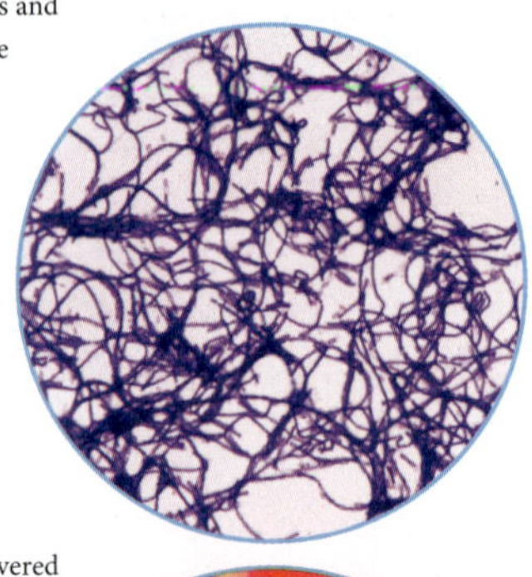

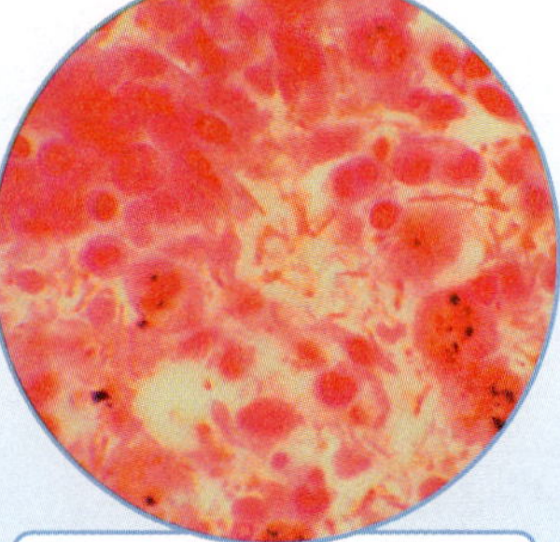

Disease Agent. These microorganisms cause the disease known as tuberculosis.

10

11

Telescopes and Microscopes

Let's see what you know about the history of telescopes and microscopes.

1. What happens when light hits an object?

2. What is the difference between a convex and a concave lens?

3. How are a refracting telescope and a reflecting telescope different?

4. How are some microscopes like a refracting telescope?

5. What are microorganisms?

Acknowledgments
Grateful acknowledgment is given to the authors, artists, photographers, museums, publishers, and agents for permission to reprint copyrighted material. Every effort has been made to secure the appropriate permission. If any omissions have been made or if corrections are required, please contact the Publisher.

Photographic Credits
Cover Design NASA/ESA/T. Megeath (University of Toledo) and M. Robberto (STScI)/Space Telescope Science Institute; 2-3 NASA/ESA/T. Megeath (University of Toledo) and M. Robberto (STScI)/Space Telescope Science Institute; 3 Corbis; 4 sciencephoto/Alamy Images, Ned Frisk/Jupiterimages; 5 SPL/Getty Images, Scala/Art Resource Inc., Stockbyte/Getty Images; 6 Roger Ressmeyer/Corbis; 7 Richard Wainscoat/Alamy Images, Corbis, Stephen Coburn/Shutterstock; 8 Deco Images II/Alamy Images; 8-9 Biodisc/Getty Images; 10 SSPL/Getty Images; 11 Gary D. Gaugler/Photo Researchers Inc., Medical-on-Line/Alamy Images, Dr. Cecil H. Fox/Photo Researchers Inc.; 12 Sandra van der Steen/Shutterstock.

Neither the Publisher nor the authors shall be liable for any damage that may be caused or sustained or result from conducting any of the activities in this publication without specifically following instructions, undertaking the activities without proper supervision, or failing to comply with the cautions contained herein.

Program Authors
Malcolm B. Butler, Ph.D., Associate Professor of Science Education, University of South Florida, St. Petersburg, Florida; Judith Sweeney Lederman, Ph.D., Director of Teacher Education and Associate Professor of Science Education, Department of Mathematics and Science Education, Illinois Institute of Technology, Chicago, Illinois; Randy Bell, Ph.D., Associate Professor of Science Education, University of Virginia, Charlottesville, Virginia; Kathy Cabe Trundle, Ph.D., Associate Professor of Early Childhood Science Education, The Ohio State University, Columbus, Ohio; David W. Moore, Ph.D., Professor of Education, College of Teacher Education and Leadership, Arizona State University, Tempe, Arizona

The National Geographic Society
John M. Fahey, Jr., President & Chief Executive Officer
Gilbert M. Grosvenor, Chairman of the Board

Copyright © 2011 The Hampton-Brown Company, Inc., a wholly owned subsidiary of the National Geographic Society, publishing under the imprints National Geographic School Publishing and Hampton-Brown.

All rights reserved. No part of this book may be reproduced or transmitted in any form or by any means, electronic or mechanical, including photocopying, recording, or by an information storage and retrieval system, without permission in writing from the Publisher.

National Geographic and the Yellow Border are registered trademarks of the National Geographic Society.

National Geographic School Publishing
Hampton-Brown
www.NGSP.com

Printed in the USA.
RR Donnelley, Johnson City, TN

ISBN-13: 978-0-7362-7727-3

10 11 12 13 14 15 16 17 18 19
10 9 8 7 6 5 4 3 2

Physical Science
PATHFINDER EDITION

It's Time to Explore on Your Own!

Good readers use multiple strategies as they read on their own. Use the four key reading comprehension strategies below:

1 PREVIEW AND PREDICT
- Look over the text.
- Form ideas about how the text is organized and what it says.
- Confirm ideas about how the text is organized and what it says.

2 MONITOR AND FIX UP
- Think about whether the text is making sense and how it relates to what you know.
- Identify comprehension problems and clear up the problems.

3 MAKE INFERENCES
- Use what you know to figure out what is not said or shown directly.

4 SUM UP
- Pull together the text's big ideas.

Remember that you can choose different strategies at different times to help you understand what you are reading.

NATIONAL GEOGRAPHIC
School Publishing

Scope This Out

PATHFINDER EDITION

By Juan Quintana

CONTENTS

The Hubble Space Telescope circles the Earth, looking deep into space. The amazing pictures it sends back help us learn more about the universe.

For about 400 years, telescopes have been used to learn about space. The telescope changed people's understanding of places beyond Earth.

But the story of telescopes really started with eyeglasses. Let's scope it out.

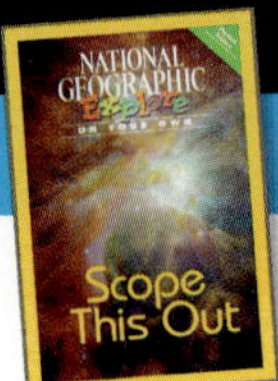

How Glasses Work

What is light? It's a form of energy that people can see. Light travels in a straight line until it hits an object. If the object is clear the light rays will travel through it, but they will be bent or **refracted**. If the object is not clear, it will absorb some of the light. The rest of the light will be reflected, or bounced off, the object.

Your eyes use light to see. Built-in lenses in the eye bend the reflected light rays. This focuses the rays, and they form a clear image at the back of the eye.

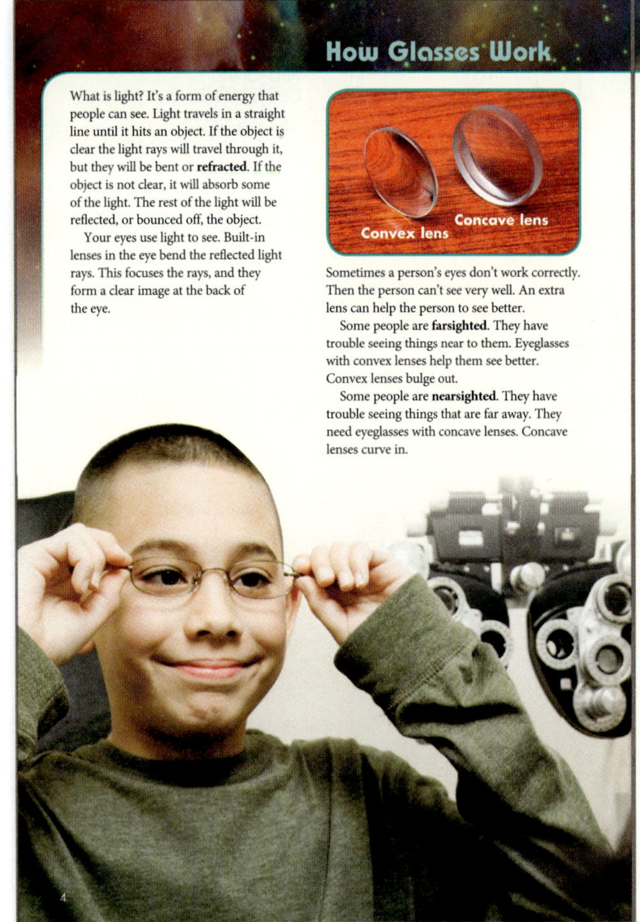

Sometimes a person's eyes don't work correctly. Then the person can't see very well. An extra lens can help the person to see better.

Some people are **farsighted**. They have trouble seeing things near to them. Eyeglasses with convex lenses help them see better. Convex lenses bulge out.

Some people are **nearsighted**. They have trouble seeing things that are far away. They need eyeglasses with concave lenses. Concave lenses curve in.

How Telescopes Were Made

The idea for the first telescope may have started with eyeglasses. About 400 years ago many European cities had shops selling eyeglasses. These shops sold concave and convex lenses.

The first telescopes were made from these lenses. They were tubes with a convex lens at one end and a concave lens near the eye. When people looked through the telescope, objects that were far away looked much closer.

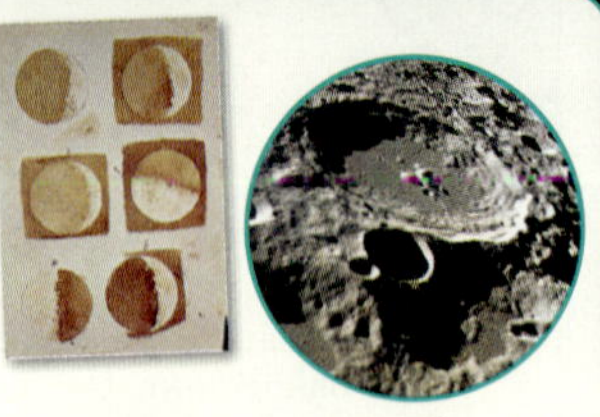

Full Moon. Galileo's drawing shows how he saw the moon. Today's telescopes show much more of the moon's surface.

Discoveries with the Telescope

Galileo Galilei was an Italian scientist. He heard about the telescope. He bought a few different types of lenses. After a while, he put a convex lens and a concave lens in a tube and made his own telescope!

Galileo was interested in what he could learn about objects in the sky using this new invention. He made his own lenses. He found the right ones to make a better telescope. Then he aimed his telescope at the sky.

What Galileo Saw

In Galileo's time, most people thought that the sun, the moon, the planets, and the stars were perfect **spheres**. They thought that these spheres all revolved around Earth.

Through his new telescope, Galileo saw that the moon was not a perfect sphere. It had bumps and holes and all sorts of scars.

Galileo also looked at the planet Jupiter with his telescope. He noticed some spots moving across the planet. These smaller spots were **satellites**, or small objects, that circled Jupiter just like the moon circles Earth. What Galileo saw helped prove that not everything revolved around Earth! His discoveries changed what people thought about the solar system.

Improved Model. These telescopes were made by Galileo. Galileo's best telescopes made things look more than 30 times bigger!

More Power

There are different types of telescopes used to view our world and beyond. Telescopes that use only lenses are called refracting telescopes. You can make a refracting telescope more powerful by using two lenses that curve in. The first lens makes a big image. That image is made even bigger by the second lens. These two convex lenses together create an image that looks upside down!

Mirrors, Not Lenses

Some telescopes don't use lenses. They use curved mirrors instead. Telescopes that use curved mirrors are called reflecting telescopes. Curved mirrors reflect, or bounce, light rather than refract, or bend, it.

Mirror, Mirror. This reflecting telescope is at the Beijing Observatory in China.

Reflectors Rule!

Today, the world's biggest and most powerful telescopes are reflectors. That's because with modern technology it is easier to make very wide mirrors than very wide and thick lenses.

The Hubble Space Telescope is a reflecting telescope. It floats above the atmosphere in space. You can get a much clearer image from above the atmosphere.

Telescopes have changed a lot over the years. Now telescopes such as the Hubble help us see farther into space and make new discoveries.

Looking Out. The Keck I and Keck II telescopes in Hawaii are reflecting telescopes.

Wordwise

farsighted: sees faraway objects better than objects up close
nearsighted: sees close-up objects better than faraway objects
refract: to bend light
satellite: smaller object that circles around a larger object in space
sphere: a perfectly round figure, like a perfect ball

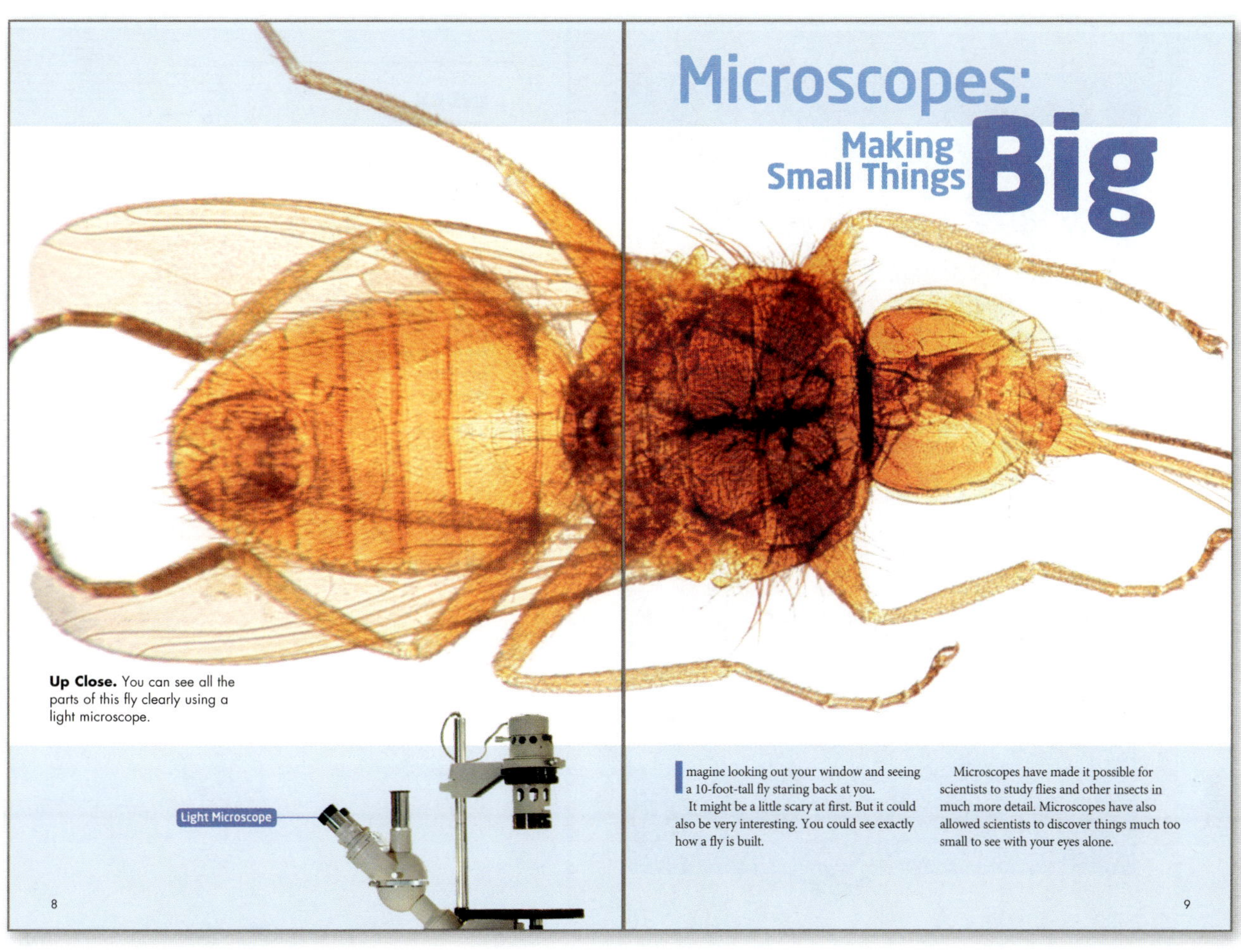

Up Close. You can see all the parts of this fly clearly using a light microscope.

Imagine looking out your window and seeing a 10-foot-tall fly staring back at you.

It might be a little scary at first. But it could also be very interesting. You could see exactly how a fly is built.

Microscopes have made it possible for scientists to study flies and other insects in much more detail. Microscopes have also allowed scientists to discover things much too small to see with your eyes alone.

8

9

The First Microscopes

Some microscopes work in the same way as a refracting telescope. They have convex lenses that work together to create a very large image.

These two lenses make small images seem bigger. The bigger image made by one lens is made even bigger by the second lens. For example, one lens makes things look 100 times bigger. The second lens makes things look 7 times bigger. So looking through both lenses will make things look 700 times bigger! Here is how the math works: $100 \times 7 = 700$.

Portrait of a Flea

The first microscopes were invented 400 years ago. They were known as "flea glasses" because some people used them to look closely at fleas and other small insects.

One scientist, Robert Hooke, looked at fleas and insects. But he also looked at other things through the microscope. He drew pictures of what he saw.

Hooke's drawings and notes were published in a book called *Micrographia*. The book was an instant bestseller. It made people interested in the world of things too small to see.

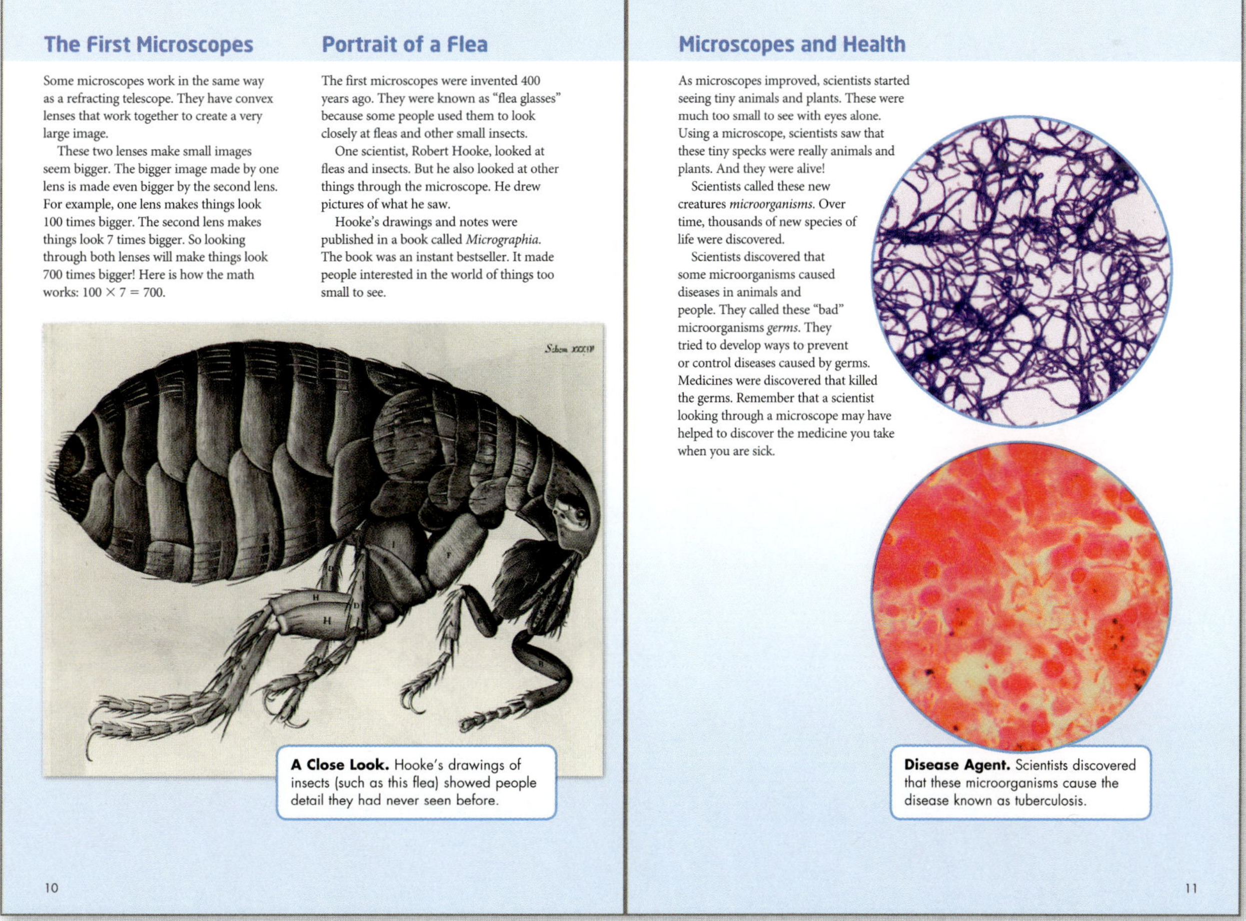

A Close Look. Hooke's drawings of insects (such as this flea) showed people detail they had never seen before.

Microscopes and Health

As microscopes improved, scientists started seeing tiny animals and plants. These were much too small to see with eyes alone. Using a microscope, scientists saw that these tiny specks were really animals and plants. And they were alive!

Scientists called these new creatures *microorganisms*. Over time, thousands of new species of life were discovered.

Scientists discovered that some microorganisms caused diseases in animals and people. They called these "bad" microorganisms *germs*. They tried to develop ways to prevent or control diseases caused by germs. Medicines were discovered that killed the germs. Remember that a scientist looking through a microscope may have helped to discover the medicine you take when you are sick.

Disease Agent. Scientists discovered that these microorganisms cause the disease known as tuberculosis.

10

11

Telescopes and Microscopes

Let's see what you know about telescopes and microscopes.

1. What happens when light hits an object?

2. What is the difference between a convex and a concave lens?

3. How are a refracting telescope and a reflecting telescope different?

4. How are some microscopes like a refracting telescope?

5. Why was the discovery of microorganisms important?

Acknowledgments
Grateful acknowledgment is given to the authors, artists, photographers, museums, publishers, and agents for permission to reprint copyrighted material. Every effort has been made to secure the appropriate permission. If any omissions have been made or if corrections are required, please contact the Publisher.

Photographic Credits
Cover Design NASA/ESA/T. Megeath (University of Toledo) and M. Robberto (STSci)/Space Telescope Science Institute; 2-3 NASA/ESA/T. Megeath (University of Toledo) and M. Robberto (STSci)/Space Telescope Science Institute; 3 Corbis; 4 sciencephotos/Alamy Images, Ned Frisk/Jupiterimages; 5 SPL/Getty Images, Scala/Art Resource Inc., Stockbyte/Getty Images; 6 Roger Ressmeyer/Corbis; 7 Richard Wainscoat/Alamy Images, Corbis, Stephen Coburn/Shutterstock; 8 Deco Images II/Alamy Images; 8-9 Biodisc/Getty Images; 10 SSPL/Getty Images; 11 Gary D. Gaugler/Photo Researchers Inc., Medical-on-Line/Alamy Images, Dr. Cecil H. Fox,/Photo Researchers Inc.; 12 Sandra van der Steen/Shutterstock.

Neither the Publisher nor the authors shall be liable for any damage that may be caused or sustained or result from conducting any of the activities in this publication without specifically following instructions, undertaking the activities without proper supervision, or failing to comply with the cautions contained herein.

Program Authors
Malcolm B. Butler, Ph.D., Associate Professor of Science Education, University of South Florida, St. Petersburg, Florida; Judith Sweeney Lederman, Ph.D., Director of Teacher Education and Associate Professor of Science Education, Department of Mathematics and Science Education, Illinois Institute of Technology, Chicago, Illinois; Randy Bell, Ph.D., Associate Professor of Science Education, University of Virginia, Charlottesville, Virginia; Kathy Cabe Trundle, Ph.D., Associate Professor of Early Childhood Science Education, The Ohio State University, Columbus, Ohio; David W. Moore, Ph.D., Professor of Education, College of Teacher Education and Leadership, Arizona State University, Tempe, Arizona

The National Geographic Society
John M. Fahey, Jr., President & Chief Executive Officer
Gilbert M. Grosvenor, Chairman of the Board

Copyright © 2011 The Hampton-Brown Company, Inc., a wholly owned subsidiary of the National Geographic Society, publishing under the imprints National Geographic School Publishing and Hampton-Brown.

All rights reserved. No part of this book may be reproduced or transmitted in any form or by any means, electronic or mechanical, including photocopying, recording, or by an information storage and retrieval system, without permission in writing from the Publisher.

National Geographic and the Yellow Border are registered trademarks of the National Geographic Society.

National Geographic School Publishing
Hampton-Brown
www.NGSP.com

Printed in the USA.
RR Donnelley, Johnson City, TN

ISBN-13: 978-0-7362-7730-3

10 11 12 13 14 15 16 17 18 19
10 9 8 7 6 5 4 3 2 1

Notes

Objectives

Students will be able to:

- Investigate through Open Inquiry (generate questions to investigate; plan, make, and compare investigations; collect and record data and observations; generate explanations and conclusions based on evidence or observations; share findings; ask questions based on observations to increase understanding; adjust interpretations based on findings and new ideas).

Science Process Vocabulary

question

PROGRAM RESOURCES

- Science Inquiry and Writing Book: *Physical Science*
- Science Inquiry and Writing Book **eEdition** at **myNGconnect.com**
- **Inquiry eHelp** at **myNGconnect.com**
- Science Inquiry Kit: *Physical Science*
- Learning Masters Book, pages 234–237, or at **myNGconnect.com**
- Inquiry Rubric: Assessment Handbook, page 216, or at **myNGconnect.com**
- Inquiry Self-Reflection: Assessment Handbook, page 227, or at **myNGconnect.com**

MATERIALS

Materials will vary based on students' choices of investigations.

❶ Introduce

Tap Prior Knowledge

- In Open Inquiry, students ask questions, develop written plans, and then design and conduct investigations.

- With students, brainstorm questions and evaluate explanations about scientific topics in this unit or in the natural world around them. Ask students to choose a sample question on page 210 or make up one of their own. They should then make a prediction or hypothesis.

- Students may use the Open Inquiry Checklist as they plan and conduct their investigations.

Do Your Own Investigation

Choose one of these questions, or make up one of your own to do your investigation.

- How do the masses of different pennies compare?
- How can you make water evaporate more quickly?
- How does the length of a ramp affect the distance traveled by a toy car?
- How does changing the tightness of a piece of fishing line affect the pitch of the sound it makes?
- How can you line up mirrors to make light reflect on a certain spot?

Science Process Vocabulary

question noun

You ask a **question** to find out about something.

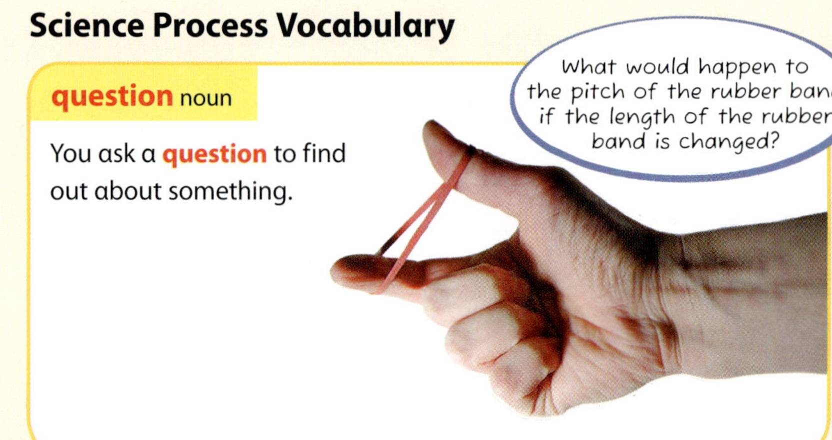

210

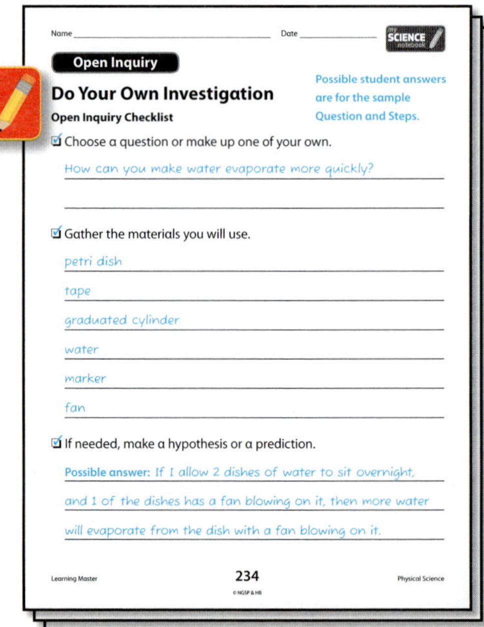

Learning Masters 234–237 or at myNGconnect.com

Open Inquiry Checklist

Here is a checklist you can use when you investigate.

- ☐ Choose a **question** or make up one of your own.
- ☐ Gather the materials you will use.
- ☐ If needed, make a **hypothesis** or a **prediction.**
- ☐ If needed, identify, manipulate, and control **variables.**
- ☐ Make a **plan** for your **investigation.**
- ☐ Carry out your plan.
- ☐ Collect and record **data. Analyze** your data.
- ☐ Explain and **share** your results.
- ☐ Tell what you **conclude.**
- ☐ Think of another question.

A siren has a high pitch.

211

Differentiated Instruction

Understand Science Process Vocabulary

Provide the following sentences for students to complete.

- *A _____ shows or tells how to do something step by step.* (plan)
- *A _____ is something you can change in an investigation.* (variable)
- *When you make a _____, you can write it as an "If … then …" statement.* (hypothesis)

❷ Build Vocabulary

Science Process Word: question

Use this routine to introduce the word.

1. **Pronounce the Word** Say **question**. Have students repeat it in syllables.
2. **Explain Its Meaning** Choral read the sentence. Ask students for another word or phrase that means the same as **question**. (something you ask to find an answer)
3. **Encourage Elaboration** Ask: **Why might you ask a question?** (to find out more about something by doing an investigation)

❸ Guide the Investigation

Have students choose steps to answer their questions. Encourage them to share their plans with other students and revise them based on feedback. Then guide students as they seek answers to their questions through careful observation or experimentation. They should use appropriate tools and techniques to collect data. See pages T180n–T180q for a model of an Open Inquiry investigation.

❹ Explain Your Results

Students should gather data and organize it in charts or graphs. They should not change their data if it differs from someone else's work. Then students should share results, inferences, and conclusions. See pages T180p–T180q for sample responses and guidance on explaining results.

❺ Find Out More

Encourage students to use observations made in their investigations to identify other problems and generate more questions. See page T180q for more on generating questions.

❻ Reflect and Assess

- Use the Inquiry Rubric on page T180q, Assessment Handbook, page 216, or at **myNGconnect.com**
- Have students use the Inquiry Self-Reflection on Assessment Handbook, page 227, or at **myNGconnect.com**

Objectives

Students will be able to:

- Investigate through Write Like a Scientist (choose a question; gather materials to use; make a hypothesis or prediction; identify, manipulate, and control variables if needed; make and carry out a plan for an investigation; collect and record data in charts, tables, and graphs; analyze data; explain and share results; tell a conclusion; think of another question).

- Identify sound as a form of energy and relate fast and slow vibrations to variations in pitch.

- Indicate materials to be used and steps to follow to conduct an investigation, and describe how data will be recorded.

- Explore and solve problems generated from school, home, and community situations, using concrete objects and materials.

Write About an Investigation

Vibrations and Pitch

- Use this writing model to help students write about their investigations. The writing model uses the Open Inquiry checklist to guide writing.

- Page 212 briefly explains the investigation that Neal did about vibration and pitch. Ask: **What did Neal want to investigate?** (He wanted to know how changing the tightness of a string affects the pitch of the sound it makes.) **What type of energy did he investigate?** (sound)

- The model begins on page 213, with Neal's question and list of materials to be used. Use the picture on page 214 to help students identify unfamiliar items, such as pegboards or fishing line.

- Neal's hypothesis is on page 214. Help students notice the "If…, then…" construction of the hypothesis. Ask: **Why did Neal include an "If… then" statement in his hypothesis?** (to clearly state how he thinks the pitch will change)

- Page 215 has Neal's variable and controls. Ask: **What variable will Neal change in this investigation?** (the tightness of the fishing line) **How will he know the effects of that change?** (by plucking the string and listening to its pitch) **What variables remain the same?** (the length of the fishing line and the way he plucks it)

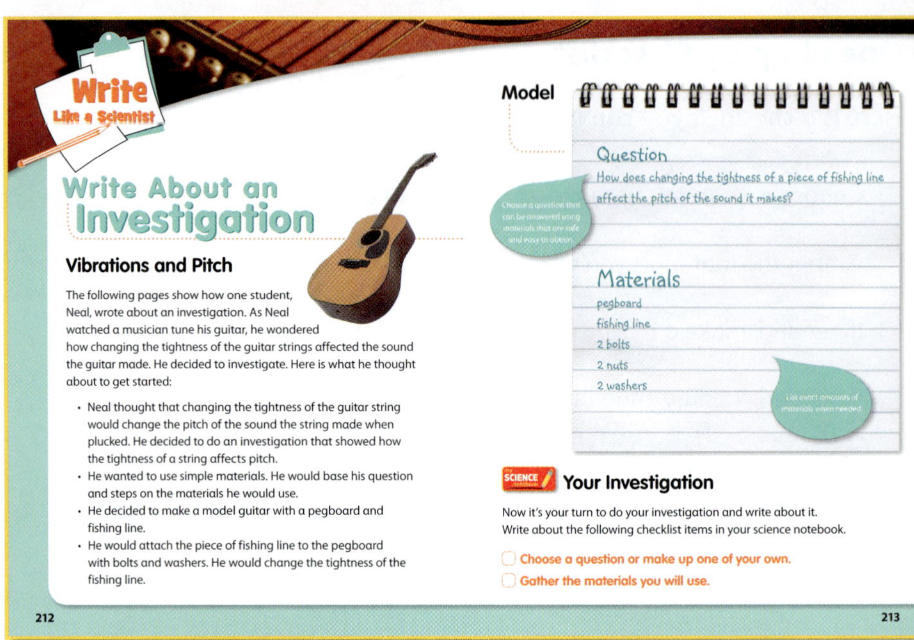

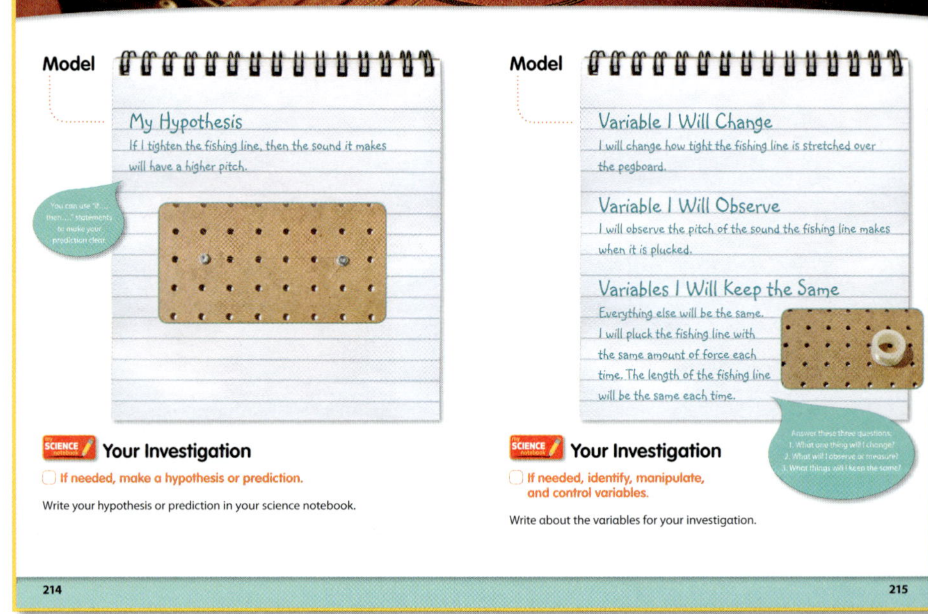

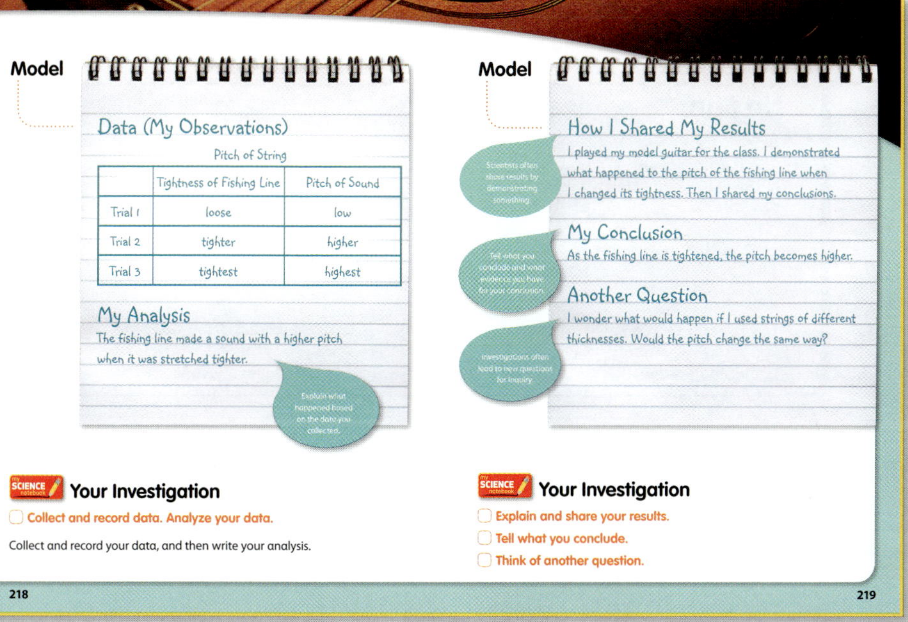

	Tightness of Fishing Line	Pitch of Sound
Trial 1	loose	low
Trial 2	tighter	higher
Trial 3	tightest	highest

Write About an Investigation, continued

- Page 216 outlines Neal's plan for the investigation. Let students know that the plan tells the steps in order, so that they, or anyone else, can carry out the investigation. The steps should be clear and complete.

- Ask: **Why is it important to clearly communicate the steps of an investigation?** (Clear steps can be understood and repeated by others.) **What could you include in your plan to make sure it is clear to others?** (Possible answers: numbered steps, specific quantities, pictures showing what to do)

- Page 217 shows a simple statement Neal wrote after carrying out the plan. Students may not always have to write a statement for this step. If students realize they need to adjust their plans, they should keep notes about any changes they make.

- Neal's data is on page 218. The data is in the form of a table with recorded observations. Ask students to decide how their observations will be recorded. Let students know that there are many ways they can record their data, such as in journals, charts, tables, or graphs.

- Discuss the importance of keeping accurate and detailed records of observations. Students should understand that Neal's observations about the pitch of a string can be used by others to draw conclusions and validate his explanations.

- After collecting and recording data, students should generate explanations as they analyze the data. Neal's analysis is below the table. Ask: **How did Neal analyze his data?** (He compared the sound of the fishing line for all three trials to see if there was a relationship between the tightness of the fishing line and the pitch.)

- Page 219 shows how results were shared, Neal's conclusion, and another question that he wondered about after completing the investigation. Point out to students that these are all important steps in an investigation.

- Extend the discussion and have students explore problems generated from school, home, or in community situations and think of ways to investigate and solve the problem using concrete objects and materials.

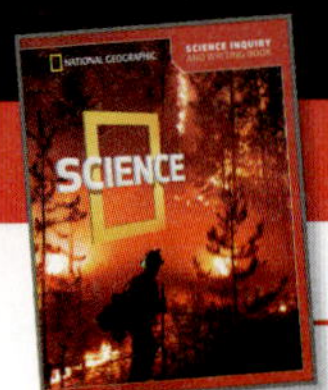

Objectives

Students will be able to:

- Investigate through Open Inquiry (generate questions to investigate; plan, make, and compare investigations; collect and record data and observations; generate explanations and conclusions based on evidence or observations; share findings; ask questions based on observations to increase understanding; adjust interpretations based on findings and new ideas).

PROGRAM RESOURCES

- Science Inquiry and Writing Book: *Physical Science*
- Science Inquiry and Writing Book **eEdition** at **myNGconnect.com**
- **Inquiry eHelp** at **myNGconnect.com**
- Science Inquiry Kit: *Physical Science*
- Learning Masters Book, pages 234–239, or at **myNGconnect.com**
- Inquiry Rubric: Assessment Handbook, page 216, or at **myNGconnect.com**
- Inquiry Self-Reflection: Assessment Handbook, page 227, or at **myNGconnect.com**

MATERIALS

Kit materials are listed in italics.

petri dish; masking tape; *graduated cylinder (50 mL);* pitcher; water; marker; *fan*

❶ Introduce

Tap Prior Knowledge

- Tell students to think about a puddle after it rains. Ask: **What happens to the puddle over time?** (It gets smaller.) **Where does the water go?** (It evaporates.) Explain to students that the water does not just disappear; it evaporates. Tell students that they will investigate evaporation in this activity.

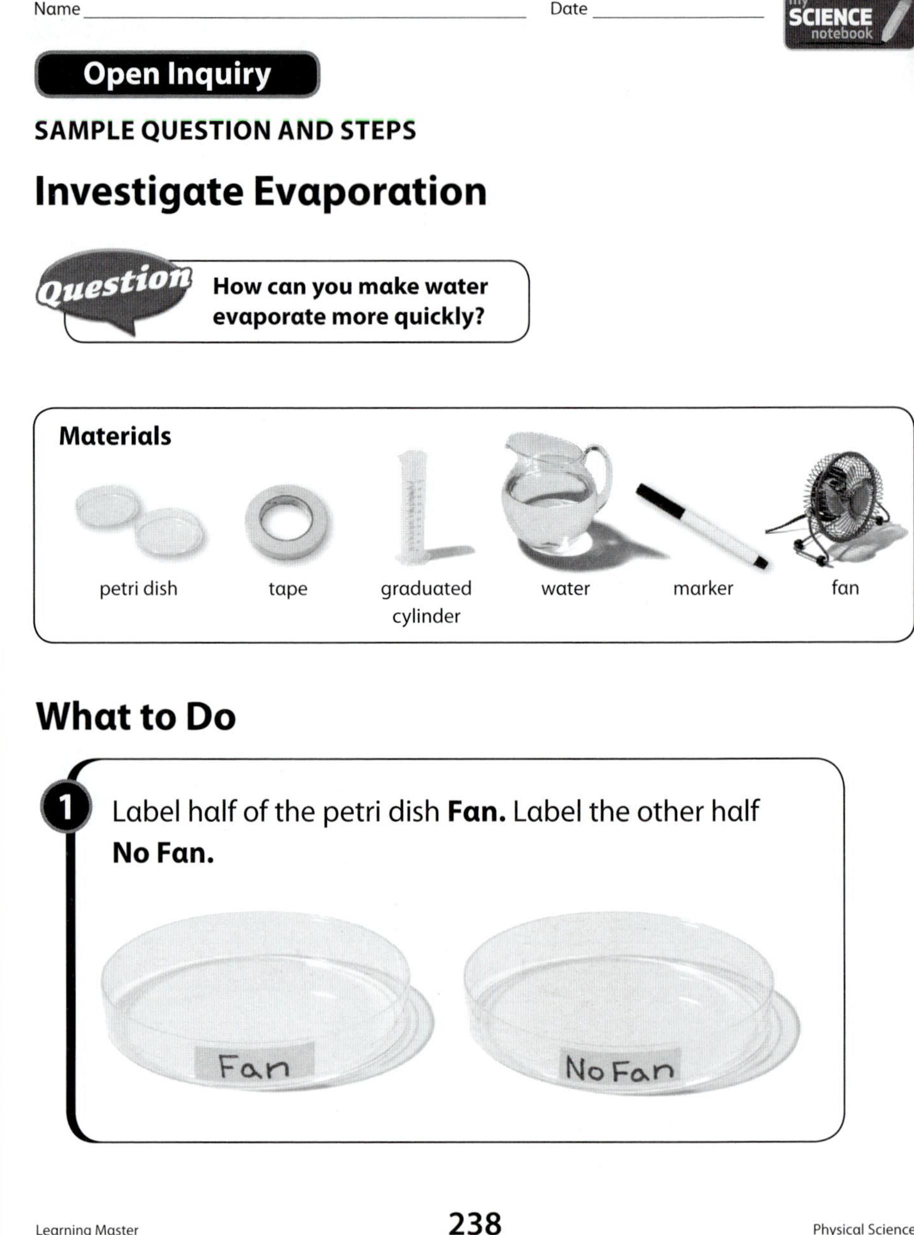

Name _______________________________   Date _______________

Open Inquiry

SAMPLE QUESTION AND STEPS

Investigate Evaporation

Question How can you make water evaporate more quickly?

Materials

What to Do

1. Label half of the petri dish **Fan.** Label the other half **No Fan.**

Learning Master   **238**   Physical Science
© NGSP & HB

MANAGING THE INVESTIGATION

Time

20 minutes

Groups

Small groups of 4

Advance Preparation

- Provide each group with enough water so they will have 30 mL of water for each petri dish.
- Clear a large, sunny area, such as on a windowsill. Make sure the area is large enough to fit all the petri dishes and will not be disturbed overnight.
- If a fan is not available for each group, have groups share an available fan or fans.

Name _______________________________ Date _______________

my SCIENCE notebook

What to Do, continued

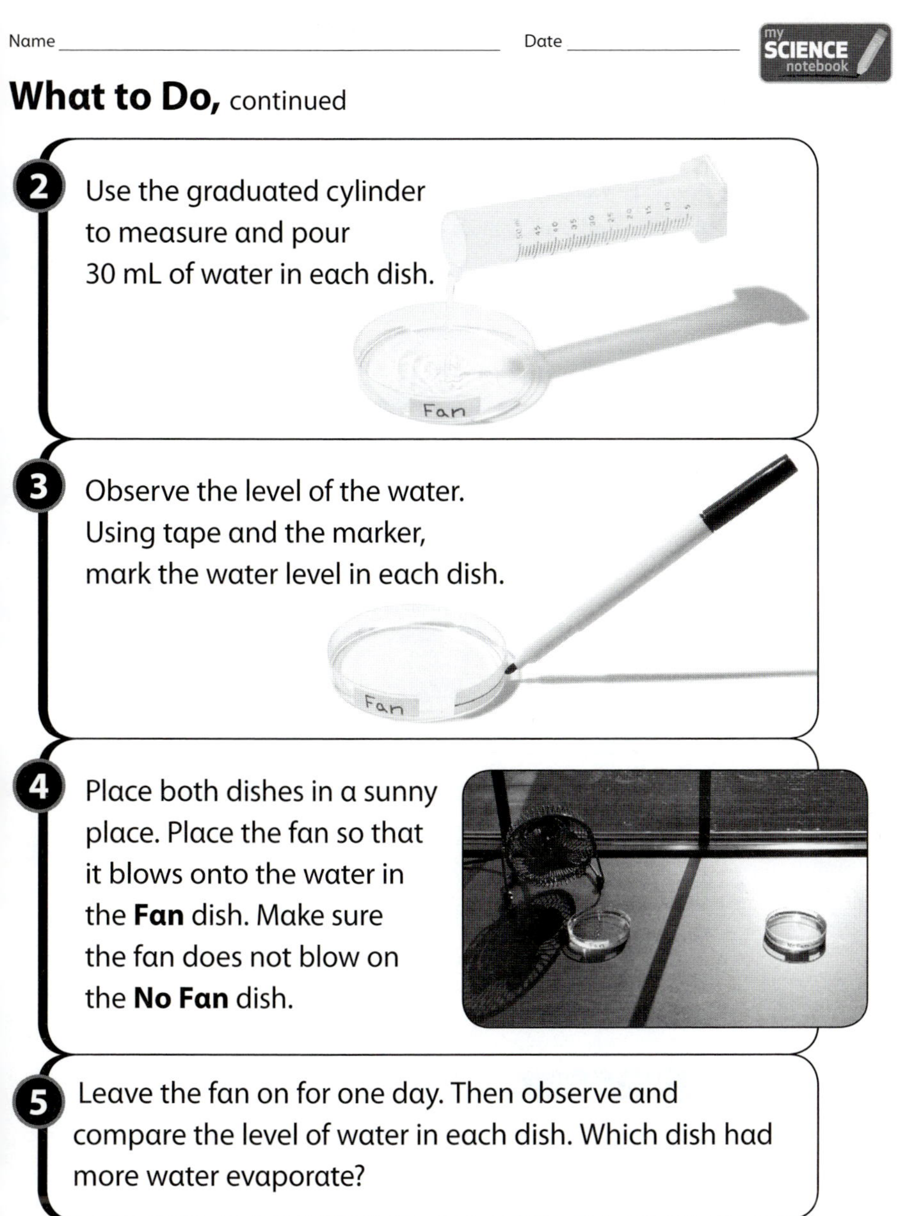

2 Use the graduated cylinder to measure and pour 30 mL of water in each dish.

3 Observe the level of the water. Using tape and the marker, mark the water level in each dish.

4 Place both dishes in a sunny place. Place the fan so that it blows onto the water in the **Fan** dish. Make sure the fan does not blow on the **No Fan** dish.

5 Leave the fan on for one day. Then observe and compare the level of water in each dish. Which dish had more water evaporate?

Learning Master **239** Physical Science

© NGSP & HB

Teaching Tips

- Tell students to use the same halves of the petri dishes so the varying size and surface area of the dishes do not become variables.
- Have students add group numbers to the labels to easily identify them the next day. If the fan cannot be left on overnight, you may cover the dishes and continue the investigation the following day.

What to Expect

- Students will observe that water will evaporate more quickly in the petri dish with the fan blowing on it.

Introduce, continued

Sample Question and Steps

- Students can use the Sample Question and Steps to practice the steps of an Open Inquiry investigation. Provide guidance as needed. Keep in mind that the goal is for students to learn how to make their own plans and to answer questions through scientific investigations.
- Have students use the Open Inquiry Checklist on Science Inquiry and Writing Book page 211. This checklist also appears on Learning Masters 234–237.

Connect to the Big Idea

- Review the Big Idea Question, *What are states of matter?* Distribute Learning Masters 234–239. Read the Question and invite students to tell what they know about evaporation. Remind students that when water evaporates, it changes from a liquid to a gas.

❷ Guide the Investigation

- Guide students in planning an experiment. Students should list what they will change, what they will observe or measure, and what they will keep the same.
- Assist students as necessary to develop a list of steps for their investigations. Stress that the steps should be able to be repeated by anyone without additional instruction. See the sample student answers on pages T180p–T180q.
- Distribute materials. Read the steps on Learning Masters 238–239. Clarify steps if necessary.
- In step 2, remind students how to measure water properly with a graduated cylinder. They should measure at eye level and look at the bottom of the meniscus. Some students may find it easier to first pour water into a plastic cup or funnel.
- Assist students in creating a table to record data.
- The following day, ask students to observe how the water levels have changed. Guide students as they compare the water levels in the two petri dishes.

⚠ Safety First

Wipe up any spills immediately.

Guide the Investigation, continued

- Ask: **How did the water change in the two petri dishes?** (Both dishes had less water.)

- Ask: **Did the water level change more in one of the dishes? Which one?** (Yes, there was a larger difference in the Fan dish.)

- If students have trouble seeing the difference between the dishes, they may use a metric ruler and measure the difference between the water levels.

❸ Explain and Conclude

- Have groups tell about the steps they followed and the data they gathered. Students should communicate scientific ideas about evaporation, using the data they collected and analyzed as evidence.

- Ask: **What changes did the water go through when it evaporated?** (It changed from a liquid to a gas.) Encourage students to describe how the water is different when it is a gas.

- Ask: **Why did the water level go down in both dishes?** (The water evaporated.)

- Ask: **Why did the water level go down more in the Fan dish?** (The moving air caused the water to evaporate more quickly.)

- Lead students to make inferences about what conditions in the world outside might lead water to evaporate faster. For example, students might infer that water would evaporate faster on a windy day, a hot day, or a dry day. Discuss how students could test some of these ideas in the classroom.

- Have each group share explanations and conclusions. Encourage students to listen actively and ask questions if they hear something they do not understand.

- Discuss possible reasons for any differences in results. Some groups may have observed a greater change in the water level because they used different halves of the petri dish, or because the fan was placed differently. Also, different areas on the windowsill or areas closer to a heater may have caused a quicker rate of evaporation.

T180p Investigation Model

Sample Student Responses

Name _______________________ Date _____________

Open Inquiry

Do Your Own Investigation
Open Inquiry Checklist

Possible student answers are for the sample Question and Steps.

☑ Choose a question or make up one of your own.

How can you make water evaporate more quickly?

☑ Gather the materials you will use.

petri dish

tape

graduated cylinder

water

marker

fan

☑ If needed, make a hypothesis or a prediction.

Possible answer: If I allow 2 dishes of water to sit overnight, and 1 of the dishes has a fan blowing on it, then more water will evaporate from the dish with a fan blowing on it.

Learning Master | **234** | Physical Science
© NGSP & HB

Name _______________________ Date _____________

Open Inquiry continued

☑ If needed, identify, manipulate, and control variables.

Variable I will change: I will change whether air blows onto the water. Variable I will measure or observe: I will observe the level of water in each dish. Variables I will keep the same: I will keep the amount and temperature of water I start with in each dish the same. I will keep the location of both dishes the same.

☑ Make a plan for your investigation.

*1. Label half of the petri dish **Fan**. Label the other half No Fan.*

2. Measure and pour 30 mL of water in each dish.

3. Observe the level of the water. Use tape and a marker to mark the water level in each dish.

*4. Place both dishes in a sunny place. Place the fan so that it blows onto the water in the **Fan** dish only.*

5. The next day, observe and compare the level of water in the dishes.

Learning Master | **235** | Physical Science
© NGSP & HB

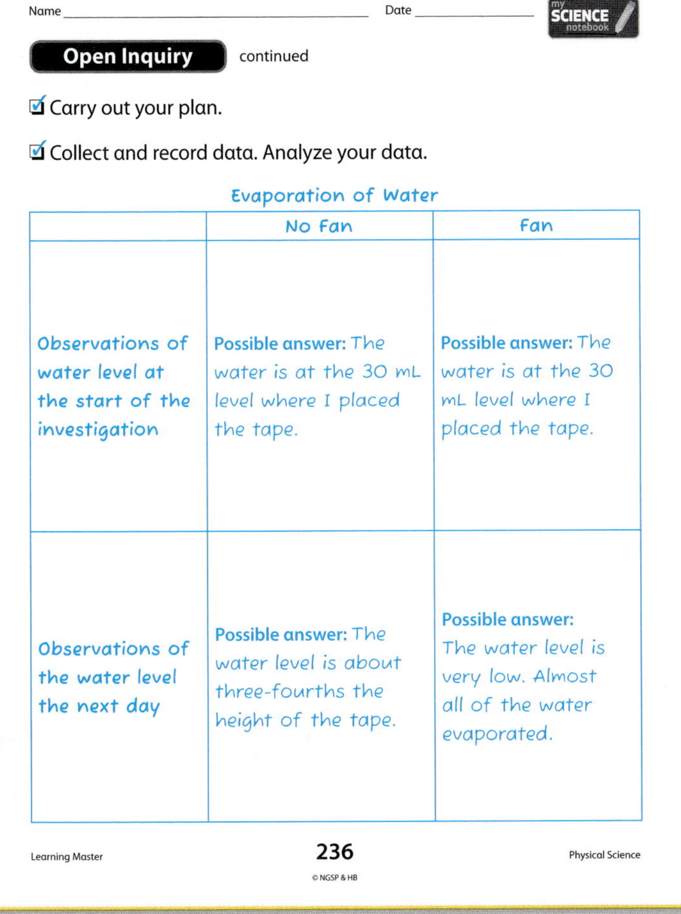

Evaporation of Water

	No Fan	Fan
Observations of water level at the start of the investigation	Possible answer: The water is at the 30 mL level where I placed the tape.	Possible answer: The water is at the 30 mL level where I placed the tape.
Observations of the water level the next day	Possible answer: The water level is about three-fourths the height of the tape.	Possible answer: The water level is very low. Almost all of the water evaporated.

❹ Find Out More

Think of Another Question

- Students should use observations made in this investigation to generate other questions that could be studied using available materials. Have them use these stems: *What happens when…? What would happen if…? How can I/we…?*

- Record student questions for possible future investigations. Discuss what students could do to find answers to the new questions.

❺ Reflect and Assess

- To assess student work with the Inquiry Rubric shown below, see Assessment Handbook, page 216, or go online at **myNGconnect.com**

- Have students use the Inquiry Self-Reflection on Assessment Handbook, page 227, or at **myNGconnect.com**

Inquiry Rubric at myNGconnect.com	Scale			
The student generated or chose a **question** to **investigate.**	4	3	2	1
The student planned an **experiment** by identifying, manipulating, and controlling **variables.**	4	3	2	1
The student **compared observations** and collected and recorded **data.**	4	3	2	1
The student formed a **conclusion** and explained results based on evidence from the collected data and observations.	4	3	2	1
The students **shared** observations and conclusions with other students.	4	3	2	1
Overall Score	4	3	2	1

Objectives

Students will be able to:

- Explain why keeping accurate and detailed records is important when doing science.
- Read and interpret data in simple tables produced by others.
- Determine the reasonableness of estimates and measurements.
- Use simple logical reasoning to develop conclusions, recognizing patterns and relationships in the environment.
- Question explanations heard from others, identifying similarities and differences, seeking clarification, and comparing them with their own observations and understandings.

PROGRAM RESOURCES

- Learning Masters Book, pages 240–241, or at
  🌐 myNGconnect.com

❶ Introduce

Tap Prior Knowledge

- Say: **Think about the investigations we have done in science class. How does each investigation begin?** (with a question) Review some of the questions that students have investigated. Say: **Scientists ask questions about the natural world. Then they investigate to find the answer.**

- Ask: **When you see something happen, which of the five senses do you use?** (sight) **You can't see thunder, but how do you know when it happens?** (You hear it.) **What other senses can be used to gather data?** (smell, touch, and taste) Discuss how the senses can be used to make observations about the natural world.

❷ Teach

Using Observations to Evaluate Explanations

- Ask for volunteers to read aloud the two paragraphs on page 220. Then ask: **What tools did these scientists use to make observations?** (graduated cylinder, microscope)

How Scientists Work

Using Observations to Evaluate Explanations

Scientists test their ideas by doing investigations. They make observations and gather information about the natural world. Then they analyze and interpret the information to explain their ideas.

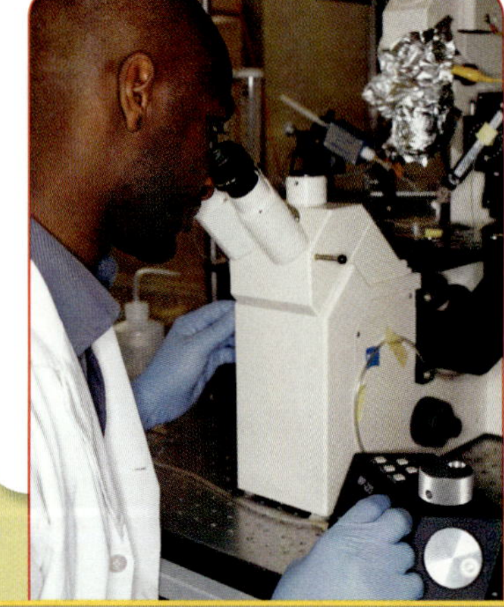

Making Observations and Collecting Data When scientists investigate, they gather information, called data. Data can be gathered by seeing, hearing, smelling, or feeling something.

220

Recording and Organizing Data Scientists might record data as written descriptions of what they observe. For example, a scientist might describe the color of an object or its motion. He or she might describe how an animal acts.

Other times data might be recorded as drawings or photographs. If scientists use measuring tools to make observations, their data might be numbers.

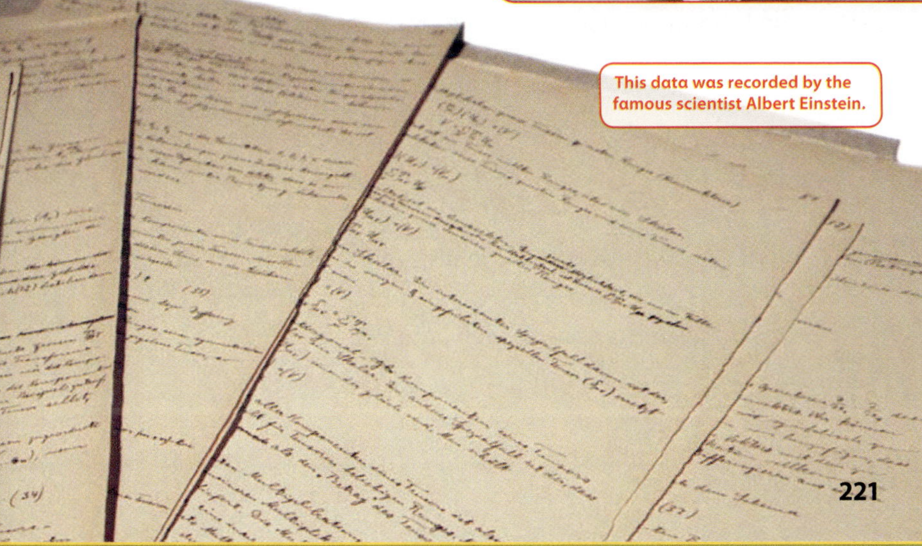

This data was recorded by the famous scientist Albert Einstein.

221

- Say: **Scientists use tools to help them test their ideas. By using their senses, scientists make observations and gather data.** Explain that scientists rarely taste things to collect data, but tasting is useful when investigating food or how taste works.

- Read page 221 aloud with students. Ask: **What is the scientist in the picture doing?** (He is looking at a test tube and writing.) **What do you think he might be writing?** (the measurement of the liquid; describing its color or how it may have changed) Point out that the scientist is recording and organizing data.

- Ask: **Why is it important to keep accurate, detailed records?** (Accurate, detailed records are easier for others to interpret, and they might help scientists when they repeat investigations.)

- Invite students to tell how they might analyze data in an investigation. For example, they might look for patterns or relationships, look for similarities or differences, and determine whether the data are reasonable. They might also produce tables, graphs, or diagrams to organize the data.

- Show students the picture of the handwritten data recorded by Albert Einstein. Tell students that Einstein was a famous scientist in the early 1900s who made new and exciting observations. The data appear to be written descriptions of what he observed. Explain that Einstein's ideas changed the way scientists think about light, heat, gravity, and matter, and that because he recorded his data, they are still used as empirical evidence today.

NATIONAL GEOGRAPHIC Raise Your SciQ!

What Is Empirical Evidence? The word *empirical* means "based in experience or observation." *Empirical evidence* is information that can be observed with the senses. Scientists use empirical evidence to support their explanations as they answer questions or test hypotheses. This is important because scientists must have their evidence reviewed by their peers before their explanation can be accepted. Empirical evidence allows scientists to replicate the studies of other scientists. Scientists question, discuss, and check the evidence and explanations of other scientists.

Teach, continued

- Read page 222 aloud with students. Say: **When scientists analyze data, they look for patterns and they see how the data fit together. The scientists in the picture are analyzing data.**

- Ask: **What if the data do not support a scientist's ideas, hypotheses, or predictions?** Explain that this does sometimes happen in a scientific investigation. Sometimes, the scientist's ideas were not completely right, or the hypothesis or prediction in that investigation was not supported. This is an opportunity for the scientist to do further investigation or test a new hypothesis.

- Ask: **What mistakes might a scientist make when doing an investigation?** (incorrect measurements, additional variables that were not measured, variations in procedure) **What mistakes might a scientist make when analyzing data?** (The scientist might draw conclusions or make inferences that are incorrect.)

- Ask: **What can scientists do when they think they may have made mistakes?** (They can perform the investigation again, record data more carefully, or discuss their ideas with other scientists to make sure they have made inferences and conclusions that are reasonable.)

- Say: **When the data do not support a scientist's ideas, the scientist may have to collect more data or try something different. Sometimes, the scientist may discover a new idea when analyzing data or doing a new investigation. That can be even more exciting than the original idea!**

- Lead a discussion about why it is important to record and keep data, even when the results don't support the original idea.

What Did You Find Out?

1. Scientists might organize their data as a written description, as a drawing or photograph, or in numbers.

2. Scientists analyze their data to find out whether the data support their ideas and explanations.

Analyzing Data Then scientists analyze their data. They look for patterns. They decide whether the data support their ideas and explanations.

Sometimes the results of an investigation may not support a scientist's ideas. Then the scientist might ask questions such as:

- Did I make mistakes?

- Do I need to collect more data?

- What else do I need to know?

- How can I change my ideas so that they are supported by the new data?

SUMMARIZE

What Did You Find Out?

1 What are three ways that scientists might organize their data?

2 Why do scientists analyze their data?

222

Learning Masters 240–241 or at

myNGconnect.com

 ## Evaluate Explanations

Rico observed two water balloons of different sizes. He thought that the bigger water balloon would have more mass because it held more water. He decided to do an investigation to find out.

First Rico used a balance to measure the mass of the small water balloon. Then he measured the mass of the big water balloon. Rico measured each balloon several times to be sure that his measurements were accurate. He organized his data in a table.

Mass of Water Balloons		
Trial	Mass of Large Balloon	Mass of Small Balloon
1	615 g	246 g
2	615 g	246 g
3	615 g	246 g

Analyze Rico's data. Write a paragraph telling whether you agree with Rico's idea that the bigger water balloon has more mass. Use Rico's data to support your argument.

Science Misconceptions

Mass and Weight Students may think that mass and weight are the same thing. Explain that *mass* is the amount of matter in something. The mass of an object remains the same whether placed on Earth or on a body with less mass than Earth, such as the moon. *Weight* is the force of gravity acting on matter. The weight of an object would be less on the moon than its weight on Earth.

Teach, continued

Evaluate Explanations

- Ask for volunteers to read aloud the first two paragraphs on page 223.

- Say: **Rico measured the mass of the balloons three times.** Have students read and interpret Rico's table. Ask: **What was Rico's measurement for the large water balloon?** (615 g) **What was his measurement for the small water balloon?** (246 g)

- Help students use simple reasoning to develop their conclusion. Ask: **What do you think of Rico's explanation? Do you think the bigger water balloon has more mass because it has more water? Explain your answer.** (Possible answer: Yes. Because the balloon is bigger, it has more water and will have more mass.)

- Have students analyze Rico's data and write a paragraph telling whether they think the data support Rico's explanation or not.

- Have students identify similarities and differences between their explanations and seek clarification about why there may be differences. Ask: **What explanations did your classmates have? How were those explanations different from yours? Why do you think there were differences?** (Answers will vary.)

❷ Assess

- Use Learning Masters Book pages 240–241. Students should be able to explain why scientists organize and analyze their data. Students should also be able to use data to evaluate and support their explanations about mass.

Create a Web Page About Matter SMALL GROUP

If students have a class Web page or a school Web site, help them create a set of pages to show how matter can be measured and described. Help students with the technology as necessary.

Steps

1. Have students use digital cameras to take photos that show examples of properties of matter. They can show different kinds of matter sorted by size, color, shape, texture, and hardness. Individual groups may focus on one property. Students may also take pictures of themselves measuring the mass and volume of different kinds of matter.

2. Upload the photos to the Web page.

3. Ask students to add captions to each image.

4. Have students share their Web pages with the class. You may also wish students to present their work to other third-grade classes.

Discuss the Big Ideas

After students present their Web pages, lead a discussion about the Big Idea Question:

• How Can You Describe and Measure Matter?

Help students see how their Web pages contribute to their understanding of the Big Idea Question.

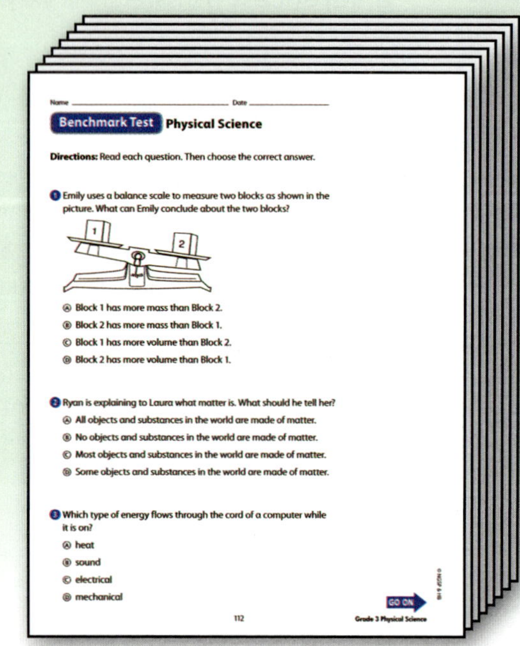

Assess Student Progress Benchmark Test

Have students complete their Physical Science Benchmark Test to assess their progress in this unit.

Benchmark Test, Assessment Handbook, pages 112–119, or at ⊘ **myNGconnect.com**

or **NGSP ExamView CD-ROM**

Make a Timeline of Water Changes SMALL GROUP

Have students draw a timeline that shows the different states of matter water goes through as its temperature changes.

Steps

1. Assign each group a different part of the timeline, showing water:
   - freezing
   - melting
   - boiling and evaporating
   - condensing

2. Make available a long roll of paper, or tape together sheets of paper.

3. Have students draw pictures showing water changing shape—for example, for the freezing section, show pouring water into an ice cube tray and then a solid ice cube.

4. Tell students to label their section with the state of matter (for example, solid), the process whereby water changes into that state (for example, freezing), and the temperature at which the change in state occurs.

5. Hang the timeline of water changes in the hallway outside your classroom.

Discuss the Big Ideas

After students complete their timeline, lead a discussion about the Big Idea Question:

- What Are States of Matter?

Help students see how their timeline contributes to their understanding of the Big Idea Question.

Demonstrate the Forms of Energy

Have students demonstrate as they explain different forms of energy.

Steps

1. Assign each group a different form of energy:
- stored, of motion, and mechanical
- sound
- electrical
- heat

2. Have students plan how they will demonstrate their assigned form of energy. Tell them to make a list of materials they will need. For the group demonstrating electrical energy, an electrical appliance will be needed. For the group demonstrating heat, a hot plate or other heat source will be needed. Explain to students in this group that you will help them use their equipment.

3. Review students' plans. Once they are approved, help students acquire the materials they need.

4. Encourage students to make notes for their explanations. Allow them time to practice their demonstrations and explanations.

5. Have students present their demonstrations to each other. Then have them give their demonstrations on a family science night.

Discuss the Big Ideas

After students complete their demonstrations, lead a discussion about the Big Idea Questions:

- What Is Energy?
- How Does Force Change Motion?

Help students see how their demonstrations contribute to their understanding of the Big Idea Questions.

Make a Movie About Light

Have students use a video camera to make a movie about the properties of light. Help them with this technology if necessary.

Steps

1. Assign each group a different topic about light to demonstrate and explain:
   - reflection
   - refraction
   - absorption and color
   - shadows

2. Have students plan how they will demonstrate their property of light. Tell them to make a list of materials they will need. Students can use flashlights for their light source.

3. Review students' plans. Once they are approved, help students acquire the materials they need.

4. Encourage students to make notes for their explanations. Allow them time to practice their demonstrations and explanations.

5. Have students use a video camera to record their demonstrations. Show their movies about light to other third-grade classes.

Discuss the Big Ideas

After students complete their movies, lead a discussion about the Big Idea Question:

- What Is Light?

Help students see how their movies contribute to the understanding of the Big Idea Question.

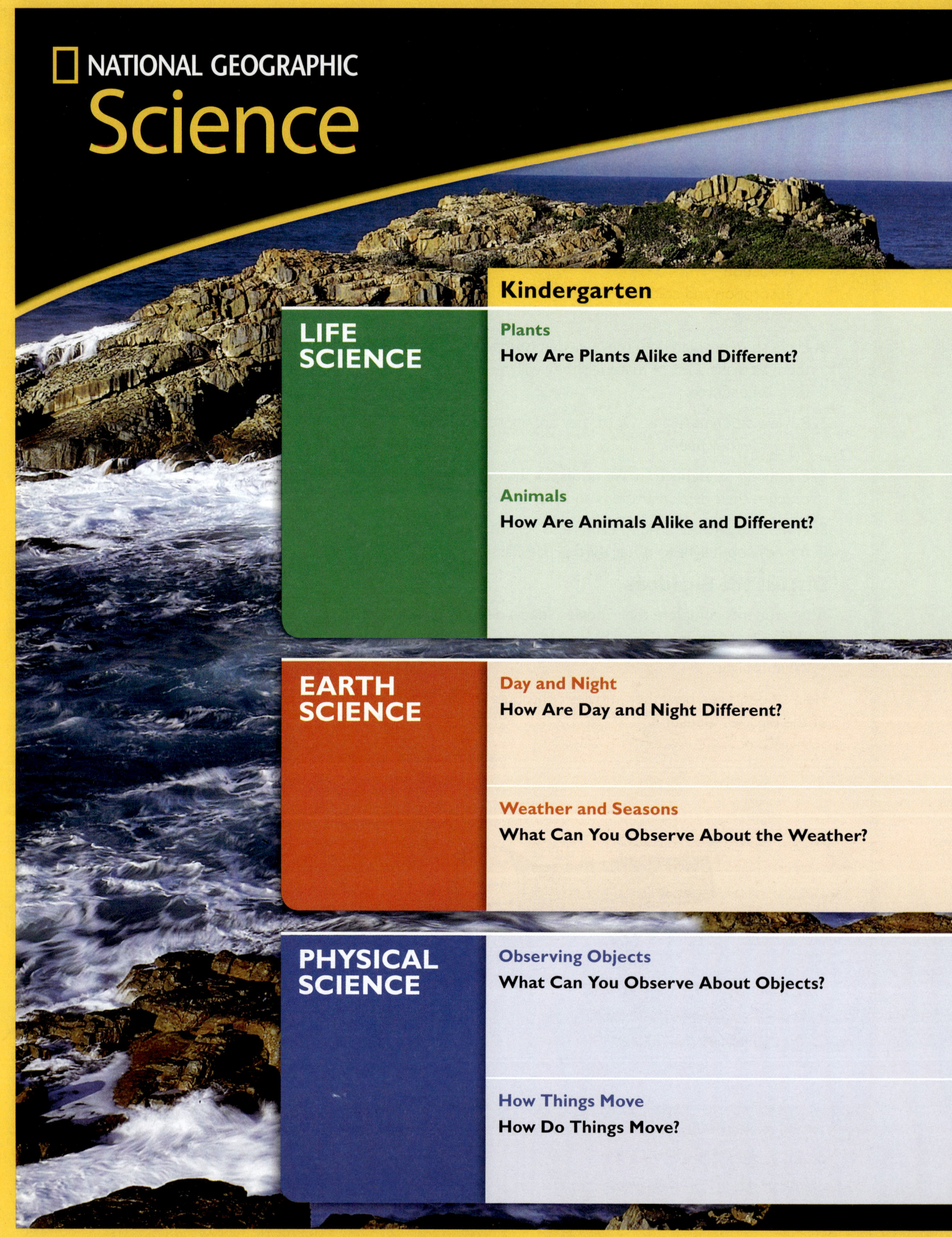

NATIONAL GEOGRAPHIC
Science

Kindergarten

LIFE SCIENCE

Plants
How Are Plants Alike and Different?

Animals
How Are Animals Alike and Different?

EARTH SCIENCE

Day and Night
How Are Day and Night Different?

Weather and Seasons
What Can You Observe About the Weather?

PHYSICAL SCIENCE

Observing Objects
What Can You Observe About Objects?

How Things Move
How Do Things Move?

K–2 Scope and Sequence

Grades 1–2

NATIONAL GEOGRAPHIC

Science

Grade 3

LIFE SCIENCE

EARTH SCIENCE

PHYSICAL SCIENCE

3–5 Scope and Sequence

Metric and SI Units

The metric system is also called the International System of Units, or SI. Scientists, as well as most countries throughout the world, use the metric system. The United States uses customary units. The chart below will help you convert units between metric measurements and customary measurements.

	METRIC/SI	CUSTOMARY	CONVERSION	
Distance	centimeter	inch	1 cm = .39 in	1 in = 2.54 cm
	meter	foot	1 m = 3.28 ft	1 ft = .30 m
	kilometer	mile	1 km = .62 mi	1 mi = 1.61 km
Volume	milliliter	teaspoon	1 mL = .20 tsp	1 tsp = 4.93 mL
	liter	pint	1 L = 2.12 pt	1 pt = .47 L
	liter	quart	1 L = 1.06 qt	1 qt = .95 L
	liter	gallon	1 L = .26 gal	1 gal = 3.79 L
Mass	gram	ounce	1 g = .035 oz	1 oz = 28.35 g
	kilogram	pound	1 kg = 2.2 lbs	1 lb = .45 kg
Temperature	Celsius	Fahrenheit		
	0° water freezes	32° water freezes	$C° = 5/9 \, (°F - 32)$	
	100° water boils	212° water boils	$F° = 9/5 \, °C + 32$	

CONVERTING AMONG CUSTOMARY UNITS

Distance	12 inches = 1 foot; 3 feet = 1 yard; 5280 feet = 1 mile
Volume	1 pint = 16 ounces; 1 quart = 32 ounces; 1 gallon = 128 ounces
Mass	1 pound = 16 ounces; 1 ton = 2000 pounds

CONVERTING AMONG METRIC UNITS

Distance	1 centimeter = .01 meter; 1 kilometer = 1000 meters
Volume	1 milliliter = .001 liter
Mass	1 milligram = .001 gram; 1 kilogram = 1000 grams

Materials List

	INQUIRY	KIT MATERIALS	SCHOOL-SUPPLIED MATERIALS
CHAPTER 1 **How Can You Describe and Measure Matter?**	**Explore Activity** *Investigate Physical Properties* How can you use tools to investigate physical properties? page T5e	burlap hand lens microscope battery wool sandpaper silk	scissors paper towel
	Directed Inquiry *Investigate Volume and Mass* How can you compare the volume and mass of solid and liquid objects? page T23a	graduated cylinder marble rock plastic cup balance gram masses	pitcher water
CHAPTER 2 **What Are States of Matter?**	**Directed Inquiry** *Investigate Water and Temperature* What happens to water as the temperature changes? page T37e	resealable plastic bags graduated cylinder	masking tape pitcher water
	Guided Inquiry *Investigate Melting* How does heating and cooling affect the properties of different materials? page T55e	plastic cups graduated cylinder salt vinegar thermometers stopwatch	masking tape water juice
CHAPTER 3 **How Does Force Change Motion?**	**Directed Inquiry** *Investigate Forces and Motion* How do forces affect motion? page T69e	table tennis ball meterstick fleece fabric	books masking tape
	Guided Inquiry *Investigate Motion and Position* How can you describe the motion of an object by observing its position? page T95a	ball straws plastic spoon string	poster board masking tape markers ruler

Materials List, continued

	INQUIRY	KIT MATERIALS	SCHOOL-SUPPLIED MATERIALS
CHAPTER 4 **What Is Energy?**	**Directed Inquiry** *Investigate Energy of Motion* How does adding more washers affect the motion of a pendulum? page T109e	string paper clip metal washers stopwatch	safety goggles masking tape ruler scissors
	Guided Inquiry *Investigate Vibrations and Sound* How does the length of a tuning fork affect the pitch of the sound it makes? page T135a	tuning forks clear plastic cup plastic wrap salt rubber band eraser	ruler
CHAPTER 5 **What Is Light?**	**Directed Inquiry** *Investigate Light and Heat* What happens to an object's temperature when light shines on it? page T149e	thermometer stopwatch	lamp
	Guided Inquiry *Investigate Light and Objects* What happens to light when it shines on different objects? page T167a	meterstick flashlight mirror foil black paper cloth	white paper masking tape paper towel scissors
UNIT INQUIRY	**Open Inquiry** *Do Your Own Investigation* Sample Question and Steps: How can you make water evaporate more quickly? page T180n	petri dish graduated cylinder fan	masking tape pitcher water marker

Notes

Glossary

A

absorption
Absorption is the taking in of light by a material.

analyze
When you analyze, you look for patterns and relationships in the data.

C

classify
When you classify, you put things in groups according to their characteristics.

compare
When you compare, you tell how objects or events are alike and different.

conclude
You conclude when you use information, or data, from an investigation to come up with a decision or answer.

condensation
Condensation is the change from a gas to a liquid.

count
When you count, you tell the number of something.

D

data
Data are observations and information that you collect and record in an investigation.

E

electricity
Electricity is energy that flows through wires.

energy
Energy is the ability to do work or cause a change.

estimate
When you estimate, you tell what you think about how much or how many.

evaporation
Evaporation is the change from a liquid to a gas.

experiment
In an experiment, you change only one variable, measure or observe another variable, and control other variables so they stay the same.

F

force
A force is a push or a pull.

friction
Friction is a force that acts when two surfaces rub together.

G

gas
A gas is matter that spreads to fill a space.

gravity
Earth's gravity is a force that pulls things to the center of Earth.

H

heat
Heat is the flow of energy from a warmer object to a cooler object.

hypothesis
You make a hypothesis when you state a possible answer to a question that can be tested by an experiment.

I

infer
When you infer, you use what you know and what you observe to draw a conclusion.

investigation
You investigate when you make a plan and carry out the plan to answer a question.

L

light
Light is a kind of energy you can see.

liquid
A liquid is matter that takes the shape of its container.

M

magnetism
Magnetism is a force between magnets and objects magnets attract.

mass
Mass is the amount of matter in an object.

matter
Matter is anything that has mass and takes up space.

measure
When you measure, you find out how much or how many.

Glossary

mechanical energy
The mechanical energy of an object is its stored energy plus its energy of motion.

model
You can make and use a model to show how something in real life works.

motion
When an object is moving it is in motion.

O

observe
When you observe, you use your senses to learn about an object or event.

operational definition
When you make an operational definition, you use your own words to tell what something can do.

P

plan
When you make a plan to answer a question, you list the materials and steps you need to take.

predict
When you predict, you tell what you think will happen.

Q

question
You ask a question to find out about something.

R

reflection
Reflection is the bouncing of light off of an object.

refraction
Refraction is the bending of light when it moves through one kind of matter to another.

S

share
When you share results, you tell or show what you have learned.

solid
A solid is matter that keeps its own shape.

sound
Sound is energy that can be heard.

speed
Speed is the distance an object moves
in a period of time.

states of matter
States of matter are the forms in which
a material can exist.

T

texture
Texture describes the surface of any
area made up of matter.

V

variable
A variable is something that could
change in an investigation.

volume
Volume is the amount of space matter
takes up.

Index

Note: EM refers to the End Matter and SN refers to the Nature of Science/ Science Notebook.

A

Absorption, T152–T153, T162–T163, T167a–T167d, T168, T176–T177

Adams, Constance, T4

Air, as a gas, T45

Analyze, T25, T97, T155, T161, T167, T169, T180l–T180m

Apply, T45, T51, T75, T79, T89, T115, T119, T129, T137, T142–T143, T169

Assess the Big Idea, T25, T57, T97, T137, T169

Assessment
assess the big idea, T25, T57, T97, T137, T169
become an expert, T30–T31, T34–T35, T62–T63, T66–T67, T102–T103, T106–T107, T139, T142–T143, T146–T147, T174–T175, T178–T179
before you move on, T15, T17, T21, T45, T51, T53, T75, T79, T85, T89, T93, T95, T99, T115, T119, T125, T129, T133, T155, T159, T161, T163, T165
reflect and assess
directed inquiry, T23d, T37h, T69h, T109h, T149h
explore activity, T5h
guided inquiry, T55h, T95d, T135d, T167d
investigation model, T180q
open inquiry, T180k

Atomic Force Microscope, T98, T99

Attraction. *See* Pull

B

Balances, comparing mass using, T16, T23a–T23d

Balls and gravity, T69e–T69h

Bath, Patricia, T170–T171

Batteries, T127

Beakers, measuring volume with, T19

Become an Expert
High-Speed Trains, T100-T101 – T108
Laser: A Special Kind of Light, T172-T173 – T180
Properties of Graphene, T28-T29 – T36
Sweden: Ice Hotel Construction, T60-T61 – T68
Use Some Energy: Have Some Fun, T140-T141 – T148

Before You Move On, T15, T17, T21, T45, T51, T75, T79, T85, T89, T93, T115, T119, T125, T129, T133, T155, T159, T165

Big Idea Question and Vocabulary, T6–T7, T8–T9, T38–T39, T40–T41, T70–T71, T72–T73, T110–T111, T112–T113, T150–T151, T152–T153

Big Idea Questions
How Can You Describe and Measure Matter? T5–T36
How Does Force Change Motion? T69–T108
What Are States of Matter? T37–T68
What Is Energy? T109–T148
What Is Light? T149–T180

Big Idea Review, T25, T57, T97, T137, T169

Biographical information on scientists. *See* Physical Science Expert; Meet a Scientist

Boiling, T49

Brice, Steve, T58–T59

Bullet trains, T102–T103

Burning, T155

C

Carbon, *chart,* T26, T28–T29, *chart,* T28–T29

Careers, science. *See* Physical Science Expert; Meet a Scientist

Cause and effect, T51, T57, T62–T63, T167, T169

Challenge, differentiated instruction, T16, T45, T48, T74, T83, T91, T124, T160

Change
of color in response to conditions, T13
energy's ability to cause, T114–T115
in position (*See* Motion)
in states of matter (*See* States of matter)

Changes in states of matter, T37e–T37h, T46–T51, T55e–T55h

Charts
forms of carbon, T28–T29
shapes of solids, T43

Classify, T23, T75

Clouds, T50

Cognates, ELL, T5f, T95b, T167b

Collision, T109e–T109h

Color
changing in response to conditions, T13
of graphene, T28–T29
as property of matter, T12–T13

Compare
science process vocabulary, T37f, T69f, T109f, T167b
thinking skill, T15, T17, T23a–T23d, T37e–T37h, T53, T55a, T55e–T55h, T59, T79, T106–T107, T108, T125, T137, T163, T180j–T180k, T180n–T180o, T180r–T180u

Compare and contrast, T25, T97

Conclude, T149f

Conclusion, T24, T56, T96, T136, T168

Condensation
about, T50–T51
definition of, T40–T41, T50, T64–T65
on fog fences, T55
observing, T37e–T37h
vocabulary review, T56

Connect to the Big Idea, T5f, T23b, T55f, T69f, T95b, T109f, T135b, T149f, T180o

Contact
collision, T109e–T109h
forces not requiring
gravity, T69e–T69h, T86–T89, T94–T95
magnetism, T90–T93, T104-T105 – T106-T107
forces requiring
friction, T69e–T69h, T85–T86, T94–T95, T106–T107, T132–T133
pushes and pulls, T70–T71, T80–T83

Contrast, T85, T102–T103, T119, T135, T174–T175

Cooling. *See* Temperature

Count, T109f

D

E

Credits

Acknowledgments
Grateful acknowledgment is given to the authors, artists, photographers, museums, publishers, and agents for permission to reprint copyrighted material. Every effort has been made to secure the appropriate permission. If any omissions have been made or if corrections are required, please contact the Publisher.

Photographic Credits
Front Cover, Tabs Chris Knapton/Photo Researchers, Inc.. **Inside Front Cover** (tl) Albert Moldvay/National Geographic Image Collection. (tr) Paul Nicklen/National Geographic Image Collection. (cl) Bill Hatcher/National Geographic Image Collection. (cr) Michael S. Lewis/National Geographic Image Collection. (bl) Taylor S. Kennedy/National Geographic Image Collection. (br) Gordon Wiltsie/National Geographic Image Collection. **x-xi** (bg) PhotoDisc/Getty Images. **xii-xiii** Radius Images/Photolibrary. **xiv-xv** (bg) Ihoko Saito/Toshiyuki Tajima/Dex Image/Photolibrary. **xiv** (tl) NASA Human Space Flight Gallery. (tr) Harvey Lowe. (cl) Mark Thiessen/National Geographic Image Collection. (crt) Bruce Bennett. (crb) Renee Fadiman. (bl) John Livzey. (br) Mark Mallchok. **xv** (tl, tr) Brella Productions. (clt) Calit2. (clb) Mark Mallchok/Brella Productions. (crt) Robert Clark/National Geographic Image Collection. (crb) Brett Hobson. (bl) Mark Thiessen/National Geographic Society. (br) University of Alaska, Fairbanks - Marketing and Communications. **T4a** (bg) Digital Vision/Getty Images. (inset) NASA. **T4b** John Foxx Images/Imagestate. **T5a-T5d, T37a-T37d, T69a-T69d, T109a-T109d, T149a-T149d** Digital Vision/Getty Images. **SN5** Shironina/Shutterstock. **EM2-EM5** Tom Mead/Photolibrary. **Back Cover** (bg) Chris Knapton/Photo Researchers, Inc..

Illustrator Credits
All illustrations by Susan T. Anderson and Lara Winegar.

Neither the Publisher nor the authors shall be liable for any damage that may be caused or sustained or result from conducting any of the activities in this publication without specifically following instructions, undertaking the activities without proper supervision, or failing to comply with the cautions contained herein.

Program Authors
Malcolm B. Butler, Ph.D., Associate Professor of Science Education, University of South Florida, St. Petersburg, Florida; Judith Sweeney Lederman, Ph.D., Director of Teacher Education and Associate Professor of Science Education, Department of Mathematics and Science Education, Illinois Institute of Technology, Chicago, Illinois; Randy Bell, Ph.D., Associate Professor of Science Education, University of Virginia, Charlottesville, Virginia; Kathy Cabe Trundle, Ph.D., Associate Professor of Early Childhood Science Education, The Ohio State University, Columbus, Ohio; David W. Moore, Ph.D., Professor of Education, College of Teacher Education and Leadership, Arizona State University, Tempe, Arizona

The National Geographic Society
John M. Fahey, Jr., President & Chief Executive Officer
Gilbert M. Grosvenor, Chairman of the Board

Copyright © 2011 The Hampton-Brown Company, Inc., a wholly owned subsidiary of the National Geographic Society, publishing under the imprints National Geographic School Publishing and Hampton-Brown.

All rights reserved. No part of this book may be reproduced or transmitted in any form or by any means, electronic or mechanical, including photocopying, recording, or by an information storage and retrieval system, without permission in writing from the Publisher.

National Geographic and the Yellow Border are registered trademarks of the National Geographic Society.

National Geographic School Publishing
Hampton-Brown
www.NGSP.com

Printed in the USA.
RR Donnelley, Menasha, WI.

ISBN: 978-0-7362-7739-6

11 12 13 14 15 16 17 18 19

10 9 8 7 6 5 4 3